Advances in Geographic Information Science

Series editors

Shivanand Balram, Burnaby, Canada
Suzana Dragicevic, Burnaby, Canada

More information about this series at http://www.springer.com/series/7712

Zhenjiang Shen • Miaoyi Li
Editors

Big Data Support of Urban Planning and Management

The Experience in China

Springer

Editors
Zhenjiang Shen
Joint International FZUKU Lab SPSD
Fuzhou University
Fuzhou City
China

Kanazawa University
Kanazawa City
Japan

Miaoyi Li
Joint International FZUKU Lab SPSD
Fuzhou University
Fuzhou City
China

ISSN 1867-2434 ISSN 1867-2442 (electronic)
Advances in Geographic Information Science
ISBN 978-3-319-51928-9 ISBN 978-3-319-51929-6 (eBook)
DOI 10.1007/978-3-319-51929-6

Library of Congress Control Number: 2017952939

Printed on acid-free paper

This Springer imprint is published by Springer Nature
The registered company is Springer International Publishing AG
The registered company address is: Gewerbestrasse 11, 6330 Cham, Switzerland

Foreword

Urban planning and governance innovation have been topics of the latest annual conferences conducted by Urban Planning Society of China (UPSC), which best represented the spirit of UPSC in terms of transformation of urban planning system in China from conventional professional planning and design to highly complex, multidisciplinary policy processes. In China, demand-oriented approach in planning could not reveal the true direction of urbanization and fails to respond to the economic challenges. We should adopt a solution-oriented approach to urban diseases after the rapid urbanization process. Many Chinese planners and decision makers now argued that a better understanding of what happens to Chinese cities is vital to planning and design for existing cities and new towns. To say simply that a reasonable data analysis should be enhanced to conventional process of urban planning and design, comprehensive evaluation of alternative plans based on urban simulation should be conducted for decision making and regular diagnosis of status quo for city development should be made prior to any revision of statutory plans. Hence, geospatial analysis and simulation techniques become very hot topics within the field of urban planning in China recently.

Accordingly, data source for analysis, simulation, and diagnosis become important for planners to handle the new challenges posed by the solution-oriented approach. While some planners are not familiar with data analysis techniques for finding planning solutions, they have difficulties in learning data-driven approach with their professional experiences. Big data collected from sensors of Internet of Things (IoT) is a great solution for planners to find data sources. Even though statistical analysis using survey data is very popular in the field of urban planning, planners failed to obtain survey data due to the shortage of open data in China. The reasons are complicated, such as data copyright and other administrative barriers. As we all know, data analysis is the first step of the planning process for current situation analysis, and an indicator system is very powerful as planning evaluation tool for making a final decision on alternative plans. Thus, the role of data-driven approach is the same as that of the conventional planning process. Hence, it is expected that big data could bring more effective analysis than conventional survey data in order to improve quality of planning and design.

In China, smart city construction is growing rapidly under the implementation of the scheme for promotion of smart city in China. For example, in mobile devices installed with social network system (SNS), GPS function can be used as data source of big data; other data sources such as traffic card, cellphones, and so on are very often to be seen nowadays in Chinese cities. Most of chapters in this book are related to mobile devices and human mobility. There are many kinds of big data that can be possibly used in the planning field, such as SNS, Points of Interest (POI), Taxi GPS tracking, and cellular signal for urban structure analysis from views of human mobility. In addition, real time big data collected from smart infrastructure, such as transportation monitoring system, environment monitoring system, and security system can be stored in database and developed as Cyber infrastructure that is a cloud-based urban management platform for urban management, as presented in this book.

There is a big demand for big data analysis in China. In the field of planning, many our colleagues, particularly young generation, are eager to handle big data for urban planning and design, including those in planning research institutes, faculties, and students in universities. It is very exciting that symposiums and conferences were organized for sharing planning experience on big data application for new approaches of urban planning in the recent years. In 2014, we hosted the first symposium of smart city and big data application in UPSC Annual Conference in Haikou city, and after that many events and forums were organized on this topic. We also invited Prof. Zhenjiang SHEN from Japan and others to give lectures on the topic. As promoted by our Society, more planners and researchers in China have shown strong interests in big data application, and best practices and case studies have been published in journals and books.

Thanks to many researchers and planners in China who have contributed to this book; it is my great honor to present this first book introducing Chinese case studies of big data application in planning. I sincerely hope that the audience will enjoy reading this book and take it as a reference for understanding big data analytics. This publication, which shares experiences on smart city and big data application, will of immense use to friends and colleagues who concern with China.

Dr. Nan SHI
Executive Vice President
Urban Planning Society of China
shinan@planning.org.cn

Foreword

It is the first book to introduce case studies of big data application for planning practices in China. This book, consisting 22 chapters, elaborates diverse planning support efforts on big data analysis in many Chinese cities, which primarily introduced how geospatial analysis using big data could be conducted and how big data will help planners and decision makers for understanding what are happening in their cities.

Using big data for spatiotemporal analysis of human mobility is a new challenge for easing the urban diseases caused by rapid urbanization in China. The Chinese Society for Urban Studies has newly launched the Urban Big Data Commission on 15 November 2016. Most of the contributors in this book are members of this community. I am happy to see that all authors explore the new potential of urban-rural planning and management using data-driven approaches from practical perspectives in this book, which is a multidisciplinary cooperation related to smart city construction in China. As mentioned above, a collection of up-to-date case studies using big data has been presented, and contributors draw a picture of the current research status of big data application in urban China.

The applications of geospatial analysis using big data in this book are described as several essential aspects of elaborating urban structure: social network system and human behaviors, POI mapping and urban space recognition, mobile device data, and urban form reflecting human activities, all these works are related to visualizing conflicts between land use and transportation in China. Authors of this book proved that big data collected from mobile devices and social network are useful for analyzing human mobility, obtaining urban space recognition, and exploring spatiotemporal urban structure. Exploring urban structure after the rapid urbanization process in China is concerned with possible solutions of urban diseases in Chinese cities. The case studies in this book will explain how big data changes the perception of planners and researchers in their practices.

Under the rapid development of information and communication technology, the emergence of big data available from various IoT sensors has presented significant opportunities for urban management. While reading these contributions in this book, we are witnessing a dramatic change of our data environment with smart infrastructure construction. IoT sensors for collecting mobile device data and system platform that uses real-time traffic monitoring data are opening a new era for urban management. I expect there would be more and more new big data sources from smart city construction, hereby promoting urban studies to a new stage.

Qizhi MAO, Professor
School of Architecture, Tsinghua University
Honor Advisor, Urban Big Data Commission
Chinese Society for Urban Studies

Foreword

Rapid urbanization has not only modernized the Chinese lifestyle, but has also led to significant challenges, including, increased energy consumption, pollution and traffic congestion, to name a few.

With the development of information-sensing technologies and large-scale computing infrastructures that have generated huge amounts of urban spatial big data of mobility, environmental quality, and energy consumption, the relationship between these issues have become more easily identifiable. This identification creates a call for smarter, more environmentally conscious urban planning decisions.

One possible answer to the problems facing a rapidly urbanized China is the "smart city concept." This approach, relying on the huge volume of data being absorbed by countless inconspicuous sensors spread across the subject city, provides hints on the direction that should be taken by modern urban planners. The implementation of the smart city concept is predicated on the ubiquity of smart technologies, namely, cellphones and other portable devices comprising the Internet of Things (IoT). These technologies augment the volume, velocity, and variety of the information that is accessible to those tasked with urban planning. In other words, there is an inverse relationship between the size of technologies and the quality and quantity of information that can be gathered. Chips are capable of gathering data on environmental quality, urban mobility, and energy consumption, thereby informing future research into the human lifestyle patterns.

It behooves us to not only recognize the usefulness of big data, but to manipulate it to our greater end in urban planning: the creation of a healthy, highly developed, and highly integrated ecosystem. This book attempts to take a multidisciplinary approach to explore new perspectives on urban planning and management through the use of case studies. In most of the case studies included in this book, big data is applied to investigate more thoroughly urban phenomena.

I wish for this book to be a platform upon which researchers might share their thoughts on computational and data-driven techniques as they relate to urban planning and design in future China. I would urge the reader to make use of the accounts held within its pages to enter the conversation on the applications and limitations of big data in China's push towards smart and sustainable urbanization.

Prof. Zhiqiang Siegfried WU
Vice President, Tongji University
Vice President, Urban Planning Society of China
Advisor, Commission of Urban Big Data
Chinese Society of Urban Studies

Editorial Review Board Members

Acknowledgment

We would like to express our deep appreciation to all the authors for their outstanding contribution to this book, to series editors of "Advances in Geographic Information Science" Prof. Suzana Dragicevic and Prof. Shivanand Balram of Simon Fraser University for their kind invitation and encouragement, and to Dr. Ron Doering and Dr. Susan Westendorf of Springer for their kind editorial work and help to have such diverse topics on *Big Data Support of Urban Planning and Management: The Experience in China* published as a book.

We are deeply indebted to General Secretary Dr. Nan Shi and his staff from the Urban Planning Society of China and Dr. Xin Yuan and his staff from Beijing Tsinghua Tongheng Urban Planning and Design Institute. Dr. Shi, Dr. Yuan, and their staff invited the first editor of this book, Zhenjiang Shen, to give talks in some symposiums related to smart city construction, such as the 2014 Annual Conference of the Urban Planning Society of China and the 17th Annual Meeting of the China Association for Science and Technology in 2015. In those symposiums, we had shared a wonderful time with our Chinese colleagues, and their stimulating suggestions and encouragement helped us in all the time of considering and editing this book project with Springer.

We also appreciate the support from Dr. Baoxing Qiu, president of the Chinese Society for Urban Studies; Prof. Zhiqiang WU and Prof. Qizhi Mao, senior advisors of Commission of Urban Big Data, Chinese Society of Urban Studies; and Dr. Chenghu Zhou, director of the Commission of Urban Big Data, Chinese Society of Urban Studies. We are enjoying our cooperation with Dr. Ying Long, who is the general secretary, and all members of the Commission of Urban Big Data, Chinese Society for Urban Studies, and thanks sincerely for their assistance and contribution to this book.

Contents

Contributors

Yuheng Cai is a researcher in Innovation Center for Technology, Beijing Tsinghua Tongheng Urban Planning & Design Institute (ICT, THUPDI). She received her master's degree in 2015 at Columbia University. Her research interest includes urban redevelopment, urban economic development, and spatial modeling.

Yanwei Chai is a professor at the College of Urban and Environmental Sciences of Peking University, China. He got the doctor of arts degree from Hiroshima University, Japan, in 1994. He is concurrently the director of the Center for Research and Planning of Smart Cities in the College of Urban and Environmental Sciences of Peking University and director of the Committee of Urban Geography in the Geographical Society of China. He is on the editorial board of the Annals of the American Association of Geographers. His research areas include urban social geography, time geography, and behavioral geography. He is interested in several research topics such as urban spatial structures from an individual activity perspective, space-time behavior in the transitional context of China, and the planning applications of space-time behavior research.

Wei Chen is a senior engineer at SUTPC. He obtained a graduate degree in road and airport engineering in Tongji University in 2010. From then, he has over 5 years' experiences in transport planning, transport policy, and intelligent transport system in Shanghai and Shenzhen.

Xin Chen is the deputy general manager of GISUni company. He has led the Location Intelligence Department in GISUni to provide location analytics and consulting services for business partners like McDonald's, Mercedes-Benz, Walmart, Starbucks, and more. Via applying GIS (geographic information system) technologies to extensive demographic and business data, his teams have delivered on-demand analysis, geo-visualization maps, and ready reports for unaccountable enterprises.

Yanyan Chen is professor at Beijing University of Technology. She received a doctoral degree in structural optimization from Harbin Institute of Technology, Heilongjiang, China, in 1997. Her areas of research include structural optimization, traffic safety, intelligent transportation, traffic planning, public transportation, road guidance, and reliability research. She has visited several universities in India and abroad as a visiting scientist and professor. She has more than 120 paper publications in refereed conferences/journals and edited 5 books. She has served as a permanent member of China Communications and Transportation Association and permanent member of China Logistics Association.

Zifeng Chen is currently a PhD candidate at the Department of Urban Planning and Design in the University of Hong Kong. His research areas include urban social geography and transport geography. He is interested in suburbanization, accessibility/mobility, and social exclusion in urban China.

Qiao Chu is an assistant urban planner at Beijing Tsinghua Tongheng Urban Planning and Design Institute. Her research interests include city-regional planning and development as well as the application of emerging analytics and technologies in urban studies. At present, she mainly focuses on urban metabolism, urban morphology, and data analysis of historical city.

Boshu Cui is the urban planning director of Beijing City Quadrant Technology Co., Ltd. She possesses abundant knowledge and experience within urban planning of both China and the United States with a focus on sustainable development and urban economics. She is currently dedicating herself to solving the issues related to urban planning and management utilizing multisource data leveraged by data science.

Yikang Du is data analyst in GISUni and graduated from the University of Waterloo in Canada with a master of arts degree in geography. His work focuses on leveraging models/data analysis to achieve the success of location-based business. His interests range from spatial analysis to machine learning to business intelligence and more.

Bingruo Duan is a researcher in Innovation Center for Technology, Beijing Tsinghua Tongheng Urban Planning & Design Institute (ICT, THUPDI). He is also a researcher at Columbia China Megacity Lab, Columbia University. His research interests include urban big data analysis, urban data mining, spatial modeling, and machine learning application on urban studies. During his master's program in Columbia University, he participated in several projects involving big data research upon the urbanization issue in China and gentrification research in the USA.

Yuli Fan is currently an undergraduate at the School of Urban Design, Wuhan University, China. His research interests include planning support system, big data application in planning, and geography.

Yong Gao is the deputy president and deputy chief engineer at the Beijing branch of Shenzhen Urban Transport Planning Center. He has taken part in 15 projects in the National High-tech R&D Program of China and 3 projects for Beijing local standard establishment. His main research fields include analysis based on cellular signaling data, individual behavior analysis based on Internet user data, congestion assessment, and analysis of characteristics of road passenger transport by taxi and/or public transport for intelligent transportation system.

Qunying Huang is an assistant professor in the Department of Geography at the University of Wisconsin-Madison. Her fields of expertise include geographic information science (GIScience), cyberinfrastructure, spatiotemporal big data mining, and large-scale environmental modeling and simulation. She is very interested in applying different computing models, such as cluster, grid, GPU, citizen computing, and especially cloud computing, to address contemporary big data and computing challenges in GIScience. She published over 80 scientific articles and edited 3 books. Dr. Huang is a fellow of the Next Generation of Hazards and Disasters Researchers and CyberGIS.

Jianhui Lai received his doctorate degree in transportation planning and management from Beijing University of Technology, Beijing, China, in 2014. His research interests include intelligent transportation and traffic planning. He has more than five paper publications in refereed conferences/journals.

Yuliang Lan graduated from the School of Urban Design, Wuhan University, China. Her research interests include urban heat island effect, urban environment and climate study, and urban thermal comfort planning.

Jay Lee is a professor in the Department of Geography at Kent State University. His interests include relating geographic events and patterns through time and space. This includes quantitatively modeling changing geographic events as diffusing spatial processes. This may be applied to many aspects of our daily life: land use, air pollution, crime, public health, and how all these interact with the environment. Some of his publications and research grants have involved digital elevation models, environmental conservation, GIS, web-based GIS, urban growth, urban sprawl, management of urban growth, and areal health disparities. His recent work includes developing an urban growth simulator, an environmental pollution simulator, and an urban crime simulator. Currently, his team is developing a simulator that uses agent-based models to simulate how neighborhoods develop disparities in public health.

Dong Li is deputy director of the newly formed Innovation Center of Technology at Beijing Tsinghua Tongheng Urban Planning and Design Institute. He received his doctoral degree in ecology at the Research Center for Eco-Environmental Sciences, Chinese Academy of Sciences, and has spent more than 7 years as a senior engineer in the China Academy of Urban Planning and Design. His researches focused on planning projects of environment, infrastructure, and disaster prevention on urban and regional level. In recent years, he adopted the trend of data-driven planning, testing new data and tools for various issues in cities. His works have been published in peer-reviewed journals and conferences. He served as reviewer for *JEMA*, *CEUS*, *IJGIS*, *GeoJournal*, etc. Dr. Li also works as one of the cofounders of the first quantitatively oriented research network Beijing City Lab (BCL).

Juan Li is a PhD candidate of urban planning at the School of Architecture, Tsinghua University. She has published several papers with topics ranging from geodesign and smart community to open data. Her previous research experiences were mostly about quantitative analysis applying emerging new data to support urban planning.

Miaoyi Li served as a researcher in the Information Center of Beijing Tsinghua Tongheng Planning and Design Institute which is affiliated with the School of Architecture, Tsinghua University. He received a PhD degree from the Graduate School of Environmental Design, Kanazawa University. His research interest includes spatial/urban planning, urban big data analysis, geospatial analysis, and geo-simulations, and he published many research papers in Chinese top journals.

Shengwen Li is an associate professor in the Department of Information Engineering, China University of Geosciences, Wuhan, China. He was a visiting scholar at the Department of Geography, Kent State University. His research interests are in big space-time data analytics, social network and flow data mining, and GIS software engineering. He received his PhD in geography from the China University of Geosciences. Dr. Li has publications in *Applied Spatial Analysis and Policy*, *Geographical Analysis*, the *International Journal of Geographic Information Science*, and the *ISPRS International Journal of Geo-Information*.

Tao Li is the director of the Department of Urban Planning in Urban Development Institute at Fudan University. He received his PhD in urban planning from Tongji University and is now a Post PhD of public administration at Fudan University. He is a national registered urban planner and urban planning evaluation expert in Shanghai. He is dedicated to the regional and urban planning research for a long time. He presided over or participated in more than one hundred urban planning projects and won 1 association award, 1 provincial award as well as 2 first prizes in Tongji Urban Planning and Design Institute. His research areas are urban and regional economy, urban system, and city network.

Wenwen Li is an assistant professor in the School of Geographical Sciences and Urban Planning at Arizona State University. Her research interest is geographic information science with a focus on cyberinfrastructure, big data, semantic interoperability, spatial information retrieval, and distributed geospatial information processing. Li's research aims to develop integrated, sustainable, and smart cyberinfrastructure to revolutionize knowledge discovery in data and computational intensive geographical sciences. She is the chair of the Association of American

Geographers' Cyberinfrastructure Specialty Group from 2013 to 2014 and a member of the Spatial Decision Support Consortium at the University of the Redlands. Before joining Arizona State University, she worked at the Center for Spatial Studies at the University of California, Santa Barbara.

Wenqi Lin is a professor of urban planning in the School of Architecture, Tsinghua University, responsible for a branch of Key Laboratory of Urban-Rural Eco Planning and Green Building of the Ministry of Education. He is the chief planner of Beijing Tsinghua Tongheng Urban Planning and Design Institute and has been working in research on sustainable city and urbanization. His recent book is *Change-Space-Scenario* (Tsinghua University Press, 2013).

Liu Liu achieved the master's degree in city planning in MIT in 2014. Currently, he is the CEO and cofounder of a startup named CitoryTech. His team is focusing on urban imagery research, which is a combination of LBS analysis and computational vision. He conducted AFC data mining and visualization while interned as a data researcher in MTA in 2013. He opened up the project named "C-IMAGE" using geo-tagged photos and deep learning to study city image. During his stay in the China Academy of Urban Planning and Design, he led the big data section in national urban system planning based on LBSN data from Tencent, he developed the method to detect mixed city districts with evaluated POIs, and as the team leader he built up a project of StreeTalk which was awarded in SODA.

Yungang Liu received his PhD (multidisciplinary sciences) in human geography at the Graduate School of Arts and Science, The University of Tokyo. Now, he is a professor of the Department of Urban and Regional Planning, Sun Yat-sen University, teaching on urban geography, political geography, and geographical thought. He conducted research on the migration, territory, and border politics in urban China, especially research on the activity space of Japanese expatriates in China, and the territorial urbanization in China. He is also the chairperson of the Working Committee for Young Geographers and vice chair of the Human Geography Professional Committee of the Geographical Society of China. He is a registered urban planner in China.

Zhuqing Liu is working as an urban planner in the China Academy of Urban Planning and Design for urban strategy planning, master planning, urban design, etc. She is also a member of the International Society of City and Regional Planners, participating in a large number of international urban planning conferences, workshops, and activities.

Ying Long is now an associate professor in the School of Architecture, Tsinghua University, China. His research focuses on urban planning, quantitative urban studies, and applied urban modeling. He has an education background of both environmental engineering and city planning. Before he joined Tsinghua University, he has been working for Beijing Municipal Institute of City Planning & Design as a senior planner for 11 years. Familiar with planning practices in China and versed in the international literature, Dr. Long's academic studies creatively integrate international methods and experiences with local planning practices. He has published over one hundred journal papers and led over 20 research/planning projects. Dr. Long is also the founder of Beijing City Lab (BCL, www.beijingcitylab.org), an open research network for quantitative urban studies. More information is available at http://www.beijingcitylab.com/longy.

Jia Lu graduated from Nanjing University, majoring in geographic information system. After graduation, she worked for the Zhejiang Provincial Surveying and Mapping and Geographic Information Bureau. In 2003, she joined Shanghai Digital Intelligence System Technology Company, specializing in urban planning and land resources management information construction. She takes charge of COE (Center of Excellence) and is responsible for technical cooperation with IBM, Amazon, and other IT companies. She led the IBM development team for large data analysis direction, and related results were repeatedly introduced at the IBM conference presentation.

Xiao Luo is an associate research director of Shanghai Tongji Urban Planning & Design Institute, Shanghai, China. He got a PhD degree in Nagoya University in 2013. His publications include 14 papers on low-carbon transport policy, renewable energy policy, urban structure, big data utilization in urban planning, etc. He also served as anonymous reviewer of many international journals, such as the *Journal of Cleaner Production*; the *Journal of Regional Science*; *Resources, Conservation and Recycling*; the *Journal of Renewable and Sustainable Energy*; etc.

Liang Ma achieved the master's degree in science in transport with business management at Imperial College London in 2014 and has joined the Innovation Center of Technology at Beijing Tsinghua Tongheng Urban Planning and Designing Institute since 2015. As an urban planner and transport planner, her main research field is big data application in transport planning and urban planning. These include quantitative studies on job-housing balance, spatiotemporal individual behaviors, public transport evaluation in fine granularity, disaggregate transport mode choice, relationship between the road traffic and land use pattern, model establishment with advanced learning methods, etc.

Yaotian Ma is graduate student from the Department of Urban Planning, Tsinghua University. His research interests include land use development and low-carbon urban agglomeration by using planning support systems such as GIS and other computer-aided planning tools.

Mingrui Mao is the founder of Beijing City Quadrant Technology Co., Ltd., as well as the secretary general of the Cloud Platform Innovation Center at the Beijing Municipal Institute of City Planning and Design. He also acts as the vice secretary general of the Urban Big Data Committee of the Chinese Society for Urban Studies. After 16 years with the Beijing Municipal Institute of City Planning and Design, he took the initiative of being responsible for innovative business at the institute in 2016, thus building City Quadrant. The company has a galaxy of city data science expert professionals with multidisciplinary backgrounds, focusing on urban management and city renewal. The company has constructed a city analysis platform covering almost all cities across China with hundreds of millions of people's spatiotemporal behavior.

Jiaju Miao received BA degree in logistics management from Shanghai Ocean University and MA degree in population, resources and environmental economics from Fudan University. He is a PhD candidate in economics at Stony Brook University. His current research interests include health economics, industrial organization, and applied econometrics.

Wanlu Ouyang is currently a graduate student at the School of Urban Design, Wuhan University, China. Since she has an interdisciplinary background, her research interests lie in the intersection of geography, spatial analysis, and urban environmental planning.

Hongbin Pei is a PhD student in the College of Computer Science and Technology, Jilin University. He also received the MS and BS degrees from Jilin University. His research lies in the field of data mining, machine learning, focusing on spatiotemporal data mining and social network analysis with applications for health informatics. In 2015, he worked as an intern in Big Data Lab, Baidu Research. He then visited the Department of Computer Science at Hong Kong Baptist University as a research assistant during 2016. His research has been published in journals, such as *TPAMI*. His research project on "Imported malaria risk estimation and control" has been deployed by the China CDC for malaria elimination in Yunnan province.

Jiandong Qiu is director of Shenzhen Urban Transport Planning Center Co., Ltd. He has 10 years' experiences in transport big data modeling, simulation, GIS database, and multimodal traffic demand modeling. More than 50 projects are achieved in China, Korea, Dubai, Australia, and Saudi Arabia. He obtained graduate degrees in South China University of Technology and Beijing Institute of Technology.

Yi Shi is an assistant professor in the Department of Urban Planning, School of Architecture, Southeast University, China. He worked as a visiting scholar in the Department of Geography and GIS, University of Illinois at Urbana-Champaign, USA. He obtained his PhD and MSc in urban planning at Southeast University. As an urban planner and researcher, he has published over 20 journal papers. His research focuses on applying space-time geography method and big data to urban morphology research.

Zhenjiang Shen is a professor in Kanazawa University and is a visiting professor in Fuzhou University, Chongqing University, and Fudan University, and he has also served as an associate instructor for the PhD education program in Tsinghua

University, China. His research interest includes policy-making support system for planning and design using GIS and VR. He is now serving as academic commissioner of the City Planning Institute of Japan and vice director of the Urban Big Data Committee of the Chinese Society for Urban Studies. He is now editor in chief of *IRSPSD International* and managing editor of *IJSSoc* and *IJSSS*. He is the founder of International Community on Spatial Planning and Sustainable Development (SPSD, http://www.spsdcommunity.org).

Jia Tang is a PhD candidate at the School of Architecture and Urban Planning in Nanjing University, Nanjing, China. She is mainly interested in ICT and travel behavior, along with big data and urban study.

Xiaoyong Tang is a senior engineer in transportation planning at Chongqing Transport Planning Institute and also the vice director of Chongqing Engineering Research Center for Big Data Analysis and Application in Urban Traffic. His research interests are the application of mobile phone data to topics in urban planning such as trip characteristic and OD flow extracting, land use planning, and layout planning of transportation infrastructures. His first PhD degree was in transportation engineering at Southeast University, China.

De Wang is a professor of the College of Architecture and Urban Planning, Tongji University, a leading professor in the Division of Planning Methodology and Technology, the director of Big Data and Urban Spatial Analytics LAB, and the director of the Digital Planning Technology Research Center in Shanghai Tongji Urban Planning and Design Institute. He graduated from the Department of Geography, Nanjing University, in 1983, and obtained a PhD in geography from Nagoya University in 1994. He had worked as UNCRD researcher from 1994 to 1998 and as associate professor from 1998 and as professor from 2003 in Tongji University. He has obtained 3 National Natural Science Fund Projects, published over 120 papers, and won the Jin Jingchang Excellence Paper Awards 4 times, 6 scientific research awards at the provincial and ministerial level, and 7 planning and design provincial excellence awards. His main research directions are urban and regional planning, urban planning methodology, big data, and space and behavior models. Professor Wang is the editorial board member of two academic journals in China, *Urban Planning Forum* and *Urban Planning International*. He is the vice director of the Committee on Population Geography board in the Geographical Society of China, the vice director of the Urban Big Data Committee of the Chinese Society for Urban Studies, a committee member of foreign countries planning board in the Urban Planning Society of China, and an Urban Planning Committee member of Shanghai, Jiaxing, and Zhenjiang in China. And he is team leader of population analysis for the new master plan of Shanghai (2040) and Wuhan (2030).

Weili Wang has many years of experience dealing with spatial big data and related business intelligence products. She leads a team in GISUni in order to provide the data products for local governments and companies, by conducting better spatial data visualization and analysis of location-based business.

Haishan Wu is a senior data scientist in Big Data Lab of Baidu Research. He got his PhD from Computer Science Department of Fudan University in China in 2011. He then joined IBM Research focusing on business data mining and analytics. Since 2012, he worked in Princeton University as a postdoc researcher. He joined Baidu in 2014 and leads a spatialtemporal data mining group. His research has been widely published in *MIT Technology Review, New Scientist, Communication of ACM, The Economist, Wall Street Journal, Bloomberg, BusinessWeek, Forbes, CNN Money, Washington Post, NPR*, and so on.

Ling Wu is an assistant professor in the Department of Sociology at Kent State University. She is a criminologist and policing scientist by training. Her research interests are in victimology, crime mapping, environmental criminology, policing, courts, and comparative criminal justice. She has been working with police departments on crime analysis and crime prevention, as well as using geographic information systems and spatiotemporal analysis tools in policing information management systems. Her ongoing studies include victimology, legal research in criminal justice, crime patterns and mechanisms, as well as crime prevention and policing strategy.

Menghe Wu is a researcher in the Innovation Center for Technology, Beijing Tsinghua Tongheng Urban Planning & Design Institute (ICT, THUPDI). She works on data analysis and analytic application development related to urban issues. She holds a master's degree in urban and regional planning from Peking University, Beijing, and a bachelor's degree in landscape architecture from Tongji University, Shanghai.

Nawei Wu is an urban planner in the Innovation Center of Technology, affiliated to Beijing Tsinghua Tongheng Urban Planning and Design Institute, and focuses on data-driven urban studies. She has a bachelor's degree in architecture with a double degree in economics at Tsinghua University and an MPhil degree in urban planning, growth, and regeneration, awarded by the Department of Land Economy, Cambridge University. She received her doctorate degree in engineering in urban planning at Tsinghua University in 2016. She has more than ten publications in journals, conferences, and books, and she is also engaged in several programs supported by the National Natural Science Foundation of China and the National Key Technology R&D Program.

Yinghui Xiao is currently an associate professor at the School of Urban Design, Wuhan University. She obtained an MSc in urban planning and land administration from the International Institute for Geo-Information Science and Earth Observation (ITC), the Netherlands, in 2001. She has been involved in more than 30 scientific research and planning projects since 1995. Her research interests include GIS, remote sensing applications in urban planning and management, planning support systems, land cover and land use classification, etc. She is the author and coauthor of over 40 scientific articles and 3 books.

Dongcan Xie is a postgraduate student in the College of Architecture and Urban Planning, Tongji University, majoring in urban and rural planning. His research interests include mobile signaling data analysis, spatiotemporal behavior analysis, and application of new technology in urban planning. He published a number of papers on mobile signaling data analysis, such as commercial center analysis, population migration analysis, and built environment assessment.

Fangcao Xu is the previous data scientist for site selection and market consulting in the Location Intelligence Department of GISUni. She obtained her master's degree in urban spatial analytics at the University of Pennsylvania in 2015. She is currently pursuing her PhD at the Penn State University.

Junyan Yang is a professor in the Department of Urban Planning, School of Architecture, Southeast University, China. He is the director of the Urban Center Institute of Southeast University, a winner of the China Urban Planning Science and Technology Youth Award, and a member of the Urban Planning Society of China Academic Committee. His research focuses on urban centers and urban spatial big data. He has published over one hundred journal papers and led over 30 research/planning projects.

Xining Yang is an assistant professor in the Department of Geography and Geology at Eastern Michigan University. Before joining EMU, he was an assistant professor at California State University, Chico. His research interests are in spatiotemporal analysis, social media, volunteered geographic information (VGI), and web GIS. He received his PhD in geography from the Ohio State University in 2015. He has scientific publications in *Applied Geography*, *ISPRS International Journal of Geo-Information*, *International Journal of Applied Geospatial Research*, and *Urban Remote Sensing*. His doctoral dissertation was featured in Columbus Monthly as one of the eight fascinating OSU doctoral dissertations in 2015.

Xinyue Ye is an associate professor in the Department of Geography at Kent State University. His major expertise is on modeling the geographical perspective of socioeconomic inequality and human dynamics. He develops and implements new methods on spatiotemporal-social network analysis/modeling/simulation for different application domains such as economic development, disaster response, land use, public health, and urban crime. His work won the national first place award of "research and analysis" from the US University Economic Development Association in 2011 and received the emerging scholar award from AAG's Regional Development and Planning Specialty Group in 2012. He has received about 4 million dollars in grants as PI or co-PI from the National Science Foundation, Department of Commerce, and Department of Energy. His open-source toolbox implementation expertise has been further utilized through combining the elements of space, time, network, and semantics. He has been very active in the dissemination of new metrics for the broader research community to facilitate the dialogue among social scientists, policy-makers, and computational scientists.

Zhenxuan Yin is an assistant researcher at the Digital Planning Technology Research Center in Shanghai Tongji Urban Planning and Design Institute. He obtained master of community planning from the Department of DAAP, University of Cincinnati, in 2015, and his research focuses on urban planning methodology and big data.

Guang Yuan is studying at Beijing University of Technology. His research interests include intelligent transportation and traffic planning.

Qingming Zhan is a professor at School of Urban Design and director of Research Center for Digital City, Wuhan University. He received his BSc and MSc from Wuhan Technical University of Surveying and Mapping (WTUSM), China, and PhD from Wageningen University – ITC, the Netherlands. He has been involved in more than 70 scientific research projects since 1985, and has been PI or Co-PI of research projects from the Natural Science Foundation of China (NSFC), the Hi-Tech Research and Development Program of China (863 Project), the National Science and Technology Support Program funded by the Ministry of Science and Technology of China, the State Bureau of Surveying and Mapping of China, UNDP, World Bank, EU Framework 7, the Netherlands development aid fund, etc. His research interests include GIS, remote sensing and big data applications in urban planning and management, planning support systems, wind path planning, 3D city modeling and applications, object-based analysis of remote sensing images, land-cover and land-use classification methods, etc. He has supervised 20 PhD students and more than 110 MSc students. He is the author and co-author of over 170 scientific articles and 6 books.

Yina Zhang is currently an associate professor in the School of Social Development and Public Policy at Fudan University. She is the secretary-general of Urban Development Institute (UDI) at Fudan. Her current works are published in English journals such as *Habitat International*. Her research areas are urban and regional economy, population migration and urbanization, and application and development of geographic information system (GIS).

Feng Zhen is a professor and doctoral supervisor at the School of Architecture and Urban Planning in Nanjing University, Nanjing, China. He earned his PhD degree in science from Nanjing University in July 2001. His major research covers information geography, big data and urban study, smart city, and ICT and behavior. He is the author and coauthor of more than 300 scientific publications in core journals in both Chinese and English. Among them, the amount of SCI(E)\SSCI publications has reached more than ten. The monographs that he completed or participated in include *Urban Studies and Innovation in Urban Planning Methods in Big Data Era*, *Regional Spatial Structure in the Information Era*, *The Economics of Urban Planning*, etc. He has hosted many important projects supported by the National Natural Science Foundation of China (NSFC), including "The Influence of ICT on Process and Mechanism of Urban Network Formation: A Case Study of the Yangtze

River Delta," "Research on Changes of China's Urban Employment and Living Space in the Information Era," etc. Now, he is in charge of a new project titled "The Influence of Online Activity on Urban Space (2016–2019)" supported by NSFC. Moreover, Prof. Zhen has also hosted and participated in more than 50 ministerial and provincial research projects, mainly including "Method Research of Urban Spatial Planning in the Big Data Era," "Methods and Technology System of Urban Planning in the Big Data Era," etc.

Weijing Zhong is a postgraduate student in the Department of Architecture and Urban Planning in Tongji University, majoring in urban and rural planning. Her research interests include big data analytics, urban dynamics analysis, and spatio-temporal data mining. He publishes some refereed papers as a leading author at prestigious journals such as *Urban Planning Forum*, *Modern Urban Research*, and *Geographical Research*. She was awarded the Outstanding Students and National Scholarship in 2016, the Excellent Postgraduate Student Scholarship in 2015, and the Outstanding Graduates and the best thesis at Nanjing University in 2014.

Jingbo Zhou is a data scientist in Big Data Lab of Baidu Research, working on machine learning and data mining problems for both scientific research and business applications. Currently, he focuses on the applications of machine learning in user behavior prediction, spatiotemporal data analysis, and urban computing. He obtained his PhD degree from National University of Singapore in 2014, and BE degree from Shandong University in 2009. Prior to joining Baidu Research, he worked as a research fellow in the Interactive & Digital Media Institute (IDMI) of the National University of Singapore in 2015. He has published several papers in the top venues, such as SIGMOD, KDD, VLDB, TKDE, and AAAI.

Yi Zhu is an assistant professor in the Department of International Shipping and Transportation Engineering at Shanghai Jiao Tong University. He received the PhD degree in urban studies and planning from MIT, Cambridge, MA, in 2014. From 2014 to 2015, he was a research associate with Singapore-MIT Alliance for Research and Technology (SMART). Since 2016, he has been an assistant professor with the Department of International Shipping and Transportation Engineering, Shanghai Jiao Tong University, Shanghai, China. His research interests include urban modeling, transportation planning, and urban information system. In recent years, he has focused on better techniques and approaches in processing, exploring, and visualizing big data to understand the longitudinal evolution of urban dynamics, so as to support planning and decision-making processes.

Chapter 1
Overview: Big Data Support for Urban Planning and Management in China

Zhenjiang Shen

Abstract This book is a collection of up-to-date practice and research using big data in the field of urban and regional planning and smart city. It aims to draw a picture of the current research status of big data application in urban China; most of the works focus on spatiotemporal analysis of human mobility and transportation for understanding relevant urban structure after the rapid urbanization process in China. The case studies in this book will explain how big data changes the perception of planners and researchers in urban planning practice. It is the first book to introduce big data-based research for planning practices in China.

Keywords Social big data • Location-based data • Mobile device data • Social network system • Cyber infrastructure and urban structure

Big data is collected from sensors of Internet of Things (IoT) in smart cities, which is being generated by everything around us at all times. Conventional survey data is an important foundation for urban planning and management, which can be defined as indicator system for feasibility study by measuring and monitoring planning implementation. Planning indicators can be formulated as objectives of local planning authority to make planning decisions and review planning application. In planning process, from monitoring and planning to implementation, statistic database is conventionally employed for developing urban model in order to analyze current situation, allocate economic demand, seek solutions of planning issues such as conflicts between land use and transportation, and establish indicator system as standards for monitoring, planning, and implementation. Big data is a new kind of data source for urban planning and management because the new era of smart city construction is based on information and communications technology (ICT) and IoT. In

Z. Shen (✉)
Joint International FZUKU Lab SPSD, Fuzhou University, Fuzhou City, China

Kanazawa University, Kanazawa City, Japan
e-mail: shenzhe@se.kanazawa-u.ac.jp

Z. Shen, M. Li (eds.), *Big Data Support of Urban Planning and Management*,
Advances in Geographic Information Science, DOI 10.1007/978-3-319-51929-6_1

the meantime, new emerging possibilities of lifestyle based on ICT and IoT are explored in the world, and big data is a tool for understanding the change of lifestyle and managing the new life of human beings.

In this book, we are interested in how big data could be employed for urban planning and management in China. What kinds of big data and what kinds of application in monitoring, planning, and management could be found in China? In this book, we provide various kinds of solutions to the issue on how to support planning practice and urban management with big data in China, which presents an adequate collection of case studies from varied institutions and companies.

1.1 Smart Infrastructure and Big Data

Smart city becomes a new trend in urban construction, that is, to develop smart infrastructure for the new lifestyle with environmentalism, creativity, and knowledge, which is transformed by the use of IoT technologies and big data.

Regarding smart infrastructure, Anthony M. Townsend (2014) gives a thorough introduction of how modern cities around the world are upgrading their infrastructure to quest for a new utopia, namely, smart city in the Internet age of generating big data. For smart infrastructure construction of cities in the world such as New York and Beijing, "city mayors are partnering with organizations like Siemens and IBM to strengthen networks, communications, and crisis-intervention tools such as monitoring flu outbreaks." Smart infrastructure construction satisfies new demands for the application of innovative and forward-looking ICT and IoT sectors across business processes, production, and services. Therefore, big data is a kind of information production that is stored in smart city database system for processing real-time security, transportation, and other information in an efficient manner. Moreover, big data based on smart infrastructure such as smart grid system can monitor energy consumption opportunely for urban energy management.

Big data serves as a database for smart living, which is possible to formulate a new kind of social services in real society. Stephen Goldsmith and Susan Crawford (2014) examine digital engagement and governance in the digital age with case studies highlighting the work of pioneers in New York, Boston, Chicago, etc. As mentioned by Goldsmith and Crawford, data-smart governance is a response to smart city construction for engaging communities. Offenhuber and Ratti (2014) use examples from MIT Senseable City Lab to illustrate the impact of real-time big data on architecture and urban planning.

Smart city database contains big data collected from Internet system such as mobile devices. Thereby big data can be utilized for monitoring human mobility via mobile devices. Big data retrieved from social network system (SNS) can reflect human activities too. As discussed by Dietmar Offenhuber and Carlo Ratti (2014), big data is expected to decode the city in the Internet age, and planners are supposed to explain the way, namely, smart living, how people work and live in the city within big data environment. In our book, we will give readers a deeper understanding of how to monitor human mobility and implement the ideas using big data for the specific planning practice in each case study in Chinese context and technical details.

Unlike the books by Goldsmith and Crawford, Offenhuber and Ratti, and Anthony M. Townsend, focusing on the experience in the development of urbanization in China related to big data analysis for urban planning and management, this book shows the contribution distinction in terms of its research scope and selection of cases.

1.1.1 Smart City in China

By using new smart database services provided by ICT and IoT sectors, planners in China are paying attention to exploring solutions of urban issues. To better understand the background of this book, here we compare the definition of smart city construction in China with that in other countries. Regarding the emerging concept of smart city, there are many other terms that have been used for similar concepts including digital city, electronic communities, intelligent city, knowledge-based city, ubiquitous city, and so on. The following is the definition of smart city in Wikipedia:

> *A smart city is an urban development vision to integrate multiple information and communication technology (ICT) and Internet of Things (IoT) solutions in a secure fashion to manage a city's assets – the city's assets include, but are not limited to, local departments' information systems, schools, libraries, transportation systems, hospitals, power plants, water supply networks, waste management, law enforcement, and other community services.* (Cited from https://en.wikipedia.org/wiki/Smart_city 2017.1.21)

Even though smart city has different concepts in the different countries, the above definition might match that in China. In this book, it is not our task to give a correct definition for smart city, which is varied within the literature. In China, smart city construction has become a new trend of urban development since IBM brought the concept of "smarter planet" in China in 2009. Until 2013, 193 pilot projects of smart cities have been approved by the Ministry of Housing and Urban-Rural Development in China. The rapid development of smart cities in China is largely attributed to support of both of the Ministry of Industry and Information Technology and the Ministry of Housing and Urban-Rural Development. In China, the National Development and Reform Commission formulates a policy on new smart city construction in 2016 in order to integrate the accomplishment of the two ministries.

In Japan, based on the three themes "Environmental-Symbiotic City," "City of Health and Longevity," and "City of New Industry Creation," smart city is considered as a place where people of all age groups, including both children and the elderly, can lead their lives safely and with ease of mind. However, smart city in

Japan is actually focusing on energy management system for decreasing CO_2 emission. "Increases in energy consumption are resulting in increases in CO_2 emissions, making it necessary to build an environment-friendly, low-carbon city called a Smart City" (Smart City project, 2017). The Japanese government believes that the size of the world's smart city market is expected to reach a cumulative total of as much as US$ 42.72 trillion over the 20 years from 2010 to 2030. For this reason, a high priority in Japanese national strategies is given to smart city industry. Smart city project has been launched by 27 leading enterprises since 2009 for the purpose of developing a smart city. Smart city is defined as five layers: real estate development (first layer), basic infrastructure (second layer), smart infrastructure (third layer), life services (fourth layer), and lifestyles, culture, and art (fifth layer). Kashiwa-no-ha Smart City Project has been developed as an ambitious Japanese model project to the world.

In Europe, the concept is wider than Japan. That is because the concept of smart city in European countries is divided into six main areas: smart living, smart governance, smart economy, smart environment, smart people, and smart mobility (Giffinger et al. 2007). There are following six initiatives for establishing EU innovative partnership: Sustainable Districts and Built Environment Action; Sustainable Urban Mobility Action; Integrated Infrastructures and Processes Action; Citizen Focus Action; Business Models, Finance and Procurement Action; and Integrated Planning, Policy and Regulation Action (EIPSCC 2017). Developing smart cities is not only a technological challenge but also a regulatory, financial, and social one (ESICS 2017).

In North America, Smart America Challenge (Obama White House, 2017) is a White House Presidential Innovation Fellows project with the goal to bring together research in cyber-physical systems (CPS) and to combine projects and activities from different sectors, such as smart manufacturing, healthcare, smart energy, intelligent transportation, and disaster response. CPS is a name for the combination of the Internet of Things and system control. Intel and the city of San Jose, CA, are collaborating on a public-private partnership project titled Smart Cities USA (2017). This will help San Jose drive economic growth, create environmental sustainability, and enhance the quality of life by installing a network of air quality, sound, and microclimate sensors, namely, "sustainability lens" for the city. The "sustainability lens" uses IoT technology to measure characteristics such as particulates in the air, noise pollution, and traffic flow. City management will use this information to demonstrate how cities can use IoT to measure and act on critical data to improve life for citizens as Smart City Demonstration Platform.

This book will study how big data generated from smart infrastructure has been used to find solutions of urban issues, particularly to find the reasons why unreasonable urban structures have been shaped in the urbanization process based on human mobility dataset for improving urban planning and management in China. Comparing with the planning practices in Japan, Europe, and the USA, planners in China pay particularly more attention to monitoring human mobility for understanding citizens' activities in urban area and finding solutions of urban issues, such as traffic crowd in Chinese cities in order to encourage sustainable urban form.

There are different data sources of big data such as social media, points of interest (POIs), and location-based data, which are introduced to establish cyber infrastructure for smart governance. This book will not discuss on the smart infrastructure construction in China; instead, we summarize Chinese experiences of big data application on urban planning and management in order to give the visions of sustainable urban form in China, which develop smart city planning and smart governance.

1.1.2 Big Data for Planning and Management in China

Big data is collected from sensors of IoT, which are ICT products. If there appears deficient smart infrastructure construction, the big data analysis could not be achieved, and it is not possible to utilize big data to support urban planning and management. In Western countries, smart city construction enables the cutting-edge intelligent technology to help cities use resources more efficiently, thereby improving quality of the air and water, energy and transportation, and communication systems. In China, smart cities promoted by the Ministry of Industry and Information Technology are more focused on the technological issues of smart infrastructure, and one objective of pilot projects launched by the Ministry of Housing and Urban-Rural Development is to use big data for sustainable urban form in the process of urbanization, as aspired outcomes by smart city planning and smart governance.

We believed that smart infrastructure construction will bring the changing of lifestyle not only in Chinese cities but also other cities in the world. With huge investments on smart infrastructure, the IoT will transform citizens' lifestyle and ameliorate their quality of life in cities. Smart infrastructure construction is an integrated platform of urban development with ICT products including sensors, cloud computing, and related facilities. Those technologies make it possible to communicate between things and things, people and things, and even people and society to realize smart governance. For the escalating demands for smart living, planners are looking at ways to become "smarter" and more flexible in response to future inhabitants on many aspects such as living, working, driving, and interacting with each other. The urban services provided by IoT have Internet connection with all inhabitants in urban area, for which urban design might face the challenge that lifestyle would be transformed by smart infrastructure construction in the entire city.

On the other hand, because of shortage of survey data and difficulties in obtaining survey data from government, big data from monitoring sensors became a substitutable data of regular survey for current situation analysis in China. For example, big data collected from cell phone and GPS devices reflect spatiotemporal characteristic of human mobility, whereby urban issues can be monitored by big data more easily than by regular survey. Thus, planners and researchers are looking for methods to collect big data using social big data of social network system (SNS) and sensors of ICT. Moreover, CPS for urban planning based on smart infrastructure is developed in China in order to support advanced data acquisition, data storage, data management, data integration, data mining, data visualization, and other computing and information processing services distributed over the Internet.

Chinese planners are excited by the opportunity of using big data for urban planning and management. For examples, because IoT smart sensors are distributed in citywide, it is possible to monitor the city situation in real time for urban management. Because of shortage of survey data, planners are using mobile phone tracking to explore planning issues in urban areas. Different from conventional survey database, big data are spatial and temporal records that can be used to monitor human mobility and explore the relationship between urban structure and human behaviors in real time. Thus in practice, Chinese planners engage in developing a computer platform, namely, cyber infrastructure in local government for collecting data from ICT sensors in order to make spatial and temporal analysis of urban issues for decision making.

Researchers as contributors to this book have accomplished their projects by using big data and relevant data mining technologies for exploring the possibilities of big data, such as cell phone, social network system, and transportation cards instead of conventional survey data for urban planning support. The book is a chance for researchers to share their experience and ideas with each other on human mobility, connectivity and recognition of places, urban structure, and accessibility of transportation modes in order to provide effective analysis and forecasting tools for urban planning and design solutions in China.

1.2 Part 1: Social Big Data for Exploring Human Behaviors and Urban Structure

Social big data from SNS are social media data, and information on users' activities and researchers can be gathered from cloud database in real time. For example, locations of SNS—users' visiting places, namely, check-in spots, comments, photos, and other media data—can be retrieved from the Internet if the application programming interface (API) of the SNS system is an open API. It is possible for SNS users to check in the system with comments anywhere and anytime if their mobile devices are connecting to the Internet. In China, 11.47 billion people are users of mobile devices; therefore, SNS users' information is one of the important data sources for planning, which reflect the moving of population in China.

If the API of a location-based social network system can be used publicly, planners and researchers can retrieve location-based service (LBS) data, such as message, photo, and location information from SNS cloud database. The SNS information are useful for social big data mining, e.g., exploring the relationship of human activities and urban structure and investigating meaning of urban space, social map of urban social structure, and so on. Baidu map by Baidu Inc. and Weibo by Sina Corporation provide an open LBS cloud-computing system. Weibo is popular in China because check-in information are social big data that deeply reflect the change of spatiotemporal patterns of human behavior in real time. On the other hand, people without mobile devices are not possible to be found in SNS system, and thereby it is not possible for us to investigate all phenomena driven by human mobility from SNS. Thus, the social big data from SNS is, to some extent, supplementary to conventional national survey data from a view of spatiotemporal analysis.

In this part, six chapters focus on application of SNS data for considering the possibilities of application on urban planning and management. *Early Warning of Human Crowds Based on Query Data from Baidu Map: A Lesson from Shanghai Stampede* by Wu, Pei, and Zhou in Chap. 2 takes a recent impressive stampede tragedy in Shanghai in 2014 as case study, where 36 people died and 49 were injured in celebration of the New Year's Eve. Due to the innately stochastic and complicated individual movement, it is not easy to predict collective gatherings, which can potentially lead to crowd events. In this chapter, with applying the social big data generated on Baidu map, Wu et al. propose a novel approach to early warning such potential crowd disasters. Based on careful analysis and deep investigation on the Baidu map data, it is able to invoke warnings for potential crowd events about 13 h in advance. A sophisticated machine learning model with heterogeneous data (such as query data and positioning data) is proposed to quantitatively measure the risk of the potential crowd disasters.

"Research on Spatial Distribution Characteristics of Residents' Emotions Based on Sina Weibo Big Data: A Case Study of Nanjing" in Chap. 3 is contributed by Zhen, Tang, and Chen. The city of Nanjing has been taken as case study in this study; Sina Weibo big data has been used to extract real-time emotions and space of residents, along with geo-location-based corresponding information. This study aims to analyze the spatial distribution characteristics of residents' emotions in the overall Nanjing and in different places by using ArcGIS software with the grid statistics and analysis method, which is a bid to provide the evidence for the optimization of urban space development. Urban planning should follow the idea of being people-oriented. With the fast development of cities, more attention needs to be focused on the perception and experiences towards space by residents.

In Chap. 4, "Measuring by Movements: Hierarchical Clustering of Cities in China Based on Aggregated Massive Positioning Data" by Li, Wu, Duan, and Cai is discussed. Given that the context of communicating technologies such as high-speed railways (physical) as well as social media over the Internet (nonphysical) is changing, the concept of "city as a system" has been changed to "systems of cities." In this study, Li et al. adopt a new type of aggregated positioning data, which are social big data retrieved from massive Internet users' migration data of the three companies, Baidu, Tencent, and 360 in China to explore the spatial patterns of cities prior to the Spring Festival in 2015. By introducing new clustering algorithm on spatial constraints, models output hierarchic results with various regional zones containing different numbers of cities. Nevertheless, the very differences suggest some hidden forces which make cities be connected intensely across the administrative boundaries such as sharing mutual regional cultures or employment markets.

These facts grounded for a general picture for the study on polycentric urban regions over the whole national territory.

In Chap. 5, "Assessment of Regional Economic Integration Based on Relational Data: The Case of the Yangtze River Delta," Li, Miao, and Zhang suggested the regional economic integration index for measuring the degree of regional economic integration of the main cities. In the regional economic integration process, it is the relation between cities instead of the hierarchy system that forms the regional urban system. Based on the database of the Baidu Search Index, the information-based network is analyzed to characterize the flow of information. Based on the railway database, the transportation-based network is analyzed to characterize the flow of mobility. Based on the database of company registration information from China, the State Administration for Industry and Commerce, the firm-based network is analyzed to characterize the flow of capital and cargo in Shanghai Chang Sanjiao regional area. These three city networks are then compared and combined using multivariate analysis and interlocking network approaches.

In Chap. 6, "The Recognition of CAZ in Shanghai Based on Evaluated POI," Liu and Liu argue that the upgrade of urban core from CBD to Central Activities Zone (CAZ) becomes a trend of many cities in China. This chapter proposes a relatively clear definition of CAZ through a comparison with CBD and then presents a method to identify the potential CAZ based on evaluated DAZHONGDIANPING dataset. Taking Shanghai as an example, a data process is proposed to calibrate the degree of functional mixture on aspects of retail, catering, office, leisure, culture, and recreation. Through the compilation of rasterized maps, the top 14 most dynamic places in downtown Shanghai are investigated, and three scenarios have been defined for the discussion of CAZ boundary.

"The Fear of Ebola: A Tale of Two Cities in China," Chap. 7, is contributed by Ye, Li, Yang, Lee, and Wu. Emerging social issues have often led to rumor breeding and propagation in social media in China. In this chapter, we use the diffusion of Ebola rumors in social media networks as a case study. The topic of rumors is identified based on Latent Dirichlet allocation method, and the diffusion process is explored using the space-time methods. By comparing Ebola rumors in the two cities, the chapter explores the relationship between the spread of rumors, user factors, and contents. Public health-related rumors will harm social stability, and such noise detrimentally affects the quality of disease outbreak detection and prediction based on social media.

In this part, social big data, such as volunteer GIS map of Baidu, Baidu Data Index, Weibo blog data of Sina, and Internet users' migration data of Baidu, Tencent, and 360, are used for exploring human behavior and urban space in order to find solution for warning crowd disaster, investigate residents' emotions on urban space, classify the system of cities, suggest economic integration index, identify the definition of urban function area, and recognize diffusion of rumors in social network system. Through these case studies, contributors are trying to understand urban structures based on social big data from SNS and finds solution of the above issues related to urban planning and management.

It is exciting for planners to study urban structure from a view of spatiotemporal analysis using individual mobility data that it is never possible by conventional survey database. We can accurately investigate social big data information regarding where, when, and who are using SNS. However, it is impossible to identify personal information of SNS users individually because social big data are protected personal information. LBS information of SNS is impossible to collect from all residents because the number of SNS users is limited, and what we can investigate is spatiotemporal analysis of individual user or patterns of mass behaviors.

1.3 Part 3: POI for Exploring Urban Space Recognition

Collaborative mapping is the aggregation of web maps and user-generated content, from a group of individuals or entities. Points of interest (POIs) are recognized as one kind of big data based on collaborative mapping, which is not the sensor data from IoT. Because of data shortage problem in China, planners are using POI data to analyze current situation in planning process. There are also commercial POI collections that are sold on a subscription basis and usually are protected by its copyright. However, there are also many websites from which royalty-free POI collections can be obtained, e.g., SPOI (Smart Points of Interest).

POI is a kind of point data that stands for building points and is inputted by users of collaborative mapping. POI data obtained from OpenStreetMap with building usage is a popular trend among planners. POI data reflects the places where mapping users are interested in. However, the POI does not cover all building data. A collaborative mapping platform is available for users to input their interested points. In addition, private companies for POI collections are active in inputting those points in their system. Even though POI is a kind of data that people are interested in, POI is recognized as substitutable data for database of urban land use in China. The benefit that planners can get from using POI is not only easy access to the POI data from the Internet but also time and money saving for planners to do field survey and use POI as some kind of complementarity to national survey database.

POI data is used to explore urban space recognition, e.g., discovery center and urban functions in urban areas. In Chap. 8, "Identifying and Evaluating Urban Centers for the Whole China Using Open Data," Ma and Long argue that the central city area, namely, urban centers, has never been well delineated for the Chinese urban system due to data unavailability. Identification and evaluation of urban centers has long been a concern of the urban planning discipline because it is the core component of urban structure. After reviewing existing identification methods of urban centers, Ma and Long propose a novel approach for identifying urban centers using increasingly ubiquitous open data, namely, POIs, in order to identify nationwide urban centers while integrating various types of open data from a view of four dimensions, which are scale, morphology, function, and vitality aspects. Thus, characteristics of development of nationwide urban centers are discussed in this chapter too.

Chen, Xu, Wang, Du, and Li are working on identifying locations of retailing market, which is commercial function in urban area in Chap. 9, "Geographic Big Data's Applications in Retailing Business Market." Location has been considered as a determinant factor of success in retailing stores. However, their traditional methods are highly likely to be too expensive and difficult to expand to the whole city or nation. In order to select the best site for each store, this chapter explores methods that based on POI data and other social media data for classifying the location to generating hierarchical trading areas. Results indicate that more than 70% of stores can be explained by business district classifications suggested in this work. Moreover, results indicate that POI aggregations can match the locations of two retail stores from 83% to 90%, respectively. Besides, the relationship of distance between POIs and stores can contribute to explaining the location of retail stores.

In Chap. 10, "Redefinition the Social Space Based on Social Atlas Analysis: A Case Study of Dongguan, China," is contributed by Liu and Su. Social area analysis is a conventional approach to study social division and social structure, which is generally incapable to discover the diversity of social space. Liu and Su used the data of the Sixth Population Census and POI data to explore sociospatial features of a Chinese city. Based on demographic properties, social facilities, and organizations, 11 types of social areas and 3 kinds of social spaces are identified. Particularly, rural community, rural-migrant-worker community, and urban community are demonstrated in Dongguan. The chapter declares that the method of social atlas relieves the limitations of social area analysis, which help to understand a very complex and diverse social space and reconstruct the research framework adapting to certain social situations in China.

Li, Wang, Li, Dang, and Song have explored four functions, namely, political center, cultural center, international communication center, and innovative center of science and technology, and the development of scientific and technological innovation in Beijing in Chap. 11, "The Spatial and Temporal Evolution of Innovative Function of Science and Technology of Beijing Based on the Analysis of Enterprise Data." This chapter further focuses on analysis of Beijing's high-tech industry and its spatial evolution based on POI enterprises data in order to conclude how the government policies influence the spatial evolution. Results show that spatial distribution of high-tech industries in Beijing spreads from city center to the periphery gradually under the guidance of government policies, in which it demonstrates that top-down policies are playing the key role in the innovation-driven stage.

In this part, POI is used to explore recognition of urban space, e.g., social spaces, urban center, commercial function, and innovative function of science and technology, which are helpful for planners to consider land use function in urban areas.

1.4 Part 4: Mobile Device Data for Integrating Land Use and Transportation Planning

Cell phone and car navigation system with GPS service are widely used in China where users carry all these devices. Besides personal desktop computer users, SNS app installed in smartphones is also carried by users. In this part, SNS is not included because social big data, such as messages and photos inputted by SNS users, has been discussed in Part I. Position signal or station information, as well as, time information recorded in mobile devices with GPS or smart cards for public transportation, which are generated by mobile devices or card reader devices automatically, are used for planning support in this part.

Position and time information are useful for spatiotemporal analysis. Hence, many planners in China are engaging to use the data for analyzing urban structure to intergrade land use and transportation planning, even though urban structure is also possible to be analyzed by conventional database of land use survey. Mobile device data is useful to understand urban structure based on human mobility at an individual level. The track and time of human mobility can be used to understand spatiotemporal characteristics of urban structure, which are not possible to be identified by conventional database.

Lin, Ma, Chu, and Gao in Chap. 12, "Spatial Development Analysis of the Southern Area of Beijing Based on Multisource Data," have explored GPS data of taxi, smart card data, and other types of data. There is an Action Plan of Promoting the Development of Southern Area of Beijing implemented by the Beijing government since 2009. For understanding how this action plan influences the southern area of Beijing, this chapter provides multidimensional analysis on existing problems of the southern area of Beijing using multisource, such as taxi GPS data, transit smart card data, second-hand property pricing data, and demographic data.

Wang, Zhong, Yin, Xie, and Luo have used cell phone data in Chap. 13, which is "Study on Population Dynamics and Mobility of Shanghai Based on Cell Phone Signaling Data." Analysis of temporal and spatial dynamic distribution of population is an important basis for recognizing people's behaviors and structure of cities, allocating urban public infrastructures, and making public safety emergency plans. Because cell phone is the most popular communication terminal equipment, the distribution of cell phone users is able to reflect the distribution of population accurately. Using datasets of cell phone signaling records and related land use data from Shanghai, this study tries to build a framework based on a relationship among population, time, and behavior, which causes changes in distribution of population, to analyze spatiotemporal and dynamic distribution of population in Shanghai. Mobile devices, such as cell phones, enable to capture large amounts of human behavioral data, which can provide information about the structure of cities and their dynamical properties too. A similar work is presented by Shi and Yang, Chap. 14, "Application of Big Data in the Study of Urban Spatial Structure." In this chapter, a behavioral density parameter is defined to measure the intensity of the individual's behaviors for highlighting the centers of urban structure. A polycentric model is proposed to test the polycentric structure of Shanghai. The result shows that Shanghai has a polycentric urban structure with different categories of urban centers.

Lai, Chen, Yuan, and Li also have employed cell phone data in Chap. 15, which is "Application of Cellular Data in Traffic Planning." Traffic planning is a very important tool for planners to figure out a solution to traffic jams. The traditional method is to estimate the travel demand according to the combination of the nature of land use and the intensity of development. In this chapter, the authors use the cellular data to extract the distribution of population and their jobs, travel and other parameters, the establishment of a city only based on job distribution, and spatial location data under the conditions of travel characteristics analysis model. Based on population density and trip distance, a predictive model of different urban areas is established by multiple linear models to find a planning solution.

Zhu has used smart card data in Chap. 16, "Extract the Spatiotemporal Distribution of Transit Trips from Smart Card Transaction Data: A Comparison Between Shanghai and Singapore," which provides a comparison of a week's transit data based on smart card transaction in Shanghai and Singapore. Because individual trips have different choices of activities and destinations on each day within a week, spatially detailed urban form is helpful to explore the purpose of individual trips. At the station level, the overall spatial distributions of the boarding and alighting passengers appear to be alike at most of time in a day.

In this part, monitoring of population location and their trips is an important step for the analytics of mobile device data to integrate land use with transportation planning. Researchers have to understand general knowledge of the data, such as the formats of the data, the types of information contained, as well as the strengths and the weaknesses of the data. The spatiotemporal analysis of mobile device data helps the researchers to form a comprehensive view of the relationship between population location and their trips. The general spatiotemporal patterns of transit trips were exposed in this part providing a good foundation for further analysis and modeling in urban area. Meanwhile, it is recognized that the interpretation of observed patterns using mobile device data is mostly superficial and hypothetical. In order to have a profound understanding of land use patterns and transportation via underlying activity-travel behaviors provided by mobile device data, it is necessary to couple with the big data such as transit smart card transaction data with other data sources.

1.5 Part 5: Cyber Infrastructure for Urban Management

Apps in smartphones are possibly to be developed as a platform, namely, cyber infrastructure for providing urban services. In China, apps in smartphones, e.g., WeChat, provide most of urban services, including parking, calling taxi, information of public transportation, registration for hospital, and most of procedure of application to local administrative offices, such as tax and public charge. In China, this kind of online urban services supports smart governance and smart living with construction of smart infrastructure.

On one hand, smart city refers to intelligent and sustainable urban development, which is for smart infrastructure construction; on the other hand, the notion of a smart city refers to urban environment with pervasive and ubiquitous computing and digitally instrumented devices for providing urban services and improving quality of life. From this point of view, Li et al. (2015) divides "smart city" into four layers, which are sensor layer, network layer, platform layer, and application layer. In this book, cyber infrastructure is a term including all the four layers. How to get big data from sensor and network? How to develop a comprehensive platform to integrate different data sources as a cloud database for supporting planning and management? How to develop software applications for urban management?

We have five chapters focusing on data collecting by IoT sensors and platform development for urban management system using cyber infrastructure. In Chap. 17, "Traffic Big Data and Its application in Road Traffic Performance Evaluation: Illustrated by the Case of Shenzhen," Qiu and Chen have established a comprehensive system platform, that is, road traffic operation evaluation system with a traffic analysis model, which has been applied for decision-making process of transportation planning and management in Shenzhen city for 5 years already. This system is possible to evaluate road traffic operating conditions through time and space dimensions, providing effectively technical assessment tools for transportation authority of local government. The database is developed to integrate multisource data with real-time traffic monitoring data and support massive data processing and GIS spatial computing in real-time for urban traffic operation and management.

In Chap. 18, "Understanding Job-Housing Relationship from Cell Phone Data Based on Hadoop," Li, Wu, Tang, and Lu have taken Chongqing city as a case study to demonstrate how the platform works on job-housing analysis. A Hadoop-based cell phone data processing platform is introduced as a solution to the storage and processing of large volumes of real-time cell phone data, which is innovative in its big data computation framework, including cell phone data processing procedures and algorithms and commuting analysis for job-housing relationships. This platform is to realize the collection, processing, storage, export, and other processing of cell phone data, in order to save all the intermediate and final computational results for reuse.

Besides cyber infrastructure construction, Chaps. 19 and 20 focus on how to conduct an investigation through IoT sensors. In Chap. 19, "Towards Mobility Turn in Urban Planning: Smart Travel Planning Based on Space-Time Behavior in Beijing, China," Chai and Chen have investigated space-time behavior of residents in a suburban new town in Beijing, which the approach has become influential in China over the last two decades and has subsequently been widely applied to empirical researches and planning practices. In this research, Chai and Chen have integrated GPS tracking technology into a web-based activity-travel diary survey with detailed activity-travel information for a framework and implementation of a smart travel planning program in Beijing. Finally, Chai and Chen have elaborated on the practices of smart travel planning in the study area, in terms of web-based real-time travel information services and the challenges and the future prospects of smart travel planning in urban China.

Another work, which is conducted by Mao, is Chap. 20, "Observing and Quantifying Vitality of Dashilanr: An Automatic Human Activity Data Collection Experiment." Dashilanr subdistrict in Beijing inner city is famous for its vitality due to its incremental urban renewal strategies, which is setting an example of vitality creation for other communities in inner city of Beijing. Vitality study is necessary, although past observation methods have been mostly carried out manually. This research observes vitality of Dashilanr with sensors to achieve automatic data collection in real time of a big event, namely, Beijing Design Week. The collected data is analyzed to quantify vitality changes in Dashilanr, identify intrinsic and extrinsic factors that have affected vitality in the community, and make suggestions to management and vitality creation of Beijing inner city.

As there are many existing researches on environmental motoring, we thus have only one chapter about application of sensor data of temperature/wind collected in weather stations. Chapter 21, "Urban Wind Path Planning Based on Meteorological and Remote Sensing Data and GIS-Based Ventilation Analysis," is contributed by Zhan, Fan, Xiao, Wang, Ouyang, Yue, and Lan. This study proposes a systematic approach to support urban wind path planning for improving heat environment, using remote sensing images, wind and temperature data from local weather stations, GIS-based building and road data, etc. Potential wind paths are determined by natural wind resources and built-up environment. With the problematic regions and potentially delineated wind paths, planning measures can be formulated correspondingly.

Finally, Ye, Li, and Huang in Chap. 22, "A Synthesized Urban Science in the Context of Big Data and Cyber Infrastructure," comment on the complexities of cyber infrastructure in a system platform and their connectivity at variously spatial, temporal, and semantic scales, which can be explained by big data during the period of ICT and IoT. This kind of cyber system has posed daunting challenges to researchers in urban sustainability field, whereby rigorous analysis of such data depicting complex socioeconomic events is likely to open up a rich context in advancing urban sciences and policy interventions. Interdisciplinary approaches that are combined with rich and complex spatial data, analysis, and models are urgently needed to ignite transformatively geospatial innovation and discover enabling effective and timely solutions to challengeable urban issues in urban planning and management.

1.6 Summary

In this book, we explore the new challenges of urban-rural planning and management from practical perspectives based on many multidisciplinary projects that are related to practice of smart city planning in China. We highlight the challenges and opportunities of a synthesized urban science based on ever-increasing amounts of large-scaled diverse data and computing power.

In China, smart city practice, big data collected from mobile devices, and social network-reflected human behaviors are useful for analyzing human mobility, obtaining urban space recognition and exploring spatiotemporal urban structure, which is preconditions for planners to integrate land use with transportation planning in China.

Construction of new smart city in China focuses on citizens' welfare and urban governance. With smart infrastructure construction, IoT sensors for collecting mobile device data and platform development that use real-time traffic monitoring are opening a new era for urban management. Due to the rapid progress of information and communications technology, the emergence of big data available from various sources has presented significant opportunities for urban research. Today's gradually connected world of virtual, perceived, and real spaces, data-driven urban computing and analytics has become increasingly essential for the understanding of coupled human and socioeconomic dynamics.

From a worldwide view, the smart living is a conceptual lifestyle in smart city era innovation. Comparing with China, the concept of smart living in European countries and Japan is used as quality of life, such as cultural life, health conditions, individual safety, housing quality, education, tourism, and social cohesion (Giffinger et al. 2007). In China, smart city is an organic connection among technology, human, and institution, whereby it is to mainly serve people's living via ICT and IoT. Thus the concept of smart living is a kind of digital life in the Internet age, which is a growing requirement supporting societies' unquenchable demand for a modern digital lifestyle. However, in this book, big data supports for urban planning and management in a Chinese context recently focuses on residents' urban space recognition for smart mobility and smart governance, and there will be a long way to go for archiving the goal of smart city planning and construction in China.

References

EU Smart Cities Information System (ESCIS). (2017). http://smartcities-infosystem.eu

European Innovation Partnership on Smart Cities and Communities (EIPSCC). (2017). https://eu-smartcities.eu

Giffinger, R., Fertner, C., & Kramar, H. (2007). *Smart cities-ranking of European medium-sized cities*. http://curis.ku.dk/ws/files/37640170/smart_cities_final_report.pdf. Accessed 23 Nov 2014.

Goldsmith, S., & Crawford, S. (2014). *The responsive city: Engaging communities through data-smart governance*. Hoboken: Wiley.

Li, Y., Lin, Y., & Geertman, S. (2015). *The development of smart cities in China*. The proceedings of CUPUM2015, MIT. N291.

Obama White House. (2017). https://obamawhitehouse.archives.gov/the-press-office/2015/09/14/fact-sheet-administration-announces-new-smart-cities-initiative-help

Offenhuber, D., & Ratti, C. (2014). *Decoding the city: Urbanism in the age of big data*. Basel: Birkhauser Verlag.

Smart Cities USA. (2017). http://smartamerica.org/teams/smart-cities-usa/

Smart City Project. (2017). http://www.smartcity-planning.co.jp/en/index.html

Townsend, A. M. (2014). *Smart cities: Big data, civic hackers, and the quest for a new utopia*. New York: W.W. Norton & Company.

Part I
Social Big Data for Exploring Human Behaviors and Urban Structure

Chapter 2
Early Warning of Human Crowds Based on Query Data from Baidu Maps: Analysis Based on Shanghai Stampede

Jingbo Zhou, Hongbin Pei, and Haishan Wu

Abstract Without sufficient preparation and on-site management, the mass-scale unexpected huge human crowd is a serious threat to public safety. A recent impressive tragedy is the 2014 Shanghai Stampede, where 36 people died and 49 were injured in celebration of the New Year's Eve on December 31st 2014 in the Shanghai Bund. Due to the innately stochastic and complicated individual movement, it is not easy to predict collective gatherings, which can potentially leads to crowd events. In this chapter, with leveraging the big data generated on Baidu Maps, we propose a novel approach to early warning such potential crowd disasters, which has profound public benefits. An insightful observation is that, with the prevalence and convenience of web map service, users usually search on the Baidu Maps to plan a routine, which reveals the users' future destinations. Therefore, aggregating users' query data on Baidu Maps can obtain priori and indication information for estimating future human population in a specific area ahead of time. Our careful analysis and deep investigation on the Baidu Maps' data on various events also demonstrate a strong correlation pattern between the number of map query and the number of positioning in an area. Based on such observation, we propose a decision method utilizing query data on Baidu Maps to invoke warnings for potential crowd events about 1~3 h in advance. Then we also construct a machine learning model with heterogeneous data (such as query data and positioning data) to quantitatively measure the risk of the potential crowd disasters. We evaluate the effectiveness of our methods on the data of Baidu Maps.

Keywords Emergency early warning • Crowd anomaly prediction • Map query • Stampede

J. Zhou • H. Wu (✉)
Baidu Research, Big Data Lab, Beijing, China
e-mail: hswu85@gmail.com

H. Pei
Baidu Research, Big Data Lab, Beijing, China

College of Computer Science and Technology, Jilin University, Changchun, China

Z. Shen, M. Li (eds.), *Big Data Support of Urban Planning and Management*, Advances in Geographic Information Science, DOI 10.1007/978-3-319-51929-6_2

2.1 Introduction

The management of human crowd in public events is significantly important for public safety. The 2014 Shanghai Stampede, where 36 people died and 49 were injured in celebration of the New Year's Eve on December 31st 2014, was blamed for insufficient preparation and poor on-site management of local officials (Investigation Report 2014). Similar to other crowd disasters, one of major reasons of this tragedy is due to the wrong estimation of the human flow and human density by management officials (Investigation Report 2014). As the potential risk is increasing rapidly with the increment of mass-scale, unexpected huge crowd, inadequate emergency response preparedness may lead to a disaster. It is clear that the primary factor in assuring a safe environment for crowd is a feasible emergency action plan for potential crowd events, which can help preventing such terrible crowd disasters from happening.

Preventing crowd disasters is a challenging task which essentially depends on accurately predicting human movements. Although many existing works has modeled and predicted individuals' trajectory, most of them focus on human daily routine trajectory (Ashok 1996; Schneider et al. 2013; Zheng and Ni 2012). However, crowd anomaly is generally caused by infrequent collective activities (e.g., celebration, religious gatherings, sports tournaments). Participating these activities is a non-routine but stochastic behavior for most people. Thus, the movement of participators is much different from their daily routine and more like a random walk (Gonzalez et al. 2008).Therefore, such kind of movement can hardly be predicted readily, and there is no existing approach to handling it well by far, to the best of our knowledge.

Traditional method for crowd anomaly detection is based on video sensor and computer vision technology (Helbing et al. 2007; Gallup et al. 2012; Ali and Shah 2008). Whereas, three limitations of this kind of methods cannot be overcome: (1) they are too sensitive to noise of visual environment, e.g., rapid illumination changing may let them fail; (2) the areas with deployed video sensors are very limited in the citywide; and (3) the detection methods cannot provide significant long enough preparation time (e.g., hours level) for emergency management before crowd anomaly happens.

In this work, we propose an effective early warning method for crowd anomaly, as well as a machine learning model for crowd density prediction. Both of them are derived from a novel prospect by leveraging query data from Baidu Maps, the largest web mapping service application in China provided by Baidu.[1] Inspired by a widespread custom that people usually plan their trip on Baidu Maps before departure, we find a strong correlation pattern between the number of map query and the number of subsequent positioning users in an area. That is to say, map query behavior in some sense is a nice leading indicator and predictor for crowd dynamics. Therefore, we divide the early warning task into two parts: firstly, detecting map

[1] http://map.baidu.com/

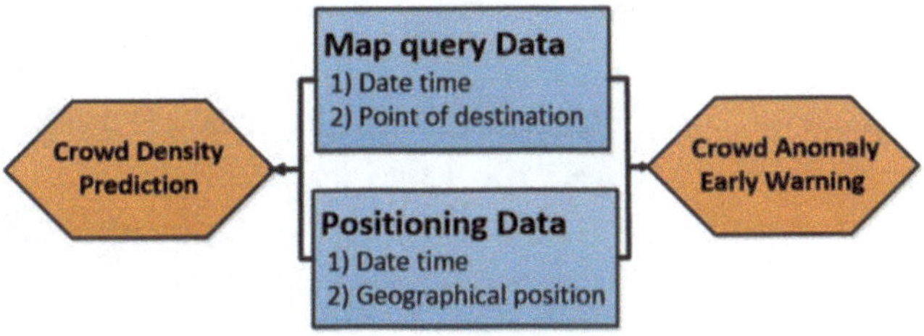

Fig 2.1 The framework of early warning for crowd anomaly based on map query data and positioning data

query dynamics of an area and sending early warning when there is a map query number anomaly, and then quantitatively predicting crowd density from historical map query data and positioning data of this area. The framework is illustrated in Fig. 2.1. Our approach can reliably provide early warning for crowd anomaly for 1–3 h which would facilitate an early intervention for administrative agency to prevent a crowd disaster from happening.

The main contributions of this work are listed as follows:

- We provide an inspiring perspective to predict the crowd anomaly according to mass map query behavior.
- We propose a simple but effective early warning method for crowd anomaly based on map query data and positioning data. Case studies and experiments on real data demonstrate that the proposed method can achieve an early warning for anomaly 1–3 h ahead.
- We develop a machine learning model for crowd density prediction based on historical map query data and positioning data. Experiments on real data show that the model can accurately predict crowd density in a specific area one hour ahead.

The rest of this chapter is organized as follows. First, we give a detailed analysis of the 2014 Shanghai Stampede based on positioning data and map query data in Sect. 2.2. Then the details of the proposed model and the experimental results are described in Sect. 2.3. Finally, we conclude this work in Sect. 2.4.

2.2 Case Study of the Shanghai Stampede

2.2.1 Observations from the Data of the Shanghai Stampede

In this section, we present several important observations from the human mobility data about the 2014 Shanghai Stampede obtained from Baidu Maps, which can provide some insights for human flow and crowd anomaly prediction. The Shanghai Stampede happened at the Bund in Shanghai on December 31st 2014. Hereafter, we use the "map query number" during a time interval (typically it is one hour) to denote the total number of queries whose destination is located in a specific area

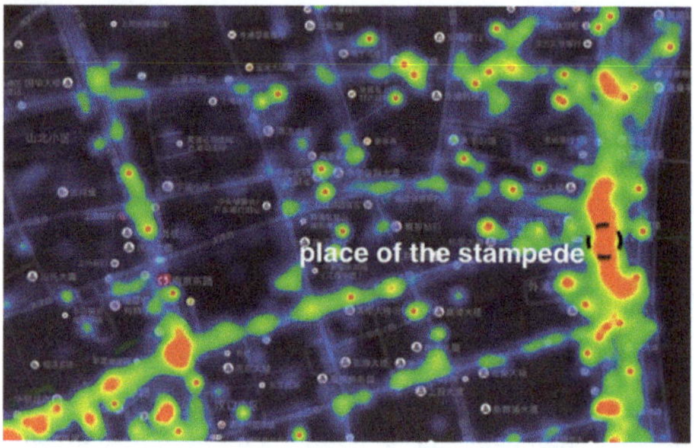

Fig. 2.2 Human population density between 23:00 and 24:00 on December 31st 2014

(like the Shanghai Bund) on Baidu Maps. Meanwhile, we denote the number of users by enabling the positioning function of Baidu Maps during a time interval in a specific area as the "positioning number". The position number can be treated as an approximation of the human population density which is the number of persons per square within an area during a fixed time interval.

Observation 1: Human Population Density of the Stampede Area Was Higher than Others
The high human population density is a necessary condition for the happening of stampede. Figure 2.2 illustrates the human density heat map between 23:00 and 24:00 on December 31st 2014 based on the positioning number. It is obvious that the Bund, which is the disaster area of 2014 Shanghai stampede, had higher human density than the rest parts of the city.

Observation 2: Human Population Density of the Disaster Area Was Higher than the Ones at Other Times
The human population density when the disaster happened was also higher than the ones at other times. Figure 2.3 shows the positioning number on Baidu Maps on the Bund over one week. During the disaster time, the positioning number on Baidu Maps was about 9 times of peak values at other times. Furthermore, Fig. 2.4 exhibits that the user positioning number on the Shanghai Bund of the New Year's Eve of 2014 was also higher than other festivals/holidays (in Fig. 2.4, the festivals/holidays are August 23rd (a common weekend of 2014), the Mid-Autumn Eve (September 7th the evening before a traditional Chinese festival), the National Day on (October 1st), and the New Year's Eve (December 31st)).

Observation 3: The Human Flow Directions of the Disaster Area Were More Chaotic
We also observe that the human flow directions in the disaster time were more chaotic than the ones at other times. Figure 2.5 illustrates the human flow directions in different festivals/holidays from 22:00 to 24:00 in the disaster area. Compared with other subfigures, Fig. 2.5(d) indicates that the disaster area had more chaos during the New Year's Eve of 2014, which implies a potential accident.

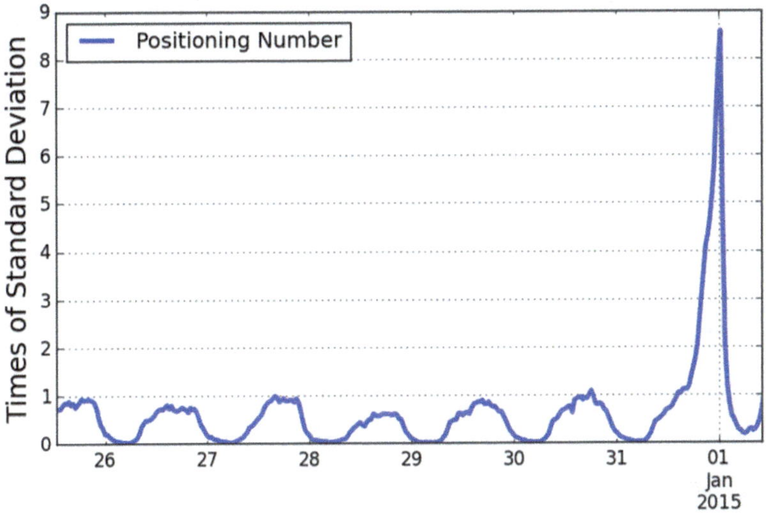

Fig. 2.3 The user positioning number in Baidu Maps on the Bund (the disaster area of 2014 Shanghai Stampede) over one week

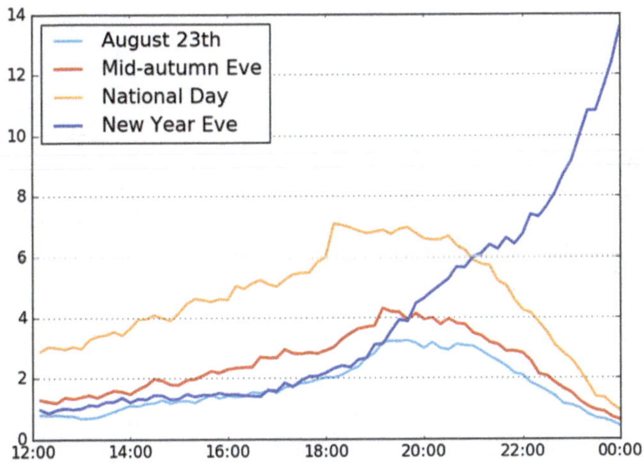

Fig. 2.4 The positioning number in Baidu Maps on the Bund (the disaster area of 2014 Shanghai Stampede) in different festivals/holidays of 2014, which are August 23rd (a common weekend of 2014), the Mid-Autumn Eve (7 September, the evening before a traditional Chinese festival), the National Day (October 1st 2014), and the New Year's Eve of 2014. All the numbers are divided by the standard deviation of the position number of Mid-Autumn Eve

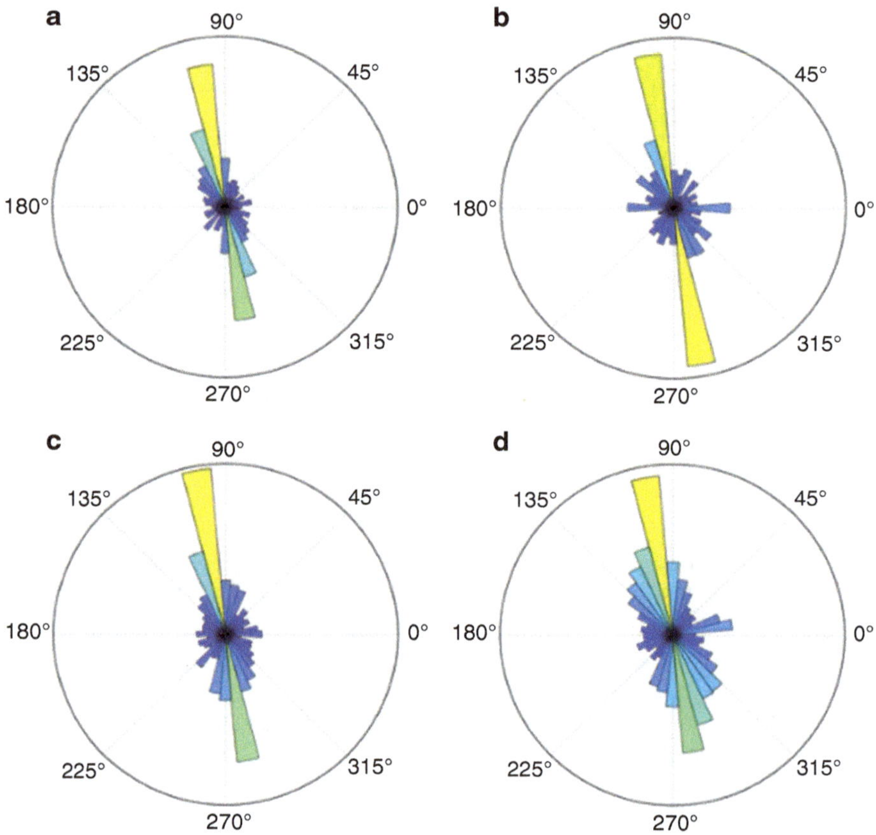

Fig. 2.5 Human flow direction distribution in Chenyi Square (the specific disaster area of 2014 Shanghai Stampede) from 22:00 to 24:00 in: (**a**), common weekend (August 23rd 2014); (**b**), the eve of the Mid-Autumn Festival (September 7th 2014); (**c**), the China's National Day (October 1st 2014); and (**d**), New Year's Eve of 2014

2.2.2 *How Baidu's Data Can Help Prevent Crowd Disasters*

Our objective is to utilize the big data, especially the query data, generated by Baidu Maps' users to avoid such crowd disasters. If without prior information, the individual movement appears as a random walk (Gonzalez et al. 2008), leading to the unavoidable impediment to predict unknown human crowd event. However, nowadays, before a user leaves for a destination, he or she usually uses a web mapping service application to plan a routine to the destination. These queries on the web map reveal some information about users' future behavior in a few of hours. Taking more than 70% market share overall in China, the Baidu Maps has innate advantage to tackle this problem. Therefore, collecting and aggregating the massive query data on Baidu Maps provides an intelligent approach to predicting crowd anomaly as well as avoiding such crowd disasters.

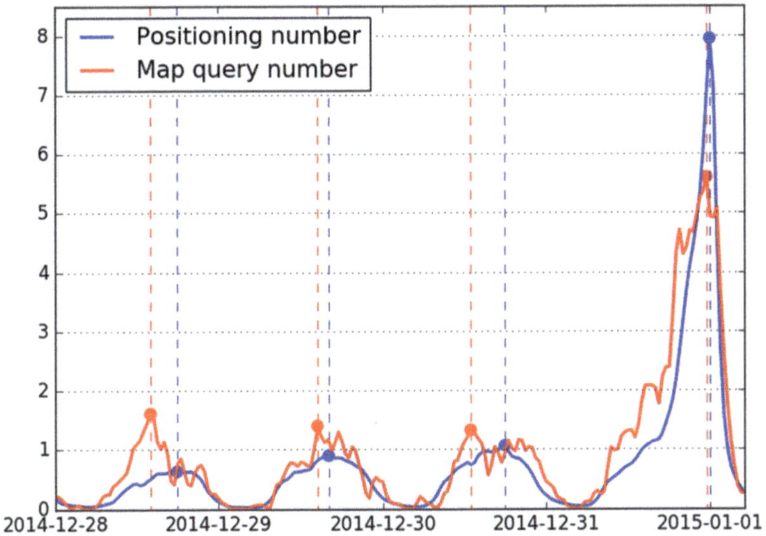

Fig. 2.6 The time series of map query number and the positioning number on Baidu Maps from December 28th 2014 to December 31st 2014 on the Shanghai Bund (both data are normalized by dividing their standard deviation). An interesting observation is that the peak values of the map query number usually appear several hours before the peak value of the positioning number in each day

After analyzing the data on the Shanghai Stampede and other crowd events (like game events and concerts) in different places, we find that there usually are an abnormally large number of queries on the Baidu Maps about the crowd event places emerging 0.5–2 h before the abnormal crowd event being perceptible to human beings.

Figure 2.6 illustrates the time series of positioning number and the one of the map query number of the Baidu Maps in the last a few of days of 2014. This figure exposures the high correlation between the positioning number and the map query number. With a more careful inspection, we can also observe that the peak value of the map query number appears, generally, several hours before the peak value of the positioning number in each day.

We expound more investigations to justify the map query number as a leading indicator of human flow from the practical and theoretical perspectives, which are introduced as follows:

First, as shown in Fig. 2.7, we exhibit the distribution of the time lags (in hours) between the peak value of the map query number and the positioning number on the Shanghai Bund in each day over one year. The negative lag means the peak value of the map query number appears ahead of the one of the positioning number. As we can see from Fig. 2.7, about 80% of days in 2014, the peak value of the map query number appears one hour before the peak value of the positioning number, and with

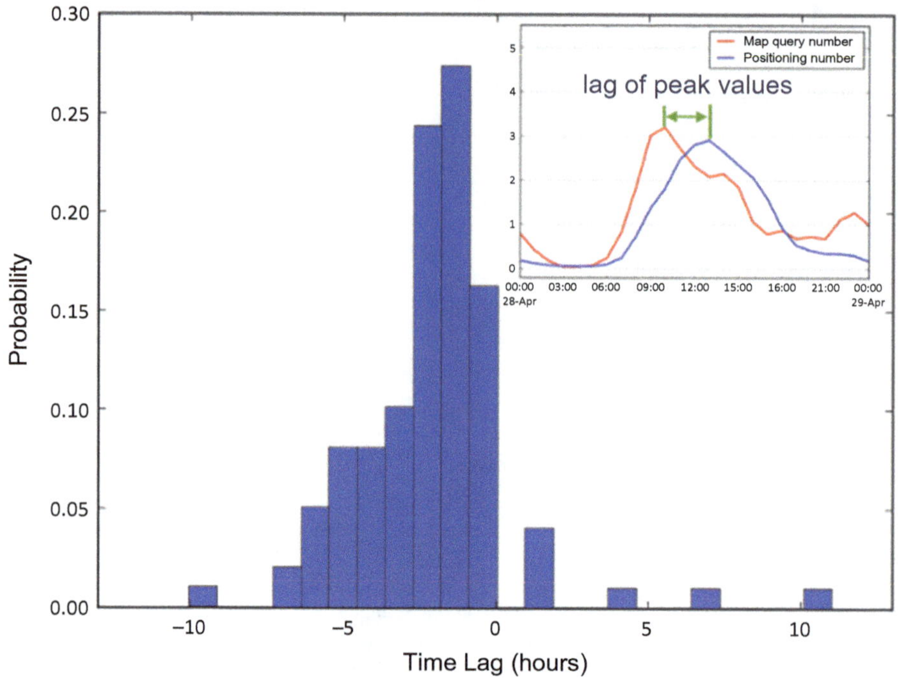

Fig. 2.7 The distribution of the lags (in hours) of the peak values between the map query number and the positioning number within the area of the Shanghai Bund in 2014. The negative lag means the peak of the map query number appears ahead of the positioning number

more than 50% of days in 2014, the peak value of the map query number appears two hours before the one of the positioning number.

Second, Fig. 2.8 theoretically demonstrates that the map query number can help predict the future human flow with the mutual information analysis. As shown in Fig. 2.8, knowing the map query number with one hour ahead can reduce the uncertainty (aka the entropy) of the positioning number, which can empower our ability to predict the future human flow in an area.

2.2.3 The General Usability of Baidu Maps' Data

We also explore the relations between the map query number and the positioning number on Baidu Maps in many different places to demonstrate the general usability of the map query data of Baidu Maps for predicting crowd anomaly. In this section, we investigate three types of such points of interest (POIs) in Beijing, which are: public event place, landmark, and transportation node. The selected representative places of them are the Beijing Workers' Stadium, the Forbidden City, and the Beijing West Railway Station, respectively.

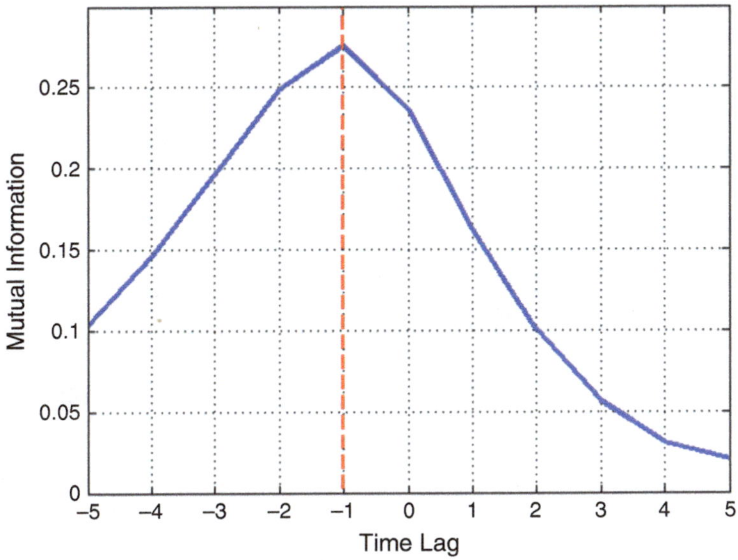

Fig. 2.8 The mutual information between the time series of the map query number and the positioning number within the area of the Shanghai Bund with different lags (in hours) in 2014

The statistical relation analysis, which is the same with the ones in Figs. 2.7 and 2.8 of Sect. 2.2.2, is exhibited in Figs. 2.9 and 2.10. All the analysis demonstrates that the map query number can help predict the crowd anomaly in an area. It is also worthwhile to note that different places may also have different characters for such relations. For example, in the Forbidden City, the peaks of map query number concentrate on two and three hours before the peaks of the positioning numbers. A more interesting observation of the Forbidden City is shown in Fig. 2.11. In the Forbidden City, for each day, there are two peaks of the time series of the map query number before the peak of the positioning number emerges. This reflects the complex dynamics of human behaviors and diverse environment conditions of different places.

2.3 Preventing Crowd Disaster with Baidu Maps' Data

In this section, we detail how to utilize the Baidu Maps' data to prevent the potential crowd disasters. We first present a qualitative decision method for crowd anomaly prediction solely based on the map query data on Baidu Maps. Then we construct a machine learning model with heterogeneous data to quantitatively measure the risk of the potential crowd disasters.

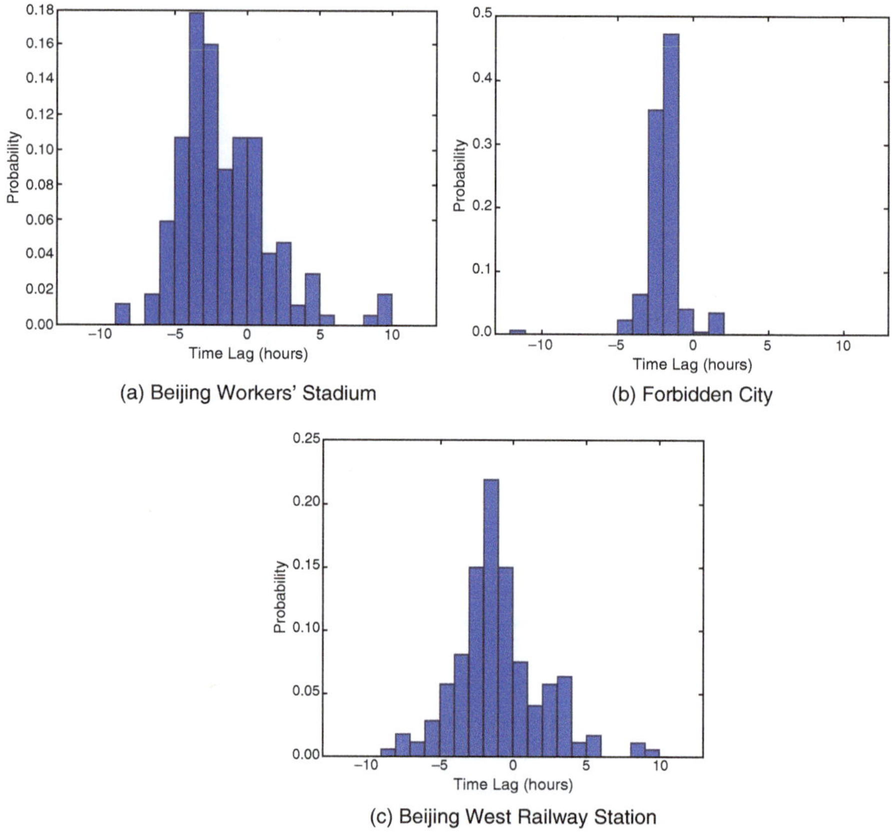

Fig. 2.9 The distribution of the lags of the peak values between the map query number and the positioning number within the area of different POIs in 2014. The negative lag means the peak of the map query number appears ahead of the one of positioning number: (**a**) Beijing Workers' Stadium, (**b**) Forbidden City, (**c**) Beijing West Railway Station

2.3.1 A Decision Method for Crowd Anomaly Prediction with Map Query Data

The general idea of the decision method for crowd anomaly prediction is that, if the number of map query per hour of a specific area is larger than a warning line (a certain threshold), we will invoke an anomaly warning about the possible crowd event in the following several hours. This decision method is based on relation analysis between the map query number and the positioning number discussed in Sect. 2.2, which indicates that a large number of map query number subsequently imply a potential large number of human population in a few of hours in a specific area.

The warning line for the map query number is statistically computed with the peak values of the map query number per hour in every day under the log-normal probability distribution model. Formally, let us denote the time series of map query

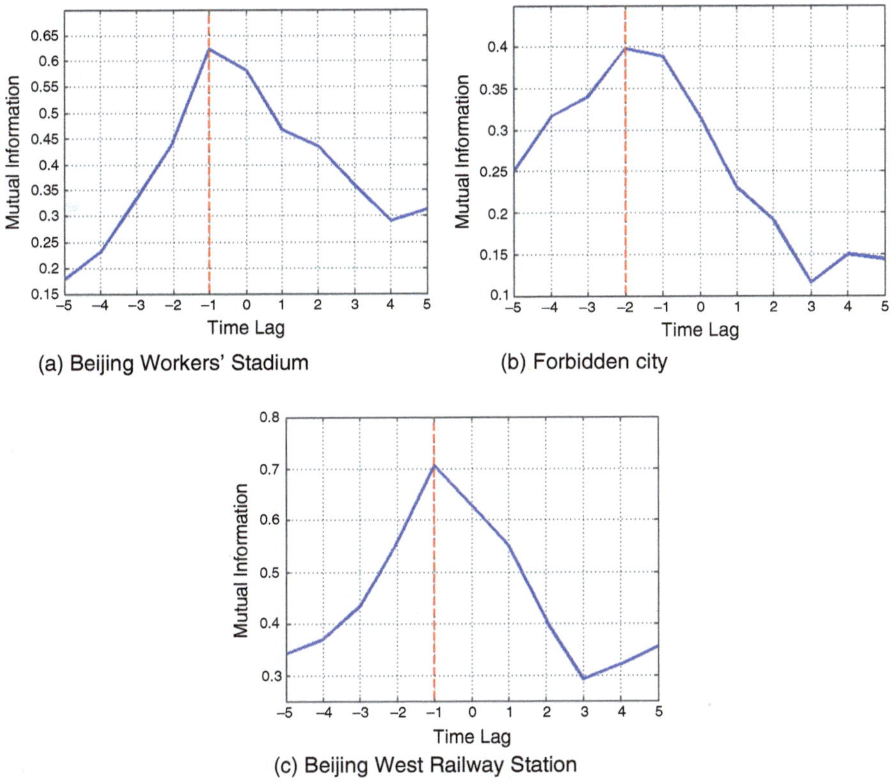

Fig. 2.10 The mutual information between the 1-year time series of the map query number and the positioning number within the area of different POIs in 2014: (**a**) Beijing Workers' Stadium, (**b**) Forbidden City, (**c**) Beijing West Railway Station

number per hour on the Baidu Maps by $M = m(t)$, $t \in \pm \mathrm{N}_0$. For a given date d, we first calculate the peak value of the map query number as pm(d):

$$pm(d) = \max_{t=24d}^{24(d+1)} m(t)$$

In our method, we consider pm(d) as random variable and assume it follows log-normal distribution, i.e.:

$$\varphi(d) = \log\left(\mathrm{pm}(d)\right) \sim N\left(\mu_{\mathrm{pm}}, \sigma_{\mathrm{pm}}^2\right) \tag{2.1}$$

We use one year time series data of map query number to obtain an unbiased estimation of μ_{pm} and σ_{pm} which are sample mean $\hat{\mu}_{\mathrm{pm}}$ and sample variance $\hat{\sigma}_{\mathrm{pm}}^2$:

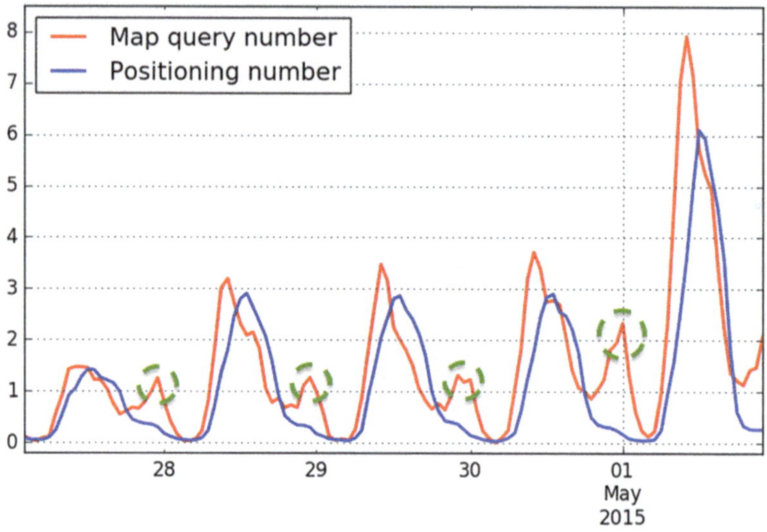

Fig. 2.11 The map query number and the positioning number of Baidu Maps' users from December 28th 2014 to December 31st 2014 in the Forbidden City (both data are normalized by dividing their standard deviation). For each day, the time series of the map query number has two peaks before the peak of positioning number

$$\hat{\mu}_{\mathrm{pm}} = \frac{1}{365}\sum_{d=0}^{364} \log(\mathrm{pm}(d)) \tag{2.2}$$

$$\hat{\sigma}_{\mathrm{pm}}^{2} = \frac{1}{365-1}\sum_{d=0}^{364} \left(\log(\mathrm{pm}(d)) - \hat{\mu}_{\mathrm{pm}}\right)^{2} \tag{2.3}$$

Accordingly, the warning line of the map query number is determined as:

$$\omega_{m} = \hat{\mu}_{\mathrm{pm}} + \alpha\hat{\sigma}_{\mathrm{pm}} \tag{2.4}$$

where α is a model parameter and $\alpha > 0$. The effect of α is evaluated in Sect. 2.3.1.2.

Similarly, we can also obtain a warning line of the positioning number. We define the time series of position number as $Q = \{q(t)\}$, and the peak value of positioning number for date d is pq(d):

$$\mathrm{pq}(d) = \max_{t=24d}^{24(d+1)} q(t)$$

We also assume that pq(d) follows log-normal distribution, i.e.:

$$\phi(d) = \log(\mathrm{pq}(d)) \sim N(\mu_{\mathrm{pq}}, \sigma_{\mathrm{pq}}^2) \tag{2.5}$$

and we have the sample estimation of $\hat{\mu}_{\mathrm{pq}}$ and $\hat{\sigma}_{\mathrm{pq}}^2$:

$$\hat{\mu}_{\mathrm{pq}} = \frac{1}{365} \sum_{d=0}^{364} \log(\mathrm{pq}(d)) \tag{2.6}$$

$$\hat{\sigma}_{\mathrm{pq}}^2 = \frac{1}{365-1} \sum_{d=0}^{364} \left(\log(\mathrm{pq}(d)) - \hat{\mu}_{\mathrm{pq}}\right)^2 \tag{2.7}$$

In this chapter, the warning line of positioning number is fixed as $\omega_q = \hat{\mu}_{\mathrm{pq}} + 3\hat{\sigma}_{\mathrm{pq}}$. Finally, our decision method can be formally described by following lemma:

Lemmma 2.1 If $m(t) \geq \omega_m$, then we have $q(t + \Delta) \geq \varpi_q$ and $1 \leq \Delta \leq T$ with high probability, where T is a limited time period.

In our following demonstrations and evaluations, we set T = 3 (hours).

2.3.1.1 Demonstration of the Decision Method

We demonstrate several examples for crowd event prediction based on Lemma 2.1 in different POIs of several China cities. Figure 2.12 illustrates such decision method on the data about the public crowd anomaly prediction of Shanghai and Shenzhen (which can be compared with Fig. 2.13). As we can see from Fig. 2.12, a warning about the abnormal crowd event can be invoked 1–3 h before the human population surpassing a warning line. Figure 2.14 demonstrates similar results for the abnormal crowd event prediction in Beijing. To sum up, with monitoring the map query number for each hour, it is possible to give a very early warning for the abnormal crowd event to avoid crowd disasters (Fig. 2.15).

2.3.1.2 Performance Evaluation for the Decision Method

We also employ a performance evaluation for the decision method. In the six POIs of Figs. 2.12 and 2.14, we first select all the time points where $q(t) > \omega_q$ and consider these points as the ground truth of crowd events. Then we use our decision method introduced in Lemma 2.1 to try to predict the possible crowd events. In other words, for a monitored POI, if the map query number is larger than a predefined warning line, we will invoke a warning that, in the following 1–3 h, there will be a crowd event. We exhibit the precision, recall, and F1-score of six POIs in Fig. 2.16. As we can see from the Fig. 2.16, there is a trade-off between the precision and recall, which is controlled by the times of the standard deviations (i.e., α in Eqn. 2.4). We also aggregate the precision, recall, and F1-score of the six POIs together and illustrate the average measures in Fig. 2.17.

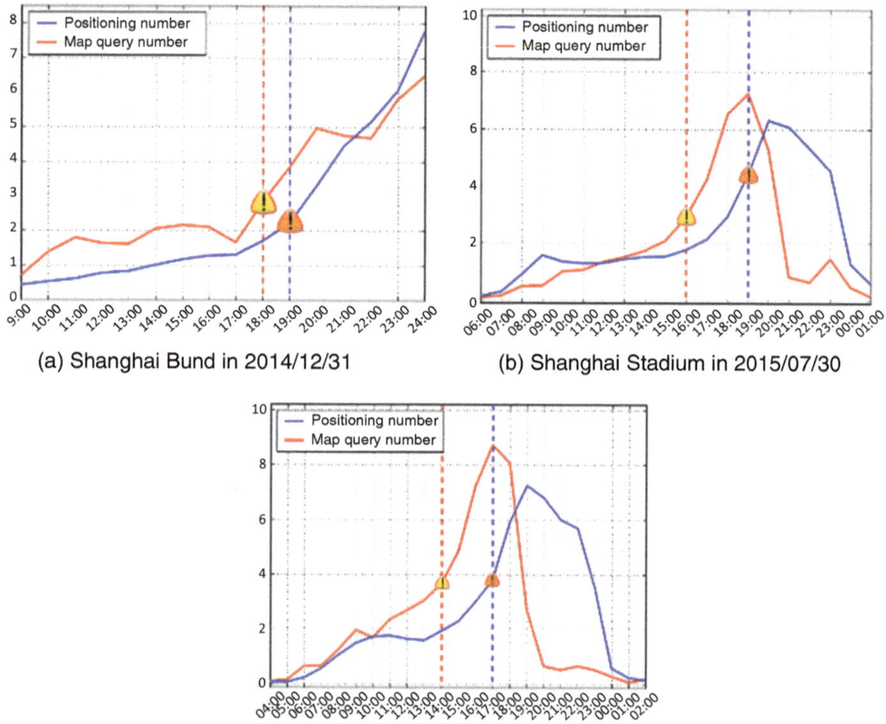

(a) Shanghai Bund in 2014/12/31

(b) Shanghai Stadium in 2015/07/30

(c) Shanghai Bay Sports Center in 2015/04/25

Fig. 2.12 An illustration of the decision method for abnormal crowd event detection with map query data on Baidu Maps in Shanghai and Shenzhen. If the map query number is larger than a warning line, we will invoke a warning for crowd event in the following several hours (see Lemma 2.1). From the figure we can see that the event warning can be sent 1–3 h before the human population surpassing the warning line. The corresponding events are (**a**) 2014 Shanghai Stampede (Shanghai Bund in 2014/12/31), (**b**) the International Champions Cup Real Madrid vs AC Milan (Shanghai Stadium in 2015/07/30) (http://www.goal.com/en-za/match/real-madrid-vs-milan/2032786/report), and (**c**) TVXQ Special Live Tour T1story in Shenzhen (Shenzhen Bay Sports Center in 2015/04/25) (http://www.springcocoon.com/html/CN/ssyc/hdgl/201504/07847.html)

2.3.2 *Machine Learning Model for Risk Control of the Potential Crowd Disasters*

We also develop a machine learning model to utilize the map query data, historical positioning data, and other information available in the Baidu Maps to predict the future human density in an area. Such predicted density can be taken as a quantitative measure to assess the potential risk of a crowd disaster. The intuition is that the probability of a crowd disaster is directly related with the human density and the human density is a necessary (though not sufficient) condition for a crowd disaster.

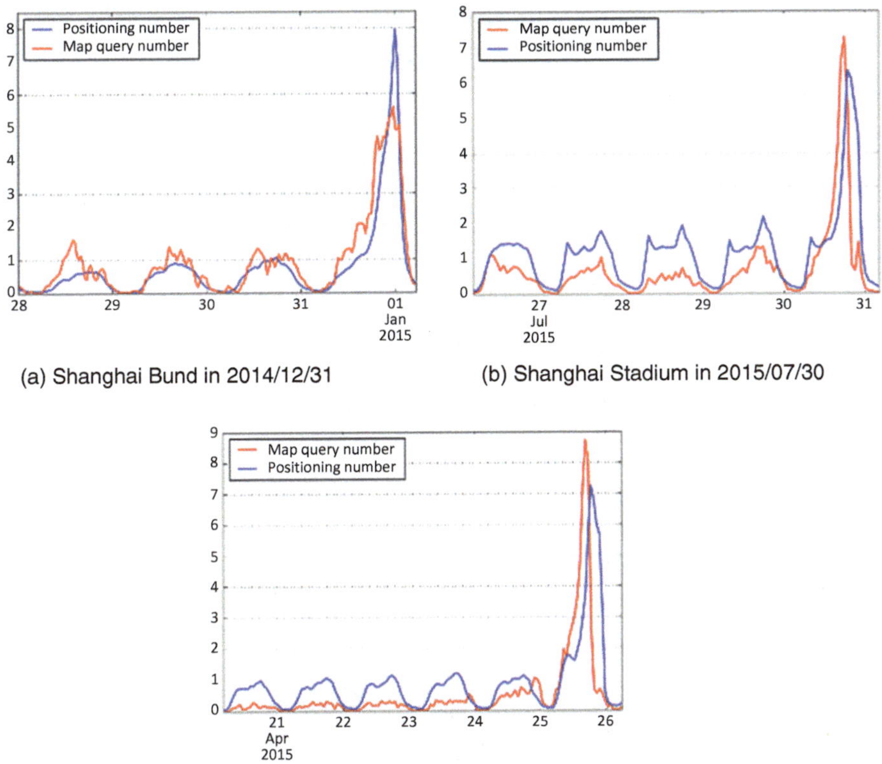

(a) Shanghai Bund in 2014/12/31

(b) Shanghai Stadium in 2015/07/30

(c) Shenzhen Bay Sports Center in 2015/04/25

Fig. 2.13 The map query number and positioning number on the Baidu Maps corresponding to the events of Shanghai and Shenzhen shown in Fig. 2.12: (**a**) Shanghai Bund on in 2014/12/31, (**b**) Shanghai Stadium on 2015/07/30, (**c**) Shenzhen Bay Sports Center in 2015/04/25

2.3.2.1 Problem Definition

Let us denote $M_{t_c} = m(t)$ and $Q_{t_c} = q(t), t \in [0,,\ldots,,t_c]$ as time series of map query number and time series of positioning number before time t$_c$ in a specific area, respectively. We set time granularity to one hour with a fine resolution and statistical significance. Let $f(M_{t_c}, Q_{t_c})$ denote the prediction model with two time series of input; then the positioning number at next time $q(t_c + 1)$ can be estimated as:

$$q(t_c + 1) \approx \hat{q}(t_c + 1) = f(M_{t_c}, Q_{t_c}) \quad (2.8)$$

where $q(t_c + 1) \approx \hat{q}(t_c + 1)$ means approximate. Thus, the problem what we are interested in is: can we train a prediction model *f*, from the historical map query and positioning data, to accurately predict the positioning number in the next time $q(t_c+1)$?

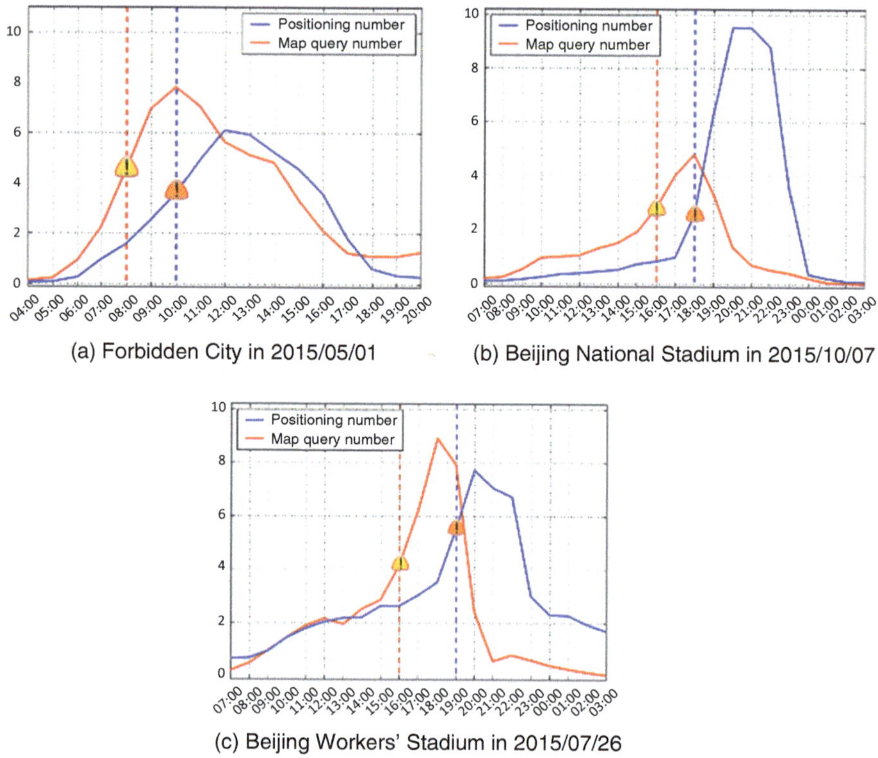

(a) Forbidden City in 2015/05/01

(b) Beijing National Stadium in 2015/10/07

(c) Beijing Workers' Stadium in 2015/07/26

Fig. 2.14 An illustration of the decision method for abnormal crowd event detection with map query data on Baidu Maps in Beijing (similar with Fig. 2.12). If the map query number is larger than a warning line, we will invoke a warning for crowd event in the following several hours (see Lemma 2.1). From the figure, we can see that the event warning can be sent 1–3 h before the positioning number surpassing the warning line. The corresponding events are (**a**) International Workers' Day (Forbidden City in 2015/05/01), (**b**) Finals of the Voice of China (Beijing National Stadium in 2015/05/01, http://www.n-s.cn/ssyc/1/), and (**c**) the Hunting Party Chinese Tour (Beijing Workers' Stadium in 2015/07/26, http://lplive.net/shows/db/2015/20150726)

2.3.2.2 Model and Feature Selection

We model the positioning number prediction task as a regression problem, where the positioning number in next hour $q(t_c+1)$ is the target value. A gradient boosting decision tree (GBDT) model, which is an effective and widely adopted ensemble machine learning technique, is employed for the task. In the GBDT model, we extract 47 features from two time series M_{tc} and Q_{tc} for the prediction task; the details of the features are described in Table 2.1.

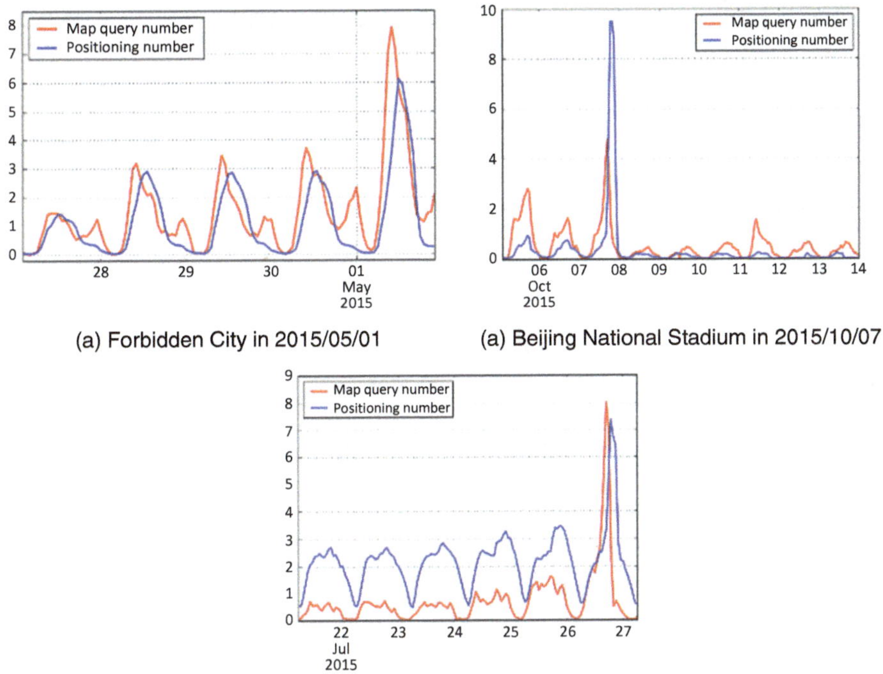

(a) Forbidden City in 2015/05/01

(a) Beijing National Stadium in 2015/10/07

(c) Beijing Workers' Stadium in 2015/07/26

Fig. 2.15 The map query number and positioning number on the Baidu Maps corresponding to the events of Beijing shown in Fig. 2.14: (**a**) Forbidden City in 2015/05/01, (**b**) Beijing National Stadium in 2015/10/07, (**c**) Beijing Workers' Stadium in 2015/07/26

2.3.2.3 Experiments

For the aforementioned six different POIs (Shanghai Bund, Shanghai Stadium, Shenzhen Bay Sports Center, Forbidden City, Beijing National Stadium, Beijing Workers' Stadium), we train six distinctive positioning number prediction models for them based on the data in 60 days before the event in each POI. As we can see from Fig. 2.18, the predicted positioning number of our model is very near to the real number. That is to say, the model provides a reliable one hour ahead prediction in the six POIs. With high quality positioning data and map query data of Baidu, we believe the positioning number prediction model can be extended to more other POIs.

Furthermore, we evaluated the importance of all features in prediction model. The top 10 important features are demonstrated in Fig. 2.19 (feature description in Table 2.1). We calculate the Gini importance score (the higher the score, the more important the feature) for each feature (L. Breiman 2001). As the figure shows, "PN1 (positioning number 1 hour ago)" and "MQ1 (map query number 1 hour ago)" are the two most important features in our model, which is consistent with our

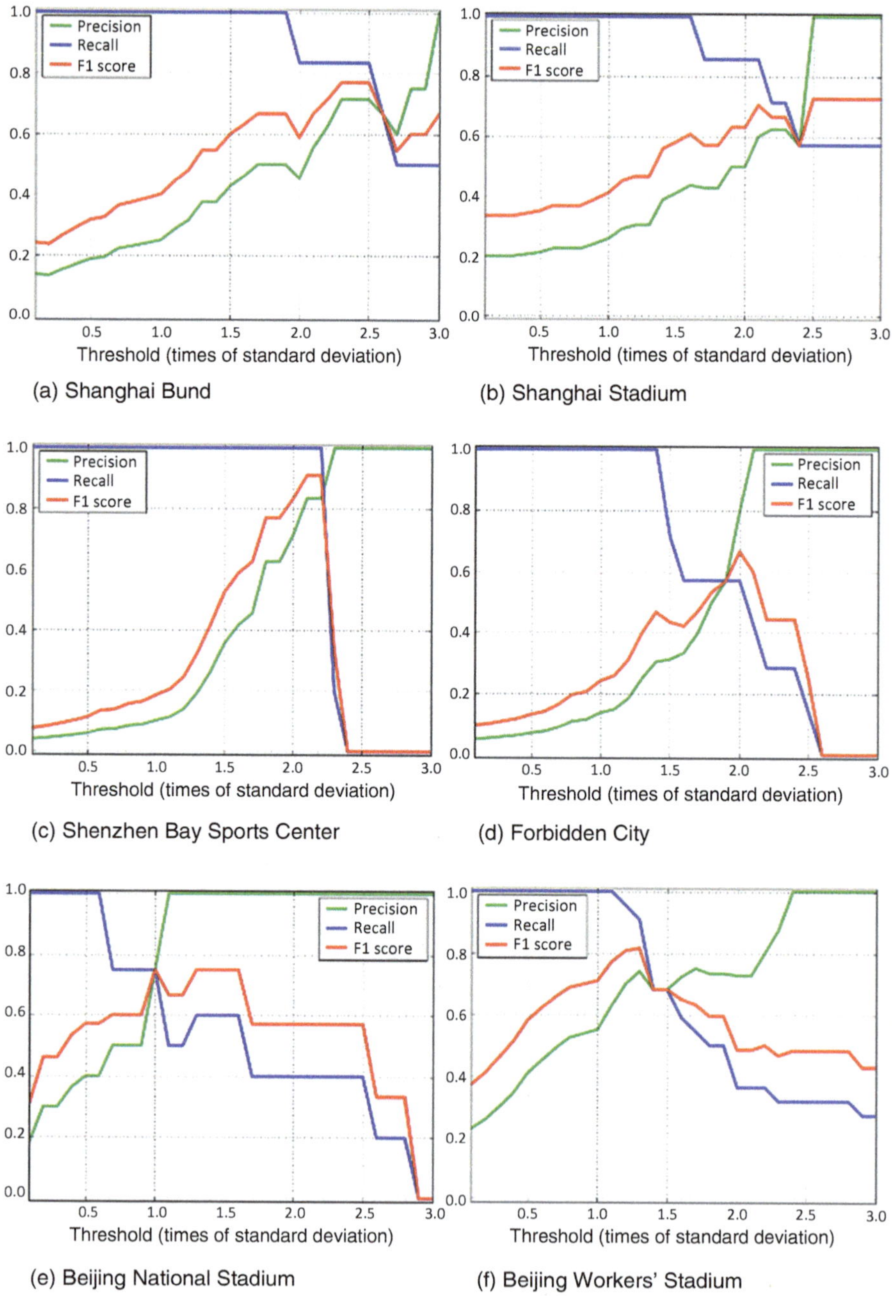

Fig. 2.16 The precision, recall, and F1-score for abnormal crowd event prediction with map query data in different places with different times (i.e., α) of standard deviations: (**a**) Shanghai Bund, (**b**) Shanghai Stadium, (**c**) Shenzhen Bay Sports Center, (**d**) Forbidden City, (**e**) Beijing National Stadium, (**f**) Beijing Workers' Stadium

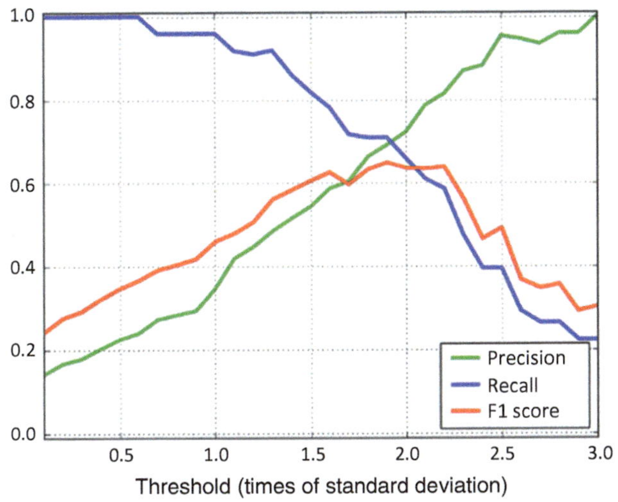

Fig. 2.17 The precision, recall, and F1-score for crowd anomaly prediction with map query data averaged in six POIs (which are the Shanghai Bund, Shanghai Stadium, Shenzhen Bay Sports Center, Forbidden City, Beijing National Stadium, and Beijing Workers' Stadium)

Table 2.1 Feature Instruction

ID	Name	Feature description
1	PN1	Positioning number 1 hago
2	PN2	Positioning number 2 h ago
....	..	...
4	PN4	Positioning number 4 h ago
5	PNS1	Positioning number at same hour 1 day ago
...	...	...
11	PNS7	Positioning number at same hour 7 day ago
12	MQ1	Map query number 1 h ago
13	MQ2	Map query number 2 h ago
14	MQY	Map query number in 20:00–24:00 yesterday
15	CT1	Current time is 1 o'clock (bool value)
...	...	...
38	CT24	Current time is 24 o'clock (bool value)
39	TW1	Today is Monday (bool value)
...	...	...
45	TW7	Today is Sunday (bool value)
46	WD	Today is weekend (bool value)
47	HD	Today is holiday (bool value)

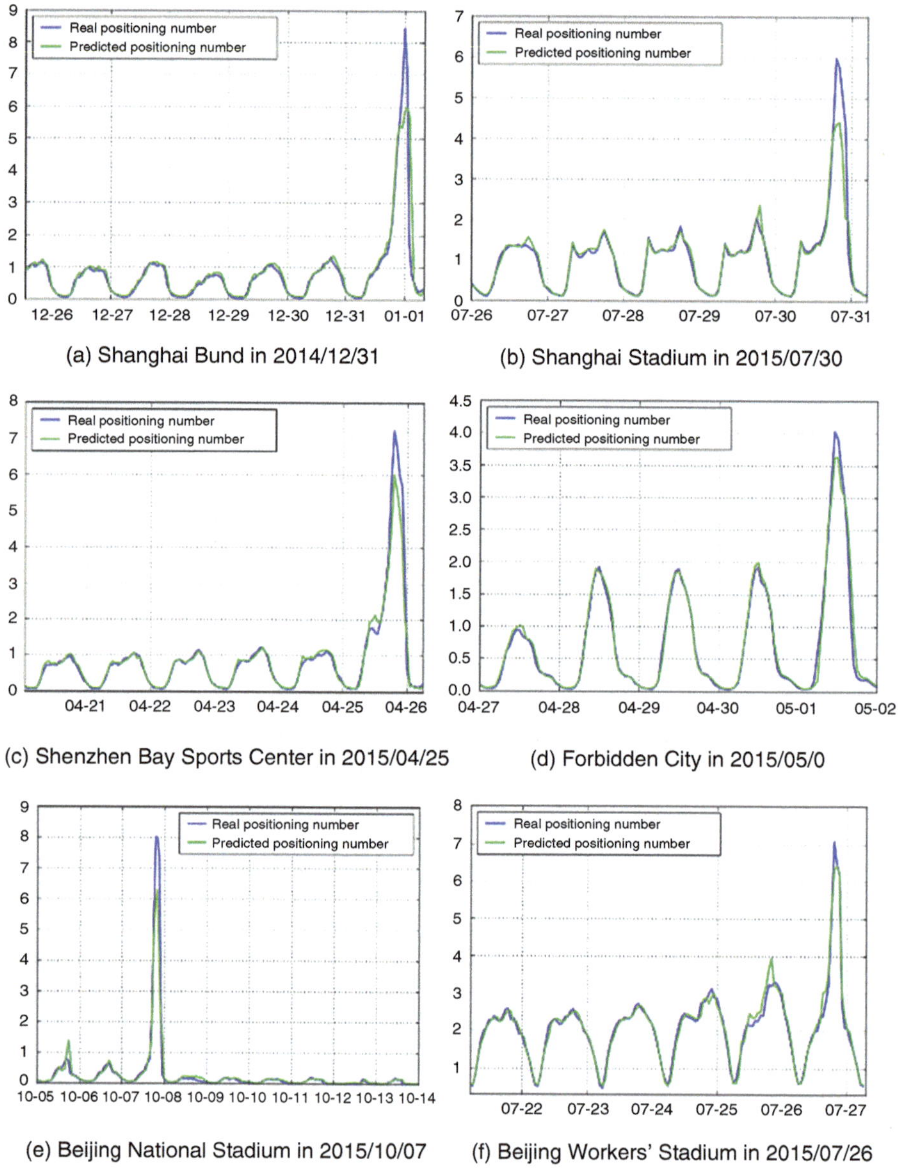

Fig. 2.18 The predicted positioning number and real positioning number on the Baidu Maps corresponding to the aforementioned events: (**a**) Shanghai Bund in 2014/12/31, (**b**) Shanghai Stadium in 2015/07/30, (**c**) Shenzhen Bay Sports Center in 2015/04/25, (**d**) Forbidden City in 2015/05/01, (**e**) Beijing National Stadium in 2015/10/07, (**f**) Beijing Workers' Stadium in 2015/07/26

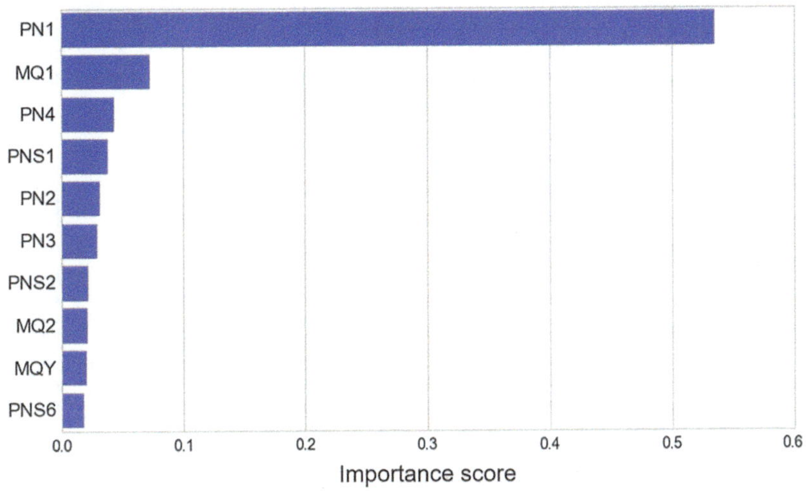

Fig. 2.19 The feature importance rankings in prediction model

intuition. It is worth noting that "MQY (map query number in 20:00–24:00 yesterday)" is also an important feature as many people prefer to plan their routine in the night before.

From the importance ranking in Fig. 2.19, we can see features from map query data are a vital part for positioning number prediction. In order to test the essentiality of map query data for prediction, we further design a comparative experiment, where two GBDT prediction models are developed: one with features from map query data and the other without. Then, we can directly capture the importance of map query data by quantity measuring the prediction error of two models. We use mean absolute error (MAE) as evaluation metrics to measure how close predictions are to the real positioning numbers. As the experiment result shown in Fig. 2.20, the performance of the model with the features is superior than the model without the features. To sum up, map query data exactly play a vital role in positioning number prediction.

2.4 Conclusion

The management of unexpected huge human crowd events is a challenging but serious problem for public health, whereas poor management of such events may lead to unfortunate stampedes. With the help of the big data generated on the Baidu Maps, we propose a solution for improving the effectiveness of the crowd management. The novelty of the solution lies in an objective fact that many people will search on Baidu Maps to plan their itineraries. This prior information indicated by anonymous users of Baidu Maps provides promising and compelling advantages for

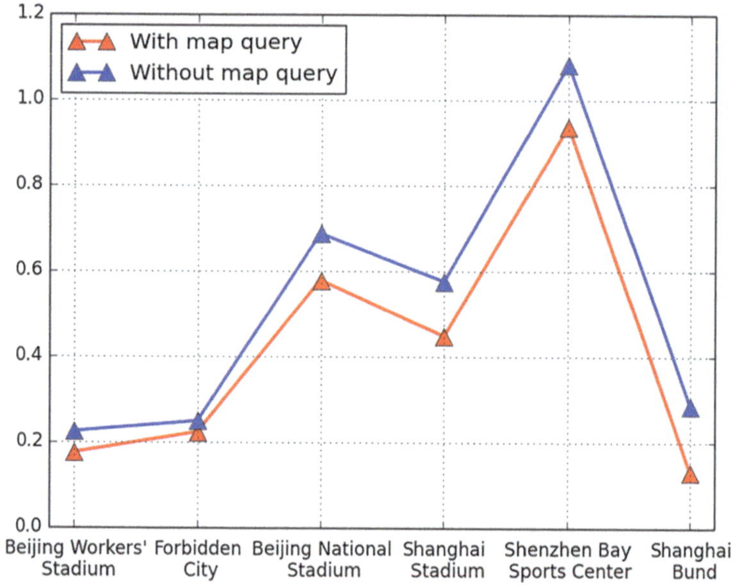

Fig. 2.20 The Mean Absolute Error (MAE) of the prediction models. *Red* line is the result of model with features in Table 2.1, and *blue* line is the one with features in Table 2.1 but without map query related features; y denotes the MAE score

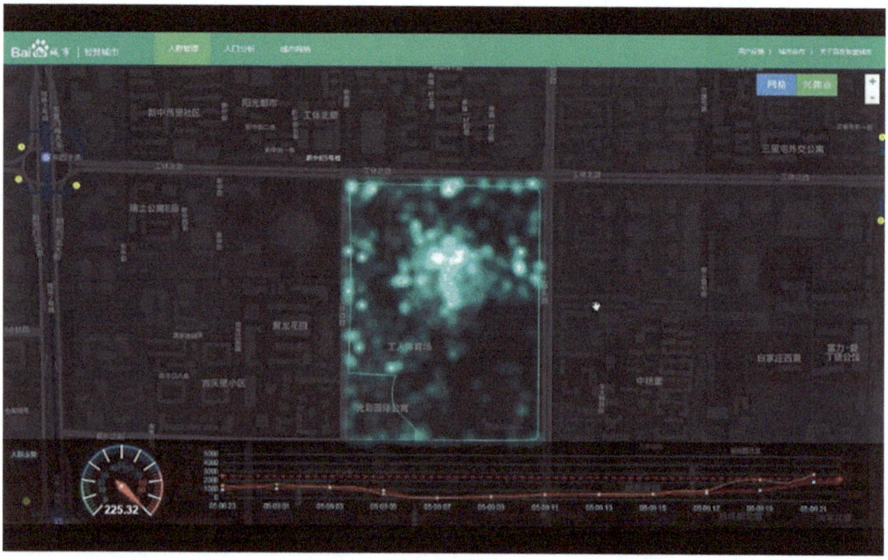

Fig. 2.21 A snapshot of the early warning system for human crowds

us to invoke early warning for possible crowd events. In this chapter, we deploy a deep study for the crowd anomaly prediction with resorting to the help of the Baidu Maps' data. Our solution includes a qualitative decision method to perceive the crowd anomaly as well as a quantitative prediction method to assess the risk of the crowd event. We believe the successful deployment of our method can bring many benefits to our society.

It is worth mentioning that we have developed an early warning system for human crowds based on the aforementioned decision method and prediction model. A completed demonstration video of the system can be visited,[2] and a snapshot is shown in Fig. 2.21.

References

Investigation report of 2014 shanghai stampede event. http://www.shjcw.gov.cn/2015jjw/n2230/n2237/ u1ai51007.html

Ali, S., & Shah, M. (2008). Floor fields for tracking in high density crowd scenes. In *Computer Vision–ECCV 2008* (pp. 1–14). Heidelberg: Springer.

Ashok, K. (1996). *Estimation and prediction of time-dependent origin-destination flows* (PhD thesis). Massachusetts Institute of Technology.

Breiman, L. (2001). Random forests. *Machine Learning, 45*(1), 5–32.

Gallup, A. C., Hale, J. J., Sumpter, D. J., Garnier, S., Kacelnik, A., Krebs, J. R., & Couzin, I. D. (2012). Visual attention and the acquisition of information in human crowds. *Proceedings of the National Academy of Sciences, 109*(19), 7245–7250.

Gonzalez, M. C., Hidalgo, C. A., & Barabasi, A.-L. (2008). Understanding individual human mobility patterns. *Nature, 453*(7196), 779–782.

Helbing, D., Johansson, A., & Al-Abideen, H. Z. (2007). Dynamics of crowd disasters: An empirical study. *Physical Review E, 75*(4), 046109.

Schneider, C. M., Belik, V., Couronn'e, T., Smoreda, Z., & Gonza'lez, M. C. (2013). Unravelling daily human mobility motifs. *Journal of the Royal Society Interface, 10*(84), 20130246.

Zheng, J., & Ni, L. M. (2012). An unsupervised framework for sensing individual and cluster behavior patterns from human mobile data. In *Proceedings of the 2012 ACM conference on ubiquitous computing* (pp. 153–162). ACM.

[2] http://www.iqiyi.com/w_19rrpwo1mt.html#vfrm=2-3-0-1

Chapter 3
Spatial Distribution Characteristics of Residents' Emotions Based on Sina Weibo Big Data: A Case Study of Nanjing

Feng Zhen, Jia Tang, and Yingxue Chen

Abstract Many urban planning approaches have emphasized the need to be people oriented. With the fast development of cities, more attention should be given to the perception and spatial experiences of the residents. Following urban planning theories and studies on environmental quality, infrastructure allocation, and planning, research on residents' emotions has drawn increasing research attention. Nanjing City of China, is used as an example for applying Sina Weibo big data, a new type of data, to extract real-time emotions and their corresponding geo-locations of residents. This study uses ArcGIS software to analyze the spatial distribution characteristics of residents' emotions in the overall city and in different types of places of Nanjing. Grid statistics are used to provide evidence to optimize urban space development.

Keywords Big data • Residents' emotions • Spatial distribution characteristics • Nanjing

3.1 Introduction

Following urban planning theories and studies on environmental quality, infrastructure allocation, and planning, research on residents' emotions has drawn increasing research attention. In recent years, emotional spatiality and emotional responses to spaces have been discussed by an increasing number of academic papers. For instance, Kwan used qualitative interviews, automatic photography, travel diaries, and collaborative 3D-GIS videography and other technologies to study the

F. Zhen • J. Tang (✉)
School of Architecture and Urban Planning, Nanjing University, Nanjing, China
e-mail: jiatang8891@163.com

Y. Chen
Shanghai Urban Planning & Verifying Center, Shanghai, China

Z. Shen, M. Li (eds.), *Big Data Support of Urban Planning and Management*,
Advances in Geographic Information Science, DOI 10.1007/978-3-319-51929-6_3

emotional trajectories of activities (Kwan 2007). Hemming analyzed children's hand drawings to identify children's feelings towards various healthy and unhealthy places and lifestyles (Hemming 2007). Heimtun illustrated feminist sentiments on spaces by investigating the emotional experiences of middle-aged single women who dine out and travel alone (Heimtun 2010). Moisi (2010) divided specific areas into three kinds of emotional spaces, hope, shame, and fear, after the September 11 terror attacks in New York (Moisi 2010). In addition, many studies have elucidated the emotional characteristics of various spaces, such as in schools (Fielding 2000), shelters (Philo and Parr 2000), workplaces (Blumen 2002), zoos (Hallman and Benbow 2007), shops (Miller 2014), and prisons (Crewe et al. 2014), for practical use in urban planning. Meanwhile, Williams et al. investigated the emotions triggered by the interactions between retail industry and consumer spaces in specific locations (Williams et al. 2001). Colls focused on the emotional spatial characteristics of the experience of women on the consumption of apparel (Colls 2006). Major cities in the UK were used by Lüscher and Weibel as examples to propose three new methods to identify urban centers based on the geographical features of urban space together with residents' emotions and their experience (Lüscher and Weibel 2013).

With the rapid development of information and communication technology (ICT), big data of residents' activities can now be collected. Big data has advantages superior to those of small sample data in terms of exploring development rules and intrinsic correlations. As a result, academics have sought to quantify the emotional characteristics of residents based on new data and emerging technologies. For instance, Hogertz attempted to identify a city's positive and negative spaces by data mining on social networking sites and geo-location tags, analyzing the emotions and feelings of residents at different locations (Hogertz 2010). Tumasjan et al. used Twitter big data to reflect the political situation in Germany as well as to predict the outcome of the election and reveal the changes in its political scene (Tumasjan et al. 2011). Stieglitz and Dang-Xuan also used Twitter big data of users' emotional perspectives to analyze their sharing behavior. They found that emotional information is often more likely to be spread out than neutral information (Stieglitz and Dang-Xuan 2013). Golder et al. produced a global emotion map by analyzing Twitter users' real-time emotions of 2.4 million tweets tracked globally from 84 nations spanning a period of 2 years.[1]

Spatial researches based on emotional analysis have generally focused on the impact of the environment on activities, the characteristics of the specific types of space, and the identification of a city's positive and negative spaces. However, restricted by the difficulty of collecting resident behavioral activity and emotional data, majority of existing research were based on the subjective responses from activity logs, in-depth interviews, surveys, narrations, and other data collection methods. In addition, most of these have been conducted on the macroscopic scale. Notably, a number of scholars have begun to explore new types of data regarding residents' sentiments, resulting in a significant increase of samples. However, previous research

[1] http://blogs.discovermagazine.com/

has not focused on the spatial characteristics of emotions and the relationship between the residents' emotions and the different places. As a result, no current studies have focused on urban space layout optimization, improvement of environmental quality, and practices.

The primary purpose of this study is to explore the spatial distribution characteristics of residents' emotions via Sina Weibo big data. Specifically, this study selects Nanjing City as an example, regards the residents' emotions as the object of study, and extracts the residents of real-time emotions and their corresponding space information based on "check-in" big data. On this basis, the general spatial characteristics of Nanjing residents' emotions and the distribution characteristics of the residents' emotions in different places were analyzed to provide a basis for the optimization of city space.

3.2 Data Collection and Research Methodology

3.2.1 Research Scope

Nanjing is one of the major cities in the Yangtze River Delta, China. Its social and economic development has been at the forefront nationwide. After its rapid urbanization process, the built-up area of Nanjing has been expanding rapidly. However, rapid urbanization also exposed the differences between areas with regard to their functional layouts, efficient use of facilities, and environmental quality. The emotions of urban residents are highly likely to reflect these differences in their daily activities. Hence, Nanjing was used as a case study. This study uses the map of Nanjing before the adjustment of administrative divisions of Nanjing in 2013 (Fig. 3.1).

3.2.2 Data Collection and Processing

3.2.2.1 Data Collection

Sina Weibo was used to collect data in this study. Sina Weibo is akin to a hybrid of Twitter and Facebook. Sina Weibo can be used by mobile terminal, which is an open data platform with comprehensive content. Meanwhile, Weibo users' emotions and geo-locations can be correlated by collecting Weibo posts with geo-tags together with emotion icons. Emotion icons used in Weibo are the actual emotional expressions of users. Hence, the relationship between residents' emotions and urban space can be analyzed by collecting emotion icons used in Weibo.

POI data were collected from 2 Mar 2015 to 15 Mar 2015 by using the geolocation service API of Sina Weibo. Data was collected by selecting coordinate points and determining the appropriate search radius in the research area. Considering the differences in the population density of urban and rural areas,

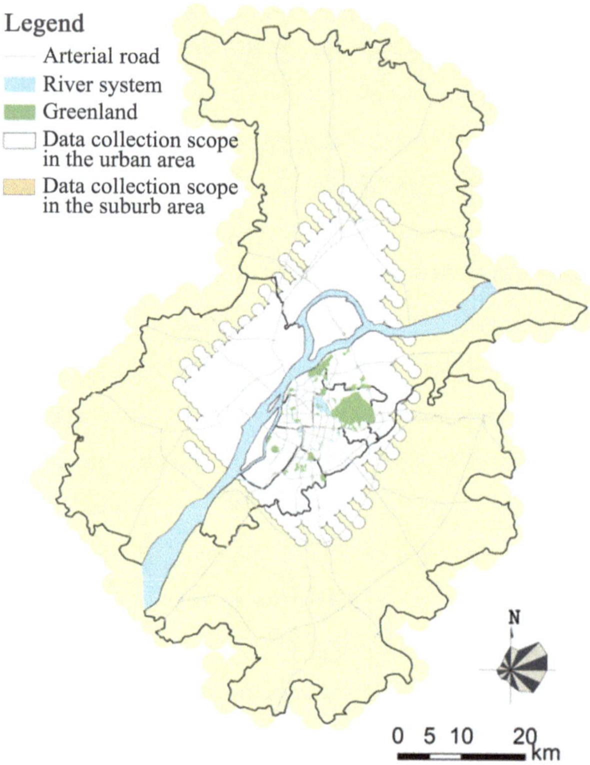

Fig. 3.1 Data collection scope

through repeated experiments, 604 POIs were chosen in the urban area with a search radius of 1 km, and 589 POIs were selected in rural area with search radius of 2 km to ensure full coverage. Finally, a total of 103,443 geo-tagged "check-in" data were collected, as shown in Fig. 3.2.

3.2.2.2 Data Processing

This study used emotion icons in Sina Weibo to quantify residents' subjective emotions (Fig. 3.3). Five emoticon icons exist: very positive, positive, neutral, negative, and very negative, with assigned values of 5, 4, 3, 2, and 1, respectively (Table 3.1). If more than one emotion icon appeared in one Weibo post, it was assigned a score based on its highest emotion icon. 34,955 Weibo posts were found with emotion icons. The numbers of Weibo posts assigned with emotion values of 5, 4, 3, 2, and 1 were 11341, 6950, 6785, 3537, and 6342 posts, respectively.

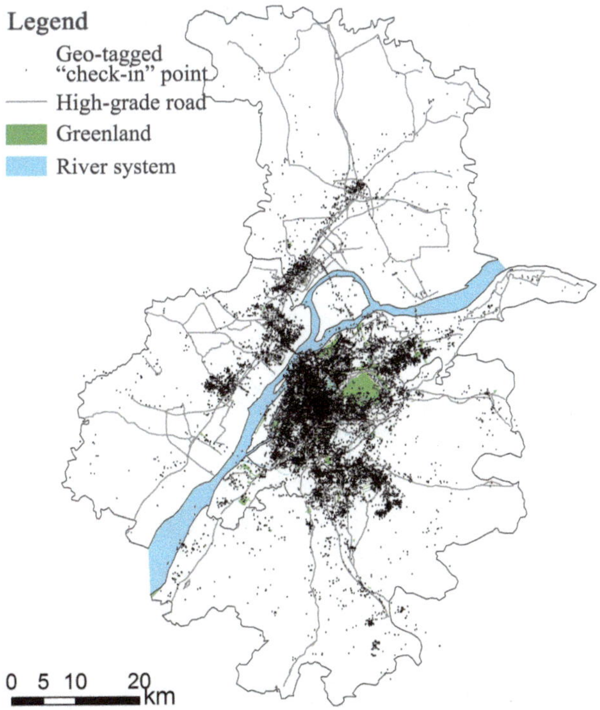

Fig. 3.2 Spatial distribution of "check-in" data

3.2.2.3 Types of Places in City

This study divided places into four types based on the functional elements in the study of city spaces and the common types of places in "check-in" Weibo posts: residence, working, traffic, and life service based on the "check-in" Weibo posts. Their detailed sub-types are presented in Table 3.2.

3.2.3 Methods

Given that more attention was given to the relationship between emotions and micro spaces, such as single buildings or a single place, the research methods were mostly based on the qualitative analysis of pictures (e.g., hand drawing, photos), the content analysis of the interview (Heimtun 2010; Miller 2014; Crewe et al. 2014; Colls 2006), and other methods. Few studies have focused on the impact of macro urban areas on the emotions of the residents or have used ArcGIS software, such as 3D-GIS (Kwan 2007) and nuclear density analysis (Lüscher and Weibel 2013), to study the emotional space of residents. Residents' emotions and their geographical

Fig. 3.3 Examples of emotion icons in Sina Weibo

locations can be expressed using ArcGIS, which is a favorable platform to study the relationship between residents' emotions and macro urban spaces.

In this chapter, the general spatial distribution characteristics and spatial distribution characteristics in different places are discussed by using grid statistics and analysis methods based on ArcGIS software. Grid statistical analysis can be used to decompose macro urban spaces to an appropriate micro spatial scale, thus facilitating the exploration of the spatial distribution characteristics of residents' emotions in urban spaces. First, based on the spatial coordinates of the user's "check-in" position, ArcGIS was used to spread the emotional value of the residents to the study area (Fig. 3.4). Second, in ArcGIS, the grids covering the research scope were set up with 1 km side length.[2] The emotional value of the residents in each grid area is the

[2] 1 km^2 is basically in accordance with the scale of urban spatial organization unit (e.g., neighborhoods).

Table 3.1 The meanings of common emotion icons in Sina Weibo and emotion value

Emotion types	Emotion value	Meanings of emotion icon examples
Very positive	5	Satisfied smile, whee, chuckle to oneself, ha ha, smile, so happy, applause, gelivable, yeah, great, love you, kiss, powerful, oh yeah, laugh, proud, happiness, toothy, enjoyable
Positive	4	Hehe, lovely, sweet, cool, heart, good, gifts, cakes, sun, breeze, make faces, cheers, infatuated, happy birthday, scattering flowers, excited, cute, airplane, mischievous, see, passing love, handsome, have a holiday, flowers
Neutral	3	Surprised, winking, hush, hug, doubt, bye-bye, think, money, what, onlookers, mutual fans, Ultraman, pig, Ok, come, clock, microphone, insidious, shy, coffee, make a wish, embarrassed, pressure, watermelon, turn back, fist, fallen leaves, boy, girl, music, recommend, movies, shake hands
Negative	2	Despise, shut up, not care you, hum, dizzy, confused, sorry, yawning, sleep, sleepy, weak, don't want to go to work, catch a cold, trapped died, nerd, ungeilivable, tragedy, tired, depressed, can't understand, vesania, dust storms, rain
Very negative	1	Pathetic, sad, tears, tears streaming down one's face, illness, vomit, anger, sorrow, disappointment, mad, losing grip, curse, injustice, miserable, collapse, crying, struggle, anger, wail, got tangled, big blow

Table 3.2 Types of places

First classification	Second classification	Third classification
Residence	Residence	–
Working	Working	Industrial park, government, office building, other organizations
Traffic	External traffic	Airport, train station, port and wharf, coach station, external rode
	Inner city traffic	Metro line/metro station, bus station, city road
Life service	Public service	Hospital, library, barbershop, automobile service places, institution of higher education, high school/middle school/vocational school, others
	Commercial sites	Supermarket, complex market store, special store
	Catering places	Ordinary restaurant, fast-food restaurant, beverage/dessert shop, take-out restaurant
	Hotel	Three-star or above hotel, economic hotel, family inn
	Entertainment places	Science and education places, theater/cinema, sports places, others
	Outdoor public places	Park, square, theme park, scenic region and spot, vacation and convalescent places

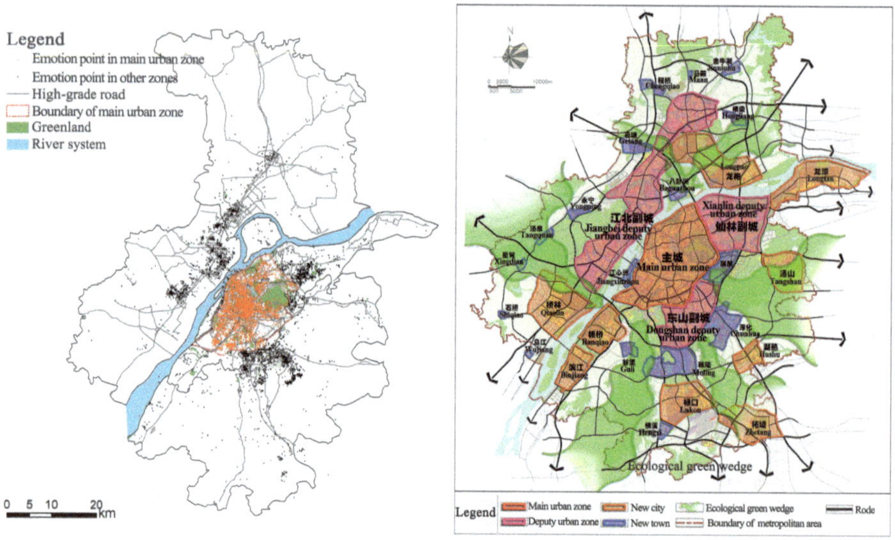

Fig. 3.4 Comparison between the "check-in" emotion point distribution (*left*) and the spatial structure planning of metropolitan area (*right*)

arithmetic mean of the emotional value of all the emotional points. The calculation formula is as follows:

$$P_i = \left(m_1 + m_2 + \ldots + m_x \ldots + m_n\right)/n$$

P_i is the mean of the residents' emotions in the *i*th space grid, m_x is the emotional value of the residents in the *x*th grid, and *n* is the total number of residential emotional points within the grid. Natural break clustering (Tang 2006) is used to divide the mean values of the emotional values of these grids into nine levels to show the differences in the spatial distribution of the residents' emotions. Finally, these were combined with the development situation of the space to identify reasons for the differences.

3.3 Overall Spatial Distribution Characteristics of Residents' Emotions

Figure 3.4 shows that the overall spatial distribution of the "check-in" emotion was highly consistent with the distribution characteristics of all the "check-in" points (Fig. 3.2). The spatial pattern of the residents' emotions was highly consistent with the urban distribution and the residents' activity space in Nanjing (Chen and Zhen 2014), which also indicated representativeness of the data.

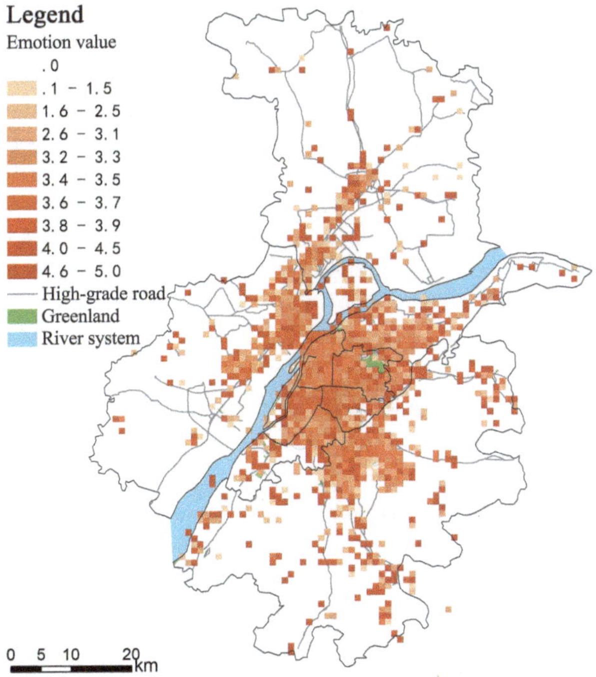

Fig. 3.5 Overall spatial distribution of residents' emotion

The overall spatial distribution of emotional space is shown in Fig. 3.5. The clustering results show that the neutral (emotional value = 3) and negative values (emotional value <3) were divided into three levels of emotional value. The other six levels of emotional value intervals were positive. In addition, the median interval of residents' emotional space was 3.4–3.5 in Nanjing.

The six continuous grids where the residents' emotional value was 3.8 and above were considered as high emotional spaces, whereas the grids where the residents' emotional value was 3.1 and below were considered as low emotional spaces.

3.3.1 Spatial Distribution Characteristics of Emotion in Urban Areas

For the main urban zone, the continuous high emotional grids are located mainly in large-scale scenic regions and their surroundings. These areas have high-grade natural, historical, or cultural landscapes. Examples include the Zijin Mountain scenic area, the Confucius Temple-Qinhuai River scenic area, the bank of the Yangtze River, and the Olympic Sports Center. These areas have a good environmental quality and function mainly for sightseeing or leisure. The remainder of the grids had emotional values ranging from 3.2 to 3.7, were characterized by neutral and positive emotions, and showed characteristics of homogenization.

The spatial distribution of residents' emotions was more obvious in the deputy urban zone than in the main urban zone. The continuous high emotion grid was mainly distributed in earlier developed areas, such as Dongshan Deputy City and Jiangbei Deputy City, which possess concentrated development and construction, superior traffic conditions, complete facilities, a pleasant living environment, and so on. The continuous low emotion grid was mainly distributed in industrial parks, and residential areas that were later developed, such as the Pukou High and New Technology Development Zone, Xianlin University City, and Jiangning University City, possess low-intensity construction, inadequate infrastructure and public service facilities, and poor living conditions.

3.3.2 Spatial Distribution Characteristics of Emotion in Suburban Areas

Overall, the proportion of the high emotional grid in the suburbs was significantly higher than that in urban areas. In the suburbs, the high emotional grid was concentrated in a region closely connected with the urban area or environments with good conditions, such as Tangshan, Lukou, Banqiao, Guli, Baguazhou, and other areas along the airport expressway. Tangshan, Gu Li, and Baguazhou are important leisure resorts in Nanjing whose leading function is special and whose environmental superiority is prominent. Lukou and the areas along the airport expressway are the bearing spaces of important transportation and related services in Nanjing. Compared with comprehensive transportation hubs inside the city, the airport has more specific function, facilities, and service levels that are often superior to those of other types of external traffic station areas.

3.4 Spatial Distribution Characteristics of Emotion in Different Places

The responses of residents' emotions to urban spaces are initially determined through the micro place environment and then through macro construction and development. Therefore, this study further studied the differences in the spatial distribution of residents' emotions in micro places of the different types of spatial function and the reasons for their formation.

3.4.1 The Emotion Distribution Characteristics in Residential Areas

As shown in Fig. 3.6, the median of residents' emotion values ranges from 3.2 to 3.8, and the grids of the emotional value higher than 3.8 accounted for more than half of the total number of grids. The number of the grids with an emotional value

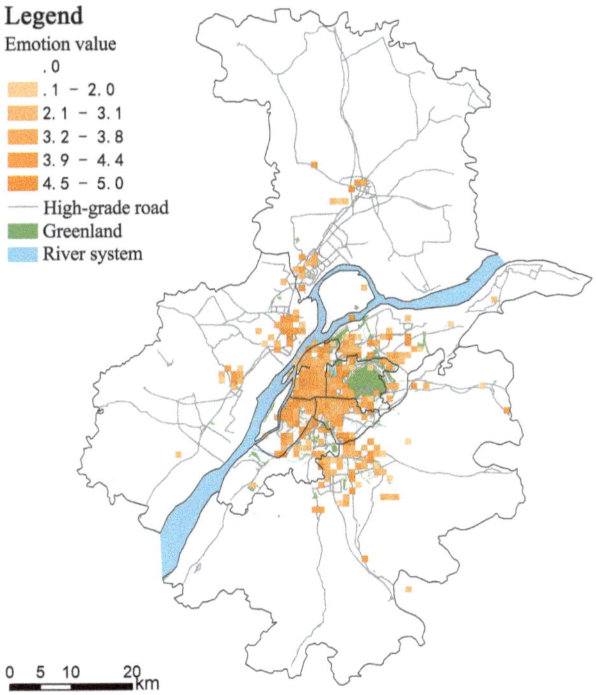

Fig. 3.6 Spatial distribution of emotion in residential areas

lower than 2.0 was 28, and those ranging from 2.1 to 3.1 was 75, all of which accounted for 37.2% of the total number of grids. These values indicate that the residents' emotion in the residential area was positive and that the overall living environment in Nanjing was good.

Furthermore, the residential areas with high emotion value were mainly distributed in the newly established residential areas with superior commercial service facilities, such as the new urban zone of Longjiang, Hexi. The residential areas with low emotion value were mainly distributed in the early established residential areas with poor living environments and poor development. Residential areas with low emotion value were mainly located in Maigaoqiao surroundings in Qixia District, Changbai Street and Chang Fu Street in Qinhuai District, and Ande Door surroundings in Yuhuatai District.

3.4.2 *The Emotion Distribution Characteristics in Workplaces*

Figure 3.7 shows that the workplaces were mainly distributed in urban areas. The workplaces with high emotions were mainly distributed in places with perfect commercial function, convenient transportation, and good internal and external

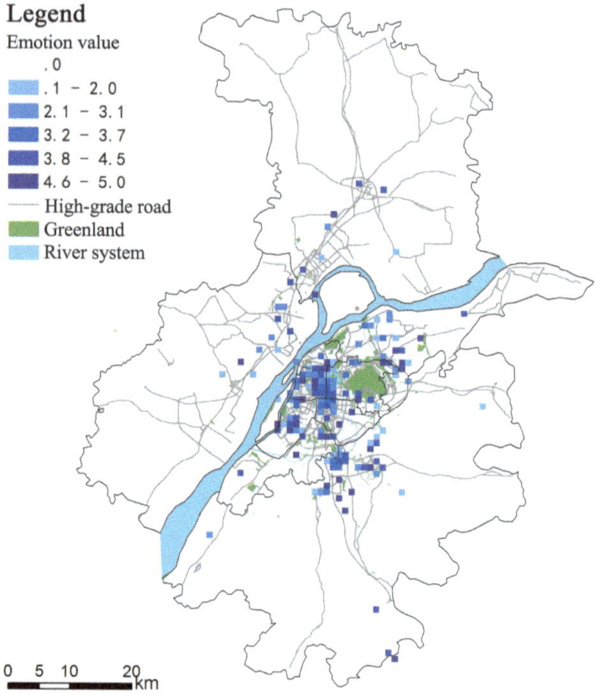

Fig. 3.7 Spatial distribution of emotion in workplaces

construction environments, such as the downtown area, the surroundings of Jiangsu provincial government, the southwest area of the Olympic Sports Center, and the northwest of Yuhuatai District. The workplaces with low emotion were mainly distributed in industrial parks located in the suburbs. The residents' emotion in the workplace was affected by the overall environmental quality of the region and depended largely on the working properties in the workplace and the type of industry. For example, a company office building was always located in a place with relatively good commerce and service functions, and the environmental quality of the industrial area was relatively poor. Two explanations for this spatial pattern exist. On one hand, the ideas of past urban planning regarding the development of such a single functional area with low land prices located far from the city center tended to focus on low prices and high economic benefit. On the other hand, given the rapid expansion and disorderly spread of urban periphery development zones and industrial areas as well as the extensive spatial organization of their interior space, elements of the material were often only able to meet production needs without considering the environmental experience of the employees.

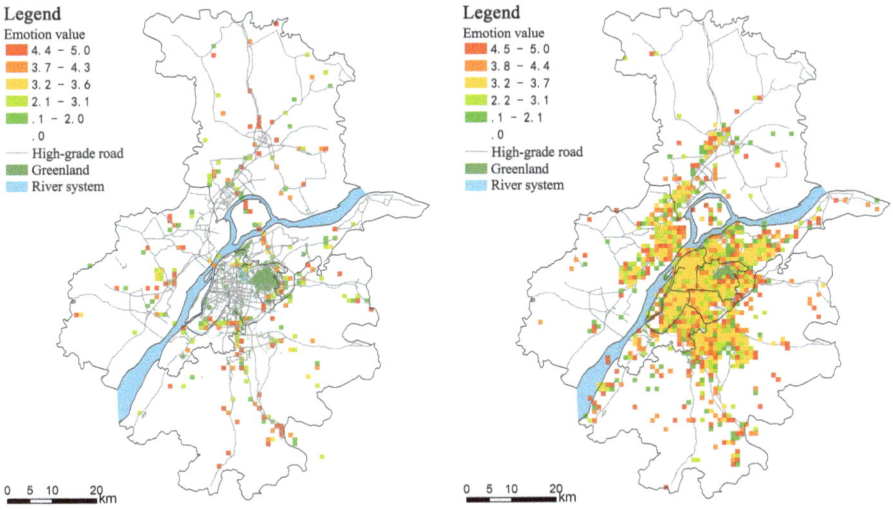

Fig. 3.8 Spatial distribution of emotion in external (*left*) and internal (*right*) traffic places

3.4.3 The Emotion Distribution Characteristics in Traffic Place

The urban transportation system comprised internal traffic and external traffic. Many internal and external traffic transfer activities happen in the urban transportation hub; thus, a comparative analysis was conducted. Figure 3.8 shows that in the external traffic areas, the continuous high emotional areas were mainly distributed in the suburbs and near the exit of the high-speed road in the urban fringe area. These areas were not directly connected with the urban internal traffic and did not need to share the internal road traffic load. Low emotional areas were located in the main urban zone and peripheral traffic connection point, such as the expressway section of Nanjing Yangtze River Bridge in the north of the Yangtze River, the expressway section of Nanjing No. 3 Yangtze River Bridge in the north of the Yangtze River, and the airport expressway entrance section. These areas linked internal and external traffic and were thus under pressure during rush hour. Therefore, urban comprehensive transportation system planning must emphasize the optimization of the traffic capacity of the road system connection between the main urban zone and their outer region.

In urban internal traffic areas, residents' negative emotion was more obvious in the subway line/subway station than along the road. This result showed that the experiences of residents on road-based transport modes were better than those regarding rail transit. The residents' negative emotion reflected the delayed construction of the Nanjing urban rail transit system, which lagged behind the needs of the residents.

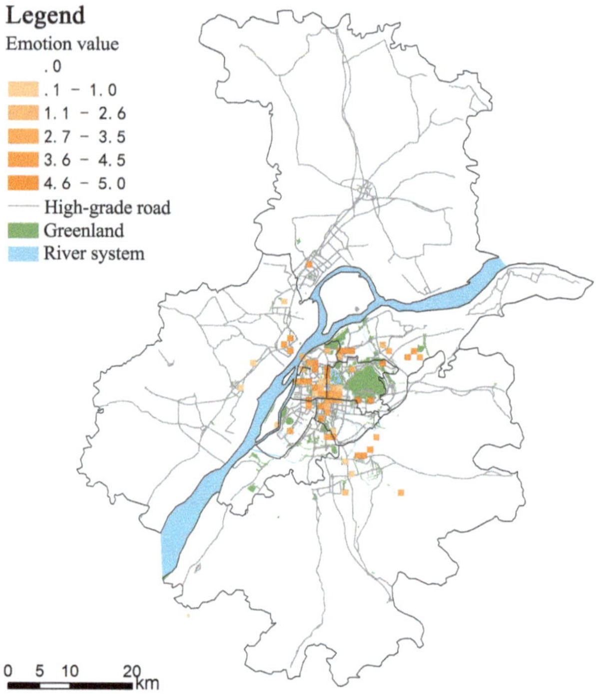

Fig. 3.9 Spatial distribution of emotion of in public service places

3.4.4 The Emotion Distribution Characteristics in the Life Service Place

3.4.4.1 The Characteristics in Public Service Place

The residents' "check-in" places mainly included education places, hospitals, libraries, barbershops, and automobile service places. The education areas will be studied individually because these are relatively independent spaces. Analyzing the other public places (as shown in Fig. 3.9), this study finds that the emotional values of residents were unevenly distributed. Most emotional spaces are concentrated in urban areas rather than in the suburb areas. Specifically, public service places with high emotional value were concentrated in areas with sufficient public service facilities, such as the northwest Gulou District and Daqiao area and the core area of deputy city. However, few distributions of emotions in the new residential area, such as the west in the main urban area, and areas with lower intensity of development of the deputy city were observed. Therefore, the study concludes that a close correlation exists between emotional value and service level and matching degree of service facilities of regional public service center.

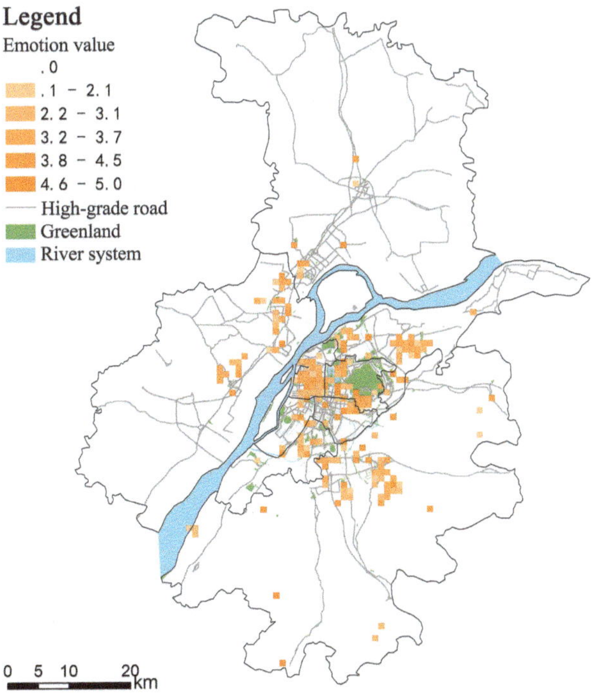

Fig. 3.10 Spatial distribution of emotion in education places

The emotional distribution of educational places is shown in Fig. 3.10. Colleges and universities accounted for 98% of the emotional points in educational places. Therefore, the spatial variation of college and university students' emotion and its relationship with the campus environment can be investigated. As seen in Fig. 3.10, the absence of a continuous high emotion or low emotion grid indicates the lack of a regional integrity of students' emotion in the colleges and universities. The reason may be that the "enclosing wall" hindered the communication between the internal material environment of the campus and the external spaces, which is "separate" from the outside and blocked from the extension of the external environment to the campus space organization and environmental quality. This results in independent functions, landscapes, and environmental organizations in the campus. Therefore, the relationship between emotion and campus space must be studied further.

3.4.4.2 The Characteristics in Commercial Sites

Commercial sites are related to the residents' shopping activities. Similar to the distribution of public service places, residents' emotions were mainly distributed in urban areas and less distributed in the suburbs (Fig. 3.11). Overall, the residents'

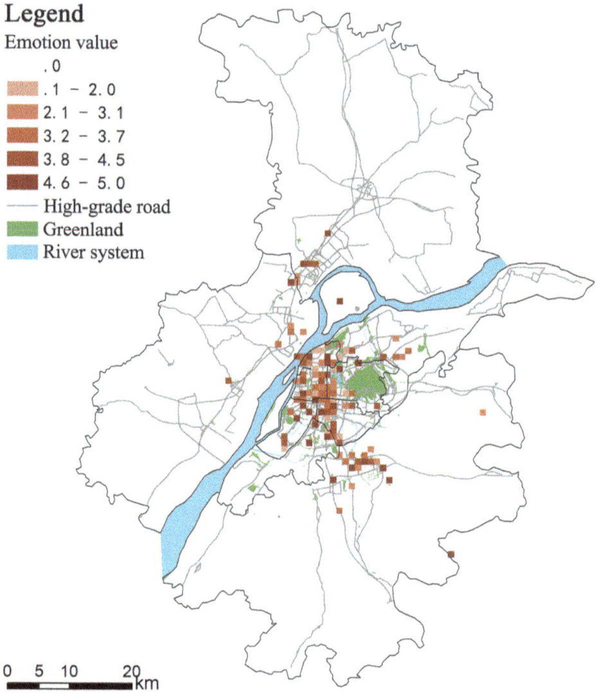

Fig. 3.11 Spatial distribution of emotion in commercial sites

emotions in these places tended to be positive, mainly due to the jovial spirit of commercial service places. The main urban zone did not have a higher emotional value in the most prosperous business centers but at the median level. In contrast, it had a higher emotional value in the business center at the edge of the main urban zone. This phenomenon may have occurred due to the aggregation of commercial sites and large costumer flow. Its management may require improvement because it was built in the early age in the urban center. In addition, it also showed a high emotional value in the better business center of the deputy city. Therefore, the residents' emotion in commercial sites was mainly affected by the building age, the overall environmental quality, the service level, the management level, and other factors. In this case, enhancing the overall quality of the environment and business services through continuous, small-scope organic renewal may help commercial sites in the urban center to meet the needs of the residents.

3.4.4.3 The Characteristics in the Catering Places and Hotels

As shown in Fig. 3.12, the spatial distribution characteristics of the residents' emotion in catering places and hotels places are similar. Specifically, the emotion value of residents was not high in the urban center because non-star hotels were not eliminated due

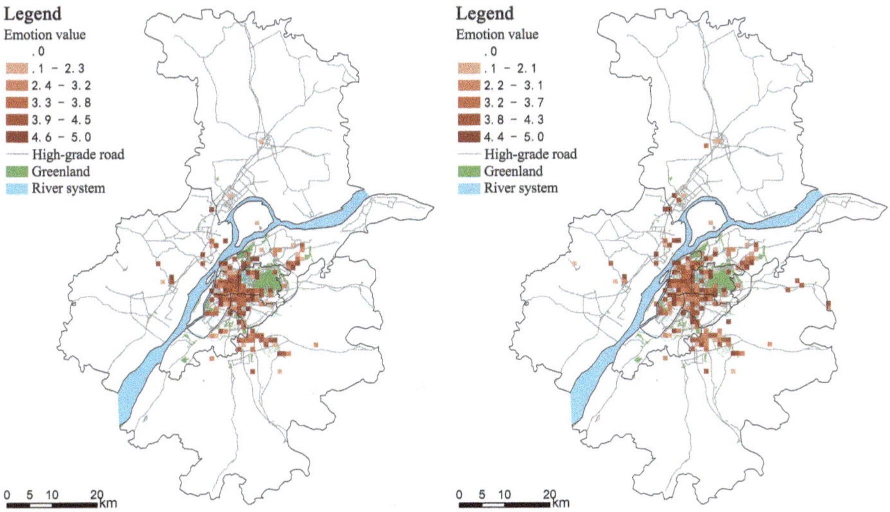

Fig. 3.12 Spatial distribution of emotion in catering places (*left*) and hotels (*right*)

to the demands of consumers, although the hotel facilities, space, and environment were relatively old and the space was cramped. Interestingly, high emotional values were distributed in the catering places and hotels at the edge of the main urban zone, although catering places and hotels had a smaller distribution. The reason may be that catering places and hotels in the urban periphery benefit from the cheap rent, the newly constructed facilities, and the quality of the environment. Therefore, the environmental quality of the catering places and hotels in the urban center should be further improved. This kind of places in the urban periphery should be matched with the size of the population and living activities to improve the residents' living service experience.

3.4.4.4 The Characteristics in Entertainment Places

For entertainment places, high emotions were mainly distributed in the west and southwest of the urban center (Fig. 3.13). The spatial distribution of high emotional value in the west of urban center is similar to that in residential areas. The recreation facilities in this area are newly constructed and are superior in quality. Compared with the basic public service facilities, the entertainment facilities in the southwest of the urban center are more mature and perfect. Therefore, the residents' emotion in entertainment places was mainly affected by the quality of construction, the allocation of service facilities, and the effect of planning and construction. Therefore, proper planning and construction adjustment should be made by planning departments according to the actual demands of the residents, including the adjustment of the scale of allocation and construction, the layout of space, and the period of construction.

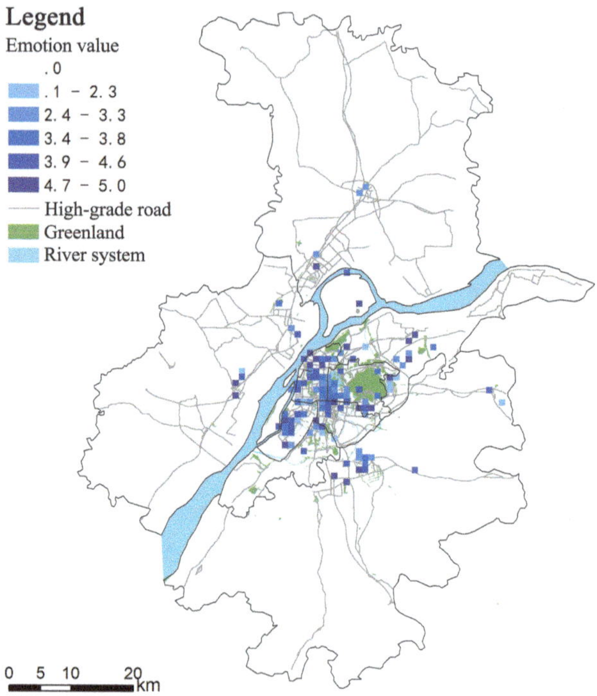

Fig. 3.13 Spatial distribution of emotion in entertainment places

3.4.4.5 The Characteristics in Outdoors Public Places

As shown in Fig. 3.14, the high emotional value of outdoor public places was mainly distributed in areas with pleasant landscapes and environments. For instance, the west of the urban center has formed a perfect green network system of parks, which had greater access to public spaces and high landscape and environmental quality. Although the per capita green area was not low at the south of the urban center, the accessibility to public spaces and the utilization rate should be improved for its unreasonable spatial organization. A high emotional value was concentrated in large scenic spots and places of leisure in the east and outside of the urban center. Therefore, the most important way to improve emotion was to optimize the spatial configuration and quality of parks and squares and to improve the rationality and accessibility of outdoor space.

3.5 Conclusions and Discussion

This study used geo-tagged "check-in" big data of Sina Weibo in Nanjing to explore the spatial distribution characteristics of residents' emotion. This study is a useful attempt at applying big data to urban planning. The mapping of emotion icons of

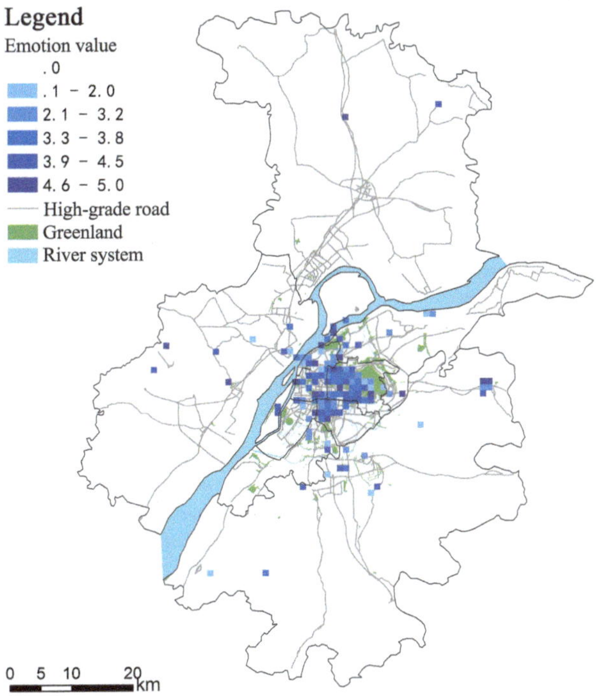

Fig. 3.14 Spatial distribution of emotion in outdoor public places

geo-tagged big data into physical geography space was found to be a feasible and innovative way to study residents' emotional responses to spaces, reflecting the development status and construction level of spatial elements. Furthermore, Sina Weibo big data can support this empirical study on the spatial distribution of residents' emotion.

Generally, high emotions in the urban center are distributed in large-scale scenic regions and their surroundings with high-grade natural, historical, or cultural landscapes. High emotions in the suburban are mainly located in areas with affinity to the urban regions and areas with a favorable environment. Additionally, residents' emotions in different places are affected by different factors. For residential areas, newly built dwellings, intensive development modes, good environments, and fewer social problems lead to high emotions. Emotions in workplaces depend on the general quality of the environment as well as on the occupation and industry features. Emotions in traffic areas are influenced by the connection between internal and external traffic. Factors influencing emotions in life service places are more complex because these places have a variety of sub-types. However, common factors included the year of construction, the quality of commercial and service facilities, the general environment quality, and the rationality of the spatial organization.

This study shows that "check-in" big data can readily capture residents' common features and depict the relationship between residents' emotions and the spaces of a city. Different spatial features of residents' emotions can be reflected in physical space. Thus, this study shows that spatial characteristics of emotions play an important role in urban space planning.

This study also has several limitations. Although ICT and smartphones are developing rapidly, the age range of residents using Sina Weibo is dominated by young people. Some older residents who do not use Sina Weibo are likely to have not been included in this study. Therefore, future studies should explore how to combine big data and survey data to remedy the weaknesses in samples' age, thereby improving analyzing methods of emotions and their application in urban planning.

References

Blumen, O. (2002). Criss-crossing boundaries: Ultraorthodox Jewish women go to work. *Gender Place and Culture: A Journal of Feminist Geography, 9*(2), 133–151.

Chen, Y. X., & Zhen, F. (2014). Further investigation into urban spatial function organization based on residents; activity data: A case study of Nanjing. *Urban Planning Forum, 5*, 72–78.

Colls, R. (2006). Outsize/outside: Bodily bignesses and the emotional experiences of British women shopping for clothes. *Gender Place and Culture, 13*(5), 529–545.

Crewe, B., Warr, J., et al. (2014). The emotional geography of prison life. *Theoretical Criminology, 18*(1), 56–74.

Fielding, S. (2000). Children's geographies and the primary school. In S. L. Holloway & G. Valentine (Eds.), *Children's geographies: Playing, living, learning* (p. 199). London: Routledge.

Hallman, B. C., & Benbow, S. M. P. (2007). Family leisure, family photography and zoos: Exploring the emotional geographies of families. *Social & Cultural Geography, 8*(6), 871–888.

Heimtun, B. (2010). The holiday meal: Eating out alone and mobile emotional geographies. *Leisure Studies, 29*(2), 175–192.

Hemming, P. J. (2007). Renegotiating the primary school: children's emotional geographies of sport, exercise and active play. *Children's Geographies, 5*(4), 353–371.

Hogertz C. (2010). WALK21 Cheltenham (UK); Emotions of the urban pedestrian: Sensory mapping. In *Pedestrians' quality needs* (pp. 31–52).

Kwan, M. P. (2007). Affecting geospatial technologies: Toward a feminist politics of emotion. *The Professional Geographer, 59*(1), 22–34.

Lüscher, P., & Weibel, R. (2013). Exploiting empirical knowledge for automatic delineation of city centres from large-scale topographic databases. *Computers Environment and Urban Systems, 37*(1), 18–34.

Miller, J. C. (2014). Malls without stores (MwS): The affectual spaces of a Buenos Aires shopping mall. *Transactions of the Institute of British Geographers, 39*(1), 14–25.

Moisi, D. (2010). *The geopolitics of emotion: How cultures of fear, humiliation, and hope are reshaping the world*. New York: Random House LLC.

Philo, C., & Parr, H. (2000). Institutional geographies: Introductory remarks. *Geoforum, 31*(4), 513–521.

Stieglitz, S., & Dang-Xuan, L. (2013). Emotions and information diffusion in social media—Sentiment of Microblogs and sharing behaviour. *Journal of Management Information Systems, 29*(4), 217–248.

Tang, G. A. (2006). *ArcGIS spatial analysis experiment course*. Beijing: Science Press.

Tumasjan, A., Sprenger, T. O., et al. (2011). Election forecasts with Twitter: How 140 characters reflect the political landscape. *Social Science Computer Review, 29*(4), 402–418.

Williams, P., Hubbard, P., et al. (2001). Consumption, exclusion and emotion: The social geographies of shopping. *Social & Cultural Geography, 2*(2), 203–220.

Chapter 4
Measuring by Movements: Hierarchical Clustering of Cities in China Based on Aggregated Massive Positioning Data

Dong Li, Menghe Wu, Bingruo Duan, and Yuheng Cai

Abstract The world is becoming linked more and more. A shift of researching focus can be observed recently, from "city as a system" to "systems of cities," given the context of fast-changing communicating technologies such as high-speed railways (physical) as well as social media over the internet (nonphysical). Flows play essential roles for a city network, indicating the trends of position and functions within the network. In this study, we adopt new type of aggregated positioning data of massive internet users in China to explore the spatial patterns of cities during the Spring Festival in 2015. By introducing new clustering algorithm highlighting spatial constraints, models output hierarchic results with vary regional zones containing different number of cities. The higher layer of results with less members is not similar to the conventional delineation according to the conditions of physical and economic geography of China. Nevertheless, the very differences suggest hidden forces driving cities connected intensely across the administrative boundaries such as sharing mutual regional cultures or employment markets. These facts grounded for a general picture for the study on polycentric urban regions over the whole national territory.

Keywords Mobility data • Spring Festival Migration • City network

4.1 Introduction

With the development of physical and cyber-communicating infrastructures such as regional transportation systems and information and communications technologies (ICT), the various connections and networks play more and more important role in current world. Flows of people, materials, and information as well as the

---

D. Li (✉) • M. Wu • B. Duan • Y. Cai
Innovation Center for Technology, Beijing Tsinghua Tongheng Urban Planning & Design Institute, Beijing 100085, China
e-mail: LIDONG@thupdi.com

Z. Shen, M. Li (eds.), *Big Data Support of Urban Planning and Management*, Advances in Geographic Information Science, DOI 10.1007/978-3-319-51929-6_4

networks they formed are supporting and reshaping the cornerstone of functions in our society. For cities, one is no longer an isolated individual, and while developing, cities are inevitably establishing connections with others located nearby or remote via different proxies, constituting regional, national, or global systems of city network.

In most cases, such network is not static but with continuous changes of external and internal conditions. There is a great significance to understand the city network as one holistic system. This trend could be leveraged on multiple levels. On the macro level, the evolving of city network provides driving force of regions and countries in current era of urbanization. On the micro level, the network is one of the most important contexts for any city influenced. A number of key issues including restructuring industries, competing and cooperating within regions, need a preliminary analysis on direct and indirect counterparts in the network, gaining the dynamic knowledge on comparative advantages.

The academic value of these practical challenges is also important. Since the complex network is an abstract expression of the interlinked world in reality, a number of nodes and edges connecting are utilized to map the graph (Easley and Kleinberg 2010). Numerous publications have focused on the theories and applications on analyzing complex network. In recent years, combining complex networks with geographic entities or saying spatially embedded networks has become one of the hot topics (Daqing et al. 2011; Expert et al. 2011). Taylor put forward the "center flow" theory as a supplementary to the traditional "center of place" theory (Taylor et al. 2010); Batty advocated a deep understanding of the city will not only be in place as a space, but it should be regarded as a network and system flow (Batty 2013). Existing studies on city network range from industries (Taylor et al. 2000), infrastructures such as aviation traffic (Smith and Timberlake 2001; Zhou and Hu 2002; Yu et al. 2008), and information such as internet domains and microblogs (Mingfeng and Yuemin 2006; Feng et al. 2012). However, few studies emphasized the network with actual mobility of population at a large scale, due to the difficulty to obtain direct observations of the population movements.

The rapid deployment of ICT, especially for mobile communication technologies, enables to capture a large number of positioning information. Since 2014 Spring Festival, several major internet companies in China have launched online visualization projects expressing the daily flows of their massive mobile users and became a normal monitoring beside holidays gradually.

The data available online is not as explicit as recording every stop of every user. It's a type of aggregated data but better than any existing statistics in terms of the level of details. Although this Internet Spring Festival Migration (hence ISFM) data still got some issues on their standards on how to aggregate just like other types of open data (Liu et al. 2015a), it does not affect the historic releasing on public-accessible massive population movement statistics at such a spatial and temporal resolution. Under this circumstance, each user becomes a part of the sensor networks nationwide. Aggregated statistics about its behavior is collected anonymously and used to demonstrate population movements every day among cities and regions within the country. In this study, the concept of social sensing has got responded in empirical life (Liu et al. 2015b).

This article aims for basic analysis on identifying clusters and ranking on importance within a network. After constructing a hierarchical structure, scenarios with different numbers of clusters are compared to conventional knowledge on regionalization. Patterns and distributions of centralities in zones are discussed. In the rest of the article, analytical methodologies utilized are described in Sect. 4.2, several data sources are compared in Sect. 4.3, and results and extended discussions including potential improvement in the future are in Sects. 4.4 and 4.5.

4.2 Method

4.2.1 Modularity and Partitioning

To understand structure of a complex network, a straightforward way is to break it down, identifying the strength of edges among nodes. Based on the differences of strength, the network would be compartmentalized into clusters, or saying communities. These steps could be repeated to obtain a result of a nested structure at multilevel finally known as hierarchical community detection. Each community holds a certain extent of intensity, i.e., the degrees of internal connection within communities are significantly higher than ones among nodes of different communities. The complexity of network has been largely simplified. This layered information extracted from community detection is an important generalization for the major pattern in a complex network in an overarching way, leading more focused and detailed explorations in subregions.

Modularity is a commonly used indicator during community detection in network based on spatial interactions, describing how well a network decomposes into modular communities. A higher modularity is a stronger community and is connected internally. Modularity is the fraction of the edges that fall within the given groups minus the expected such fraction. It is positive if the number of edges within groups exceeds the number expected on the basis of chance. For a given division of the network's vertices into some modules, modularity reflects the concentration of edges within modules compared with random distribution of links between all nodes regardless of modules. There are different methods for calculating modularity. To avoid the size effects among cities and to emphasize the links of migrating intentions, followed by Newman (2006), here we defined our modularity as the difference between actual flows and expected flows, as in Eq. (4.1). That is, little actual flows between big cites (supposed to produce huge expected flows) will prove lower migrating intentions, vice versa:

$$\text{Modularity} = \text{ActualFlows} - \text{ExpectedFlows} \tag{4.1}$$

where ExpectedFlows of nodes *i* and *j* is assumed only to be conditioned on the total marginal flows between these two nodes in the flow matrix, defined as in Eq. (4.2):

$$\text{ExpectedFlows}_{ij} = \text{Inflow}_i \times \text{Outflow}_j \Big/ \text{TotalFlow} \tag{4.2}$$

where TotalFlow is the sum total of origin-destination (OD) flow volume in the whole network.

After calculating the modularity for each pair of locations, we obtain a weighted and directed graph of cities' network. For the existing studies in nonspatial domains, there are a number of methods for partitioning (Reichardt and Bornholdt 2006; Lambiotte et al. 2008). Given the modularity calculation, the partition cuts at weak boundaries where the actual flow is much lower than the expected flow. However, those methods may not be considering spatial constraints such as spatial contiguity well. This kind of information on spatially contiguous communities (regions) is inspiring for geographers and planners.

Chi et al. (2014) adopted hybrid clustering methods to achieve a two-level hierarchical structure of a provincial mobile phone communication network. Guo (2007) and (2009) proposed a series of partitioning methods based on spatially constrained hierarchical clustering. Considering the contiguity matrix as an input, a spatially contiguous tree is constructed by iteratively merging the most connected clusters. Then the spatially contiguous tree is partitioned by finding the best edge to remove. Any removal of edge achieves a maximum gain of within-region modularity while retaining the constraints. By repeating this step for each new region, a hierarchy of regions is constructed.

4.2.2 Centrality

Another interesting issue is to determine the degree of difference of each node comparing with other ones. For the city network, people want to know the ranking of power within the communities and the whole network. The concept of centrality is closely related. There are a variety of centrality designed to evaluate the ranking of nodes listed in Table 4.1, including betweenness, degree, eigenvector, and so on (Brandes 2001; Diestel 2005). The eigenvector centrality was selected for analyzing city network due to the situation of data and the aims of study.

Eigenvector centrality (EC) is a measure of the influence of a node in a network. It assigns relative scores to all nodes in the network based on the concept that connections to high-scoring nodes contribute more to the score of the node in question than equal connections to low-scoring nodes. The approach of EC is to find the most central actors in terms of the "global" or "overall" structure of the network and to pay less attention to "local." The definition and calculation of EC values could be referred to Hanneman and Riddle (2005) and Newman (2008). Indeed, the relative importance is also concerned. Using the adjacency matrix to find eigenvector centrality, for a given graph, the eigenvector equation could be written as Eq. (4.3):

Table 4.1 Selections on indicators of centrality (Wikipedia 2016)

Name	Explanation
Betweenness centrality (BC)	How often a node appears on shortest paths between nodes in the network
Closeness centrality (CC)	The average distance from a given node to all other nodes in the network
Degree centrality (DC)	Nodes that have more ties have greater opportunities make them less dependent on any specific other node, hence more powerful
Eigenvector centrality (EC)	Node importance in a network based on node's connections
Katz centrality (KC)	The number of all nodes that can be connected through a path, while the contributions of distant nodes are penalized

$$A^{\mathrm{T}}x = \lambda x \tag{4.3}$$

When $\lambda_1$is the largest value of eigenvalues of adjacency matrix A^{T}, the EC could be obtained by Eq. (4.4):

$$\mathrm{EC} = {}^{1}\!/_{\lambda_1}\, a^{\mathrm{T}} \tag{4.4}$$

The value of EC of a city is determined by volume of flows. Information on ranking of eigenvector centrality in each community was used to evaluate, while the scenarios of clustering numbers vary at different levels. Regarding to the geographical space, community detection for a city network means to find groups of cities closely linked. Furthermore, the ranking of importance is corresponding to the polarization of population and economy in the real world. Based on various and complicated factors in practice such as location, social, economic, and cultural, many of which cannot be measured evenly; cities within one group own higher complementary in terms of supplies and demands. For areas less developed, it is crucial to know their accessible cities, measured by strong and short paths in the network rather than geographic distances, with higher ranks as resources and driving forces in a same community. Regional planning could also leverage this driving force and guide the cooperation among cities.

4.3 Data

This article used data crawled from the internet. Covering the time since 1 week before the Spring Festival to the end of February (2015-02-09 to 2015-02-28). Since there are several sources of Internet Spring Festival Migration (ISFM) data (Table 4.2), comparison has been done to choose a better one. The 360 Migration shows the statistical result based on its train ticket-purchasing software, revealing some of the interesting features of railway migration during the season. However, its

Table 4.2 Descriptive comparison on various ISFM data[a]

Data names	Main functions	Interval	Granularity of unit	OD matrix coverage
Baidu Migration http://qianxi.baidu.com/	Population migration	Daily	365 prefecture-level city units	Top ten inflow and outflow for each city, 365×10
Tencent Migration http://heat.qq.com/qianxi.php	Population migration, proportion of transportation (air, train, and highway)	Daily	365 prefecture-level city units	Top ten inflow and outflow for each city, 365×10
360 Migration http://vis.360.cn/open/traffic/	Population migration, railway statistics	Daily	31 province-level units	31×31

[a]The URLs are obtained on March 2016. They may change without notifying

portrait of migration only focuses on the provincial level, which is inferior to the rest of two. The main comparison was taken between Baidu and Tencent.

Attributes of data from Baidu and Tencent were similar. During the demonstration, both of the two sources offered their top ten origin-destination (OD) volume for each city pair. The sum of population migration among pairs of cities per day was considered as the volume of OD, as well as the strength of edges in the network.

By comparing their overall temporal and spatial patterns, we concluded that Tencent's data is more reliable. One could found the Baidu data only count the flows across borders of provinces, ignoring a large part of city pairs within in a same province. There is a clear difference between the daily trends of total inflows and total outflows between Tencent and Baidu's. Since the reason of symmetry, these two types of temporal trends supposed to be as similar as Tencent's rather than as Baidu's (Fig. 4.1).

The spatial pattern is also checked (Fig. 4.2). Cumulative flows between city pairs are classified by neutral break and encoded by width and gray scale. Overall pattern of Tencent's network is more like a "diamond" in shape (Beijing, Guangzhou-Shenzhen, Chengdu, Shanghai as four ends linked), which has been mentioned in previous studies based on different types of data such as flights and social media (Zhou and Hu 2002; Feng et al. 2012). A certain part of long trips were ignored in Baidu's network.

Overall, this chapter used summarized (sum total) ISFM data from Tencent during the whole week prior to the Spring Festival, when most of the people started to go to their hometown from the cities they were working and staying. The descriptive indicators of Tencent top ten network dataset are listed in Table 4.3.

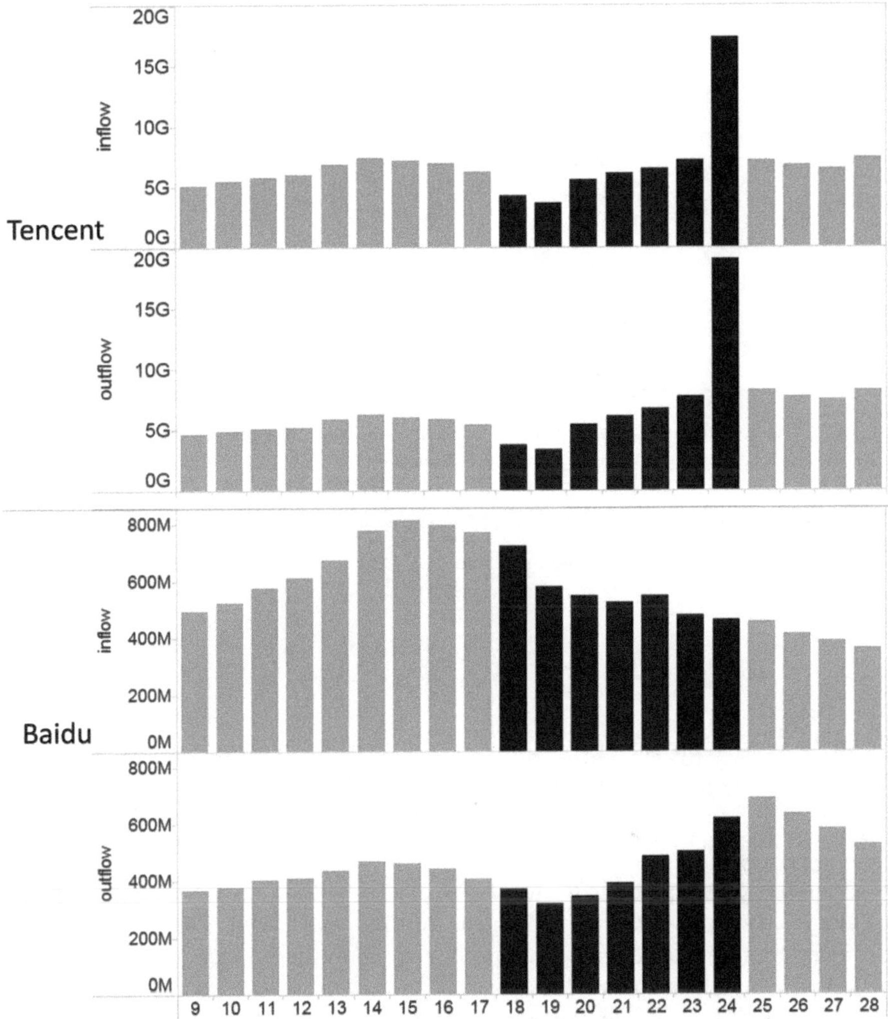

Fig. 4.1 Temporal comparison between Tencent and Baidu data in Feb. 2015. Spring Festival is marked by dark bars

4.4 Results

After calculation, a delineation of clusters based on ISFM 2015 data appeared, along with sorting of cities based on their relative EC based on Eq. (4.4) in each cluster. Two cases of 4-cluster and 7-cluster were chosen to demonstrate the credibility of new delineation, respectively, comparing with different conventional knowledge on subregions in China (Figs. 4.3 and 4.4).

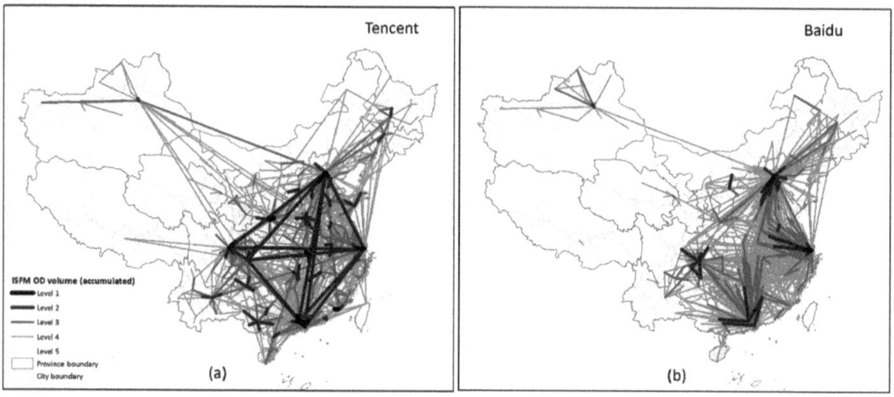

Fig. 4.2 The cumulative spatial pattern of Tencent and Baidu data in Feb. 2015. OD flows are classified according to their volumes by natural breaks

Table 4.3 Statistics descriptive indicators of Tencent dataset

Observation	Minimum	Maximum	Sum	Mean	Standard deviation
7818	1	5136263	624209129	79842.559	248550.054

4.4.1 Case of 4-Cluster

From the view of physical geography as Fig. 4.3(b), China is regionalized into four zones in accordance with the indicators of climate and terrain such as elevation, rainfall, and temperature, i.e., the northern (NR), northwestern (NW), southern (SU), and Tibetan Plateau (TP). Boundaries are defined as follows:

- Boundary-A is separating the territories of zone NR and zone NW along with the mountain ridges of Greater Khingan, Yinshan, and Helan. The main reason is regarding to the difference of monsoon and non-monsoon area, also being consistent with the isohyet of 400 mm.
- Boundary-B is separating the territories of zone NR and zone SU along with the mountain ride of Qinling as well as Huaihe River. It is generally coinciding with the isotherm of zero degree Celsius in January and isohyet of 800 mm. Thus, the dominant factor is about climate.
- Boundary-C is separating the zone TP with other three zones, roughly same as the line between the first and the second relief ladder of China along with the mountain ridges of Kunlun and Qilian. The dominant factors are related to precipitation and terrain topography.

Using the methodology mentioned above, the result of clusters divided by ISFM data with the same number of four members, named cluster N4, W4, S4, and E4 in Fig. 4.3(a), implies that some of the factors of different physical geography cast impacts on people's migration on a yearly base, while others do not. The boundary-A merely plays any cutting role from the view of collective mobility. The zones of NR and NW merge into a greater northern one, cluster N4.

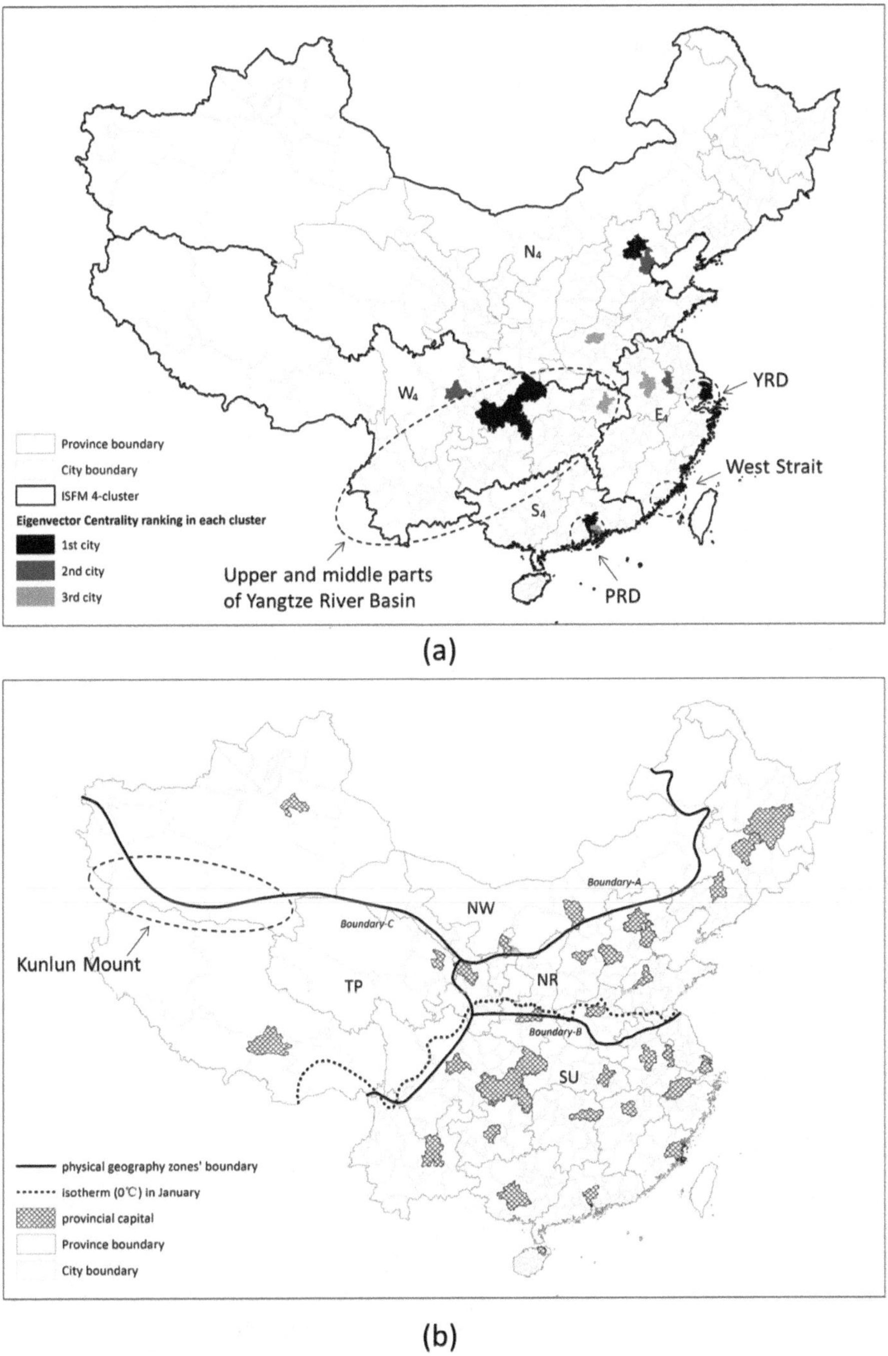

Fig. 4.3 Comparison between ISFM 4-cluster (**a**) and physical geography 4-zone (**b**). To simplify, only the top three cities with higher EC value in each cluster are shown in the map

The boundary-B separating China from north to south is retained much better, although the components inside southern China changed a lot. The southern clusters S4 and E4 defined by mobility have been reduced and pressed closer to the coast. Urban agglomerations in this region include the Yangtze River Delta (YRD), the Pearl River Delta (PRD), and the west side of the Taiwan Strait in Fujian.

The boundary-C works partially. The western end along Kunlun Mountains (Fig. 4.3 b) is blocking the movement of people. However, the middle and eastern parts are changed greatly. In the middle part, the effects of the Tibetan Plateau to the northern zone are moved south to the border between Qinghai and Tibet. While the eastern end is completely opened, the upper and middle parts of the Yangtze River Basin, roughly covering the provinces of Sichuan, Yunnan, Chongqing, Guizhou, Hubei, and Hunan, have been merged into the TP zone, constituting the big cluster W4.

Overall, based on the observation on migration, the separating effects along latitudes, or saying vertical, are much stronger than the ones along longitudes, or saying horizontal in China. Comparing to another similar study on migration-based regionalization, a stronger longitudinal segmentation could be found at a higher level (fewer clusters) in the USA (Guo 2009).

4.4.2 *Case of 7-Cluster*

The 7-cluster zoning of China is well known as economic geography major. It is shown that similarity between the ISFM 7-cluster and the partition of seven zones based on economic activities is much higher than that to the above case of four zones of physical geography (Fig. 4.4). The overall pattern is similar, for example, the dividing of northern, western, central, and southern parts. However, the biggest difference still exists in the northern region. The northeastern and northwestern economic EG zones are merged into a bigger cluster NN7 from the perspective of mobility. Furthermore, since the important cities with higher value of centrality in this cluster are not located in the northwestern zone, this entire part has been seized by the northeastern zone and disappeared at a higher views on the whole country.

The smallest difference is observed in Southern China. ISFM cluster S7 is almost entirely consistent with the southern EG zone, except three cities Hengyang, Yongzhou, and Chenzhou located in South Hunan of central EG zone which has been absorbed into their southern neighbor economically dominated by the PRD.

Besides these two clusters, more or less some local differences could be found in other partitions. Fewer differences exist within the developed Eastern China as well as in the less developed Western. The two types of boundaries located in the eastern part fit well, except Shandong Province which is merged into the northern cluster N7. While this eastern EG zone was subdivided further into two clusters, one, cluster EN7, is constituted by Shanghai, Jiangsu, and Anhui, while the southern cluster ES7 composition is containing Zhejiang, Jiangxi, Fujian, and Taiwan. For the less developed area, the western cluster W7 and the relevant western EG zone match exactly except Guizhou Province which just moves into the central cluster C7.

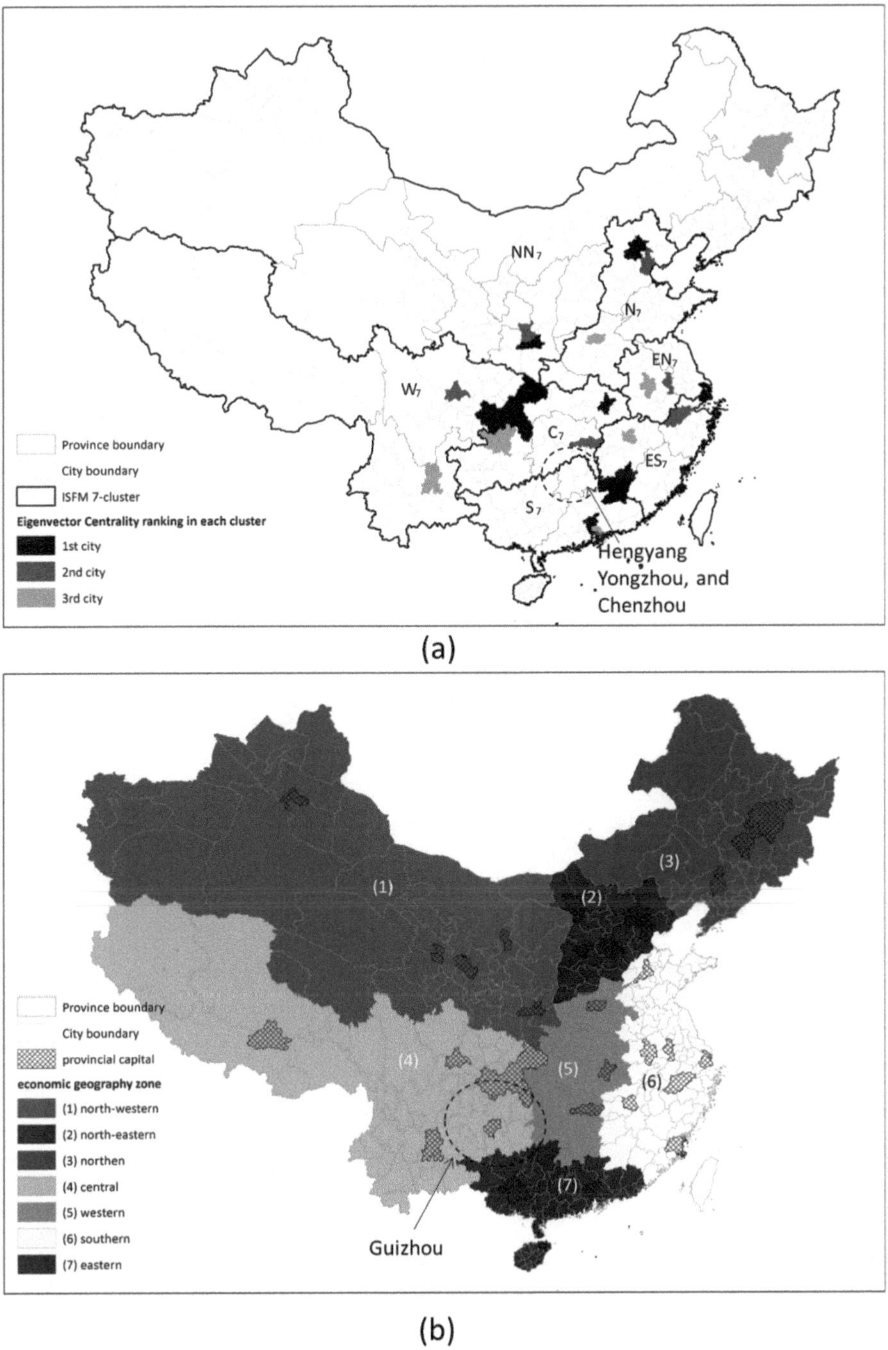

Fig. 4.4 Comparison between ISFM 7-cluster (**a**) and economic geography 7-zone (**b**)

The middle developed central and northern areas have more differences in their boundaries comparing to ISFM clusters. For the northern cluster N7 and C7, Shandong and Guizhou were included, while Shanxi was lost. Adjacent province such as Henan Province between two EG zones has been redefined into a northern one, for its stronger links with north cities.

To wrap up, comparing with the pattern of physical geography, the economic impacts on migration are much more remarkable.

4.4.3 Regional Central Cities

The top three cities in each cluster according to their values of eigenvector centrality within the whole network were listed in Table 4.4. These cities work as central and important hubs in the whole and regional network, in spite of their types. Generally, three types of advantages could be potentially relied on for ranking top as a higher centrality city in network, i.e., political (provincial capitals), economic (cities with higher GDP), and locational (border cities on cluster boundaries).

Table 4.4 Centrality cities in each cluster

Cluster	Relevant provincial capitals (province)	Top three centrality cities	Types
NN7	Urumqi (Xinjiang), Xining (Qinghai), Lanzhou (Gansu), Yinchuan (Ningxia), Hohhot (Inner Mongolia), Xi'an (Shaanxi), Taiyuan (Shanxi), Shenyang (Liaoning), Jilin (Jilin), Harbin (Heilongjiang)	Xi'an	Political
		Xianyang	Economic
		Harbin	Political
N7	Beijing, Tianjin, Shijiazhuang (Hebei), Ji'nan (Shandong), Zhengzhou (Henan)	Beijing	Political
		Tianjin	Political
		Zhengzhou	Political
W7	Lhasa (Tibet), Chengdu (Sichuan), Kunming (Yunnan), Chongqing	Chongqing	Political
		Chengdou	Political
		Kunming	Political
C7	Guiyang (Guizhou), Wuhan (Hubei), Changsha (Hunan)	Wuhan	Political
		Changsha	Political
		Zunyi	Locational
EN7	Hefei (Anhui), Nanjing (Jiangsu), Shanghai	Shanghai	Political
		Nanjing	Political
		Hefei	Political
ES7	Nanchang (Jiangxi), Hangzhou (Zhejiang), Fuzhou (Fujian), Taipei (Taiwan), Shanghai	Ganzhou	Locational
		Hangzhou	Political
		Nanchang	Political
S7	Nanning (Guangxi), Guangzhou (Guangdong), Haikou (Hainan), Hong Kong, Macao	Guangzhou	Political
		Shenzhen	Economic
		Dongguan	Economic

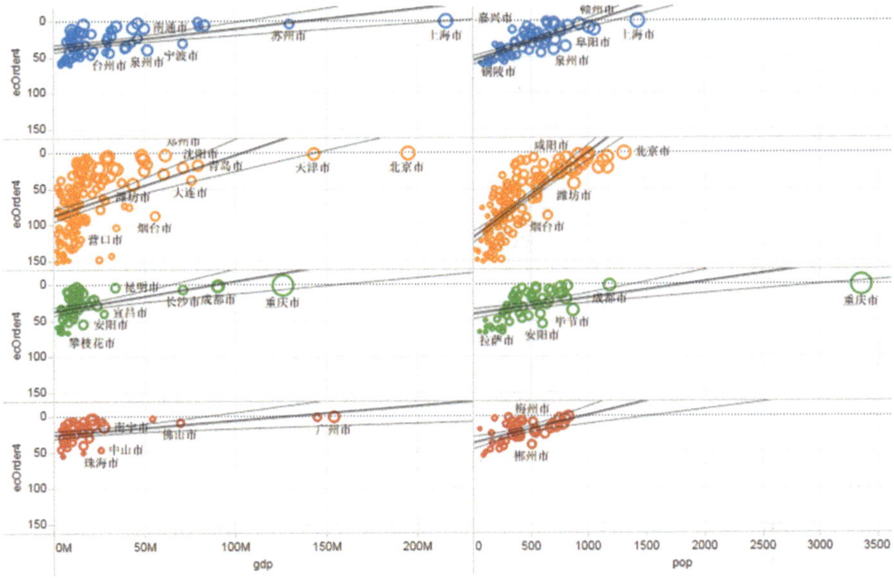

Fig. 4.5 Relationships between EC values (y axis) and GPD, population (x axis)

This table could be explained from another point of view: some of the capitals in these provinces didn't work as a real capital. Along with indicators as population and GDP, the value of EC could be used for evaluating the status of any cities in its natural cluster (Fig. 4.5).

For example, as the capital of Guizhou in cluster C7, Guiyang is not listed as a centrality city. Instead, Zunyi, a city located between Guiyang and Chongqing along the border between C7 and W7 is ranked highly. The external influence introduced by Chongqing is much powerful in Guizhou, also suggesting a weak centralized pattern in that province. Similarly, S7 is dominated by economy. Two of the top three centrality cities are due to its economy position; the rest of the capital cities in the cluster are left behind. Comparing to its political power, some hubs, in terms of economy or migration effects, play important roles and introduce a balanced structure in that province.

4.5 Conclusion

This research proposes a new analytical approach to identify subregions as well as their central cities based on the new migration data online. Results were compared with traditional geographic zonings. Common indicators of network analysis such as modularity and centrality are adopted, and some new methodologies on compartmentalizing considering modularity as well as spatial constrains are attached to obtain better outputs for geographic research. Results are promising.

Different scenarios of migration clusters are selected to compare with conventional subregions on physical and economic geography. Comparing among migration clusters and traditional zones originates new knowledge, such as the tendencies of connecting and the roles of cities in reality besides their political or economic positions in regions.

In practice, this new founding on spatial pattern provides valuable insights and can potentially support planning and policy-making, optimizing national and regional development measures, guiding projects of major infrastructures, and so on. In the era of the internet, it is possible to monitor any changes on boundaries of clusters and rankings of cities, reconsider their relative competitiveness in a greater urban system, and promote a more effective cooperation between related cities.

There is still a need for future work, mainly on improving data quality and methodology. For the data quality, Tencent is among the largest internet companies on social media and messaging in China or even on the global. The monthly active users (MAU) of its two major products named QQ and WeChat are more than 820 million and 600 million, respectively (Tencent 2016). However, limitation on the data sampling still exists. The dataset used in this study is a sort of side product of a visualizing project. Few descriptions about the quality of this aggregated dataset were released, while researchers have merely no ways to validate it. Furthermore, only top ten inflow and outflow pairs of each city were recorded in this dataset. According to estimation (Li 2015), although the ratio of top ten city pairs accounts for more than 64% of the total traffic due to the long-tail distribution of OD data among city networks, the coverage of spatial units is imperfect. Situations outside the connections of top ten pairs and their effects on the whole network remain unknown. How to reconstruct the whole OD connections for the whole city network by combining available official statistic data and new data online has a great significance for future research.

For the part of methodology, as a typical cross-disciplinary area, a number of researches focus on the topic of community detection in a complex network, from majors of mathematics, physics, computer sciences, social sciences, geography, and so on. Based on the theme, the method proposed by Guo (2009) was used in this article, emphasizing the condition of spatially constrained when partitioning the network and generating a hierarchical result. More methods should be compared for their performances.

Last but not least, the information of time has not been fully utilized. The raw data obtained was a type of aggregated sum total of the movements among cities at an interval of one day. However, only the summed total value before the Spring Festival was used in this article. Method suiting for analyzing an evolving series of temporal complex geographic network, saying a dynamic spatial network, is yet to be developed.

Acknowledgment The authors would like to thank Dr. Xingjian Liu at Hong Kong University for his suggestive comments.

References

Batty, M. (2013). *The new science of cities*. London: Mit Press.

Brandes, U. (2001). A faster algorithm for betweenness centrality. *Journal of Mathematical Sociology, 25*(2), 163–177.

Chi, G., et al. (2014). *Uncovering regional characteristics from mobile phone data: A network science approach*. Papers in Regional Science.

Daqing, L., et al. (2011). Dimension of spatially embedded networks. *Nature Physics, 7*(6), 481–484.

Diestel, R. (2005). Graph theory. Graduate texts in Mathematics 101.

Easley, D., & Kleinberg, J. (2010). *Networks, crowds, and markets: Reasoning about a highly connected world*. Cambridge: Cambridge University Press.

Expert, P., et al. (2011). Uncovering space-independent communities in spatial networks. *Proceedings of the National Academy of Sciences, 108*(19), 7663–7668.

Feng, Z., et al. (2012). China's city network characteristics based on social network space: An empirical of sina micro-blog. *Acta Geographica Sinica, 67*(8), 1031–1043.

Guo, D. (2007). Visual analytics of spatial interaction patterns for pandemic decision support. *International Journal of Geographical Information Science, 21*(8), 859–877.

Guo, D. (2009). Flow mapping and multivariate visualization of large spatial interaction data. *Visualization and Computer Graphics, IEEE Transactions, 15*(6), 1041–1048.

Hanneman, R. A., & Riddle, M. (2005). *Introduction to social network methods*. Riverside: University of California.

Lambiotte, R., et al. (2008). Laplacian dynamics and multiscale modular structure in networks. arXiv preprint arXiv:0812.1770.

Li, D. (2015). *Which Spring Festival migration data is better: A comparision*, from http://www.thepaper.cn/newsDetail_forward_1321590

Liu, X., et al. (2015a). Understanding urban China with open data. *Cities, 47*, 53–61.

Liu, Y., et al. (2015b). Social sensing: A new approach to understanding our socioeconomic environments. *Annals of the Association of American Geographers, 105*(3), 512–530.

Mingfeng, W., & Yuemin, N. (2006). The network advantage of cities: An analysis of spatial structure and node accessibility of internet backbones in China. *Geographical Research, 25*(2), 193–203.

Newman, M. E. (2006). Modularity and community structure in networks. *Proceedings of the National Academy of Sciences, 103*(23), 8577–8582.

Newman, M. E. (2008). The mathematics of networks. *The New Palgrave Encyclopedia of Economics, 2*(2008), 1–12.

Reichardt, J., & Bornholdt, S. (2006). Statistical mechanics of community detection. *Physical Review E, 74*(1), 016110.

Smith, D. A., & Timberlake, M. F. (2001). World City networks and hierarchies, 1977–1997 an empirical analysis of global air travel links. *American Behavioral Scientist, 44*(10), 1656–1678.

Taylor, P. J., Beaverstock, J. V., & Walker, D. R. F. (2000). Introducing GaWC: researching world city network formation. In S. Sassen (Ed.), *Cities and their Cross-Border Networks*. Oxford: Blackwell Publishers.

Taylor, P. J., et al. (2010). External urban relational process: Introducing central flow theory to complement central place theory. *Urban Studies, 47*(13), 2803–2818.

Tencent. (2016). *Product list of Tencent*, from http://www.tencent.com/zh-cn/content/ir/fs/attachments/ProductlistofTencent.pdf

Wikipedia. (2016). *Centrality*, from https://en.wikipedia.org/wiki/Centrality

Yu, T., et al. (2008). China's urban systems in terms of air passenger and cargo flows since 1995. *Geographical Research, 27*(6), 1407–1418.

Zhou, Y., & Hu, Z. (2002). Looking into the network structure of Chinese urban system from the perspective of air transportation. *Geographical Research, 21*(3), 276–286.

Chapter 5
Assessment of Regional Economic Integration Based on Relational Data: The Case of the Yangtze River Delta

Tao Li, Jiaju Miao, and Yina Zhang

Abstract Regional economic integration is a hot topic in regional and urban development. As the network society rises, the space of flows instead of the space of spaces is shaping regional urban system. Here, the space of flows represents the flow of capital, cargo, information, and mobility, whereas the space of spaces represents the cities in the network that act as hubs or nodes. In the regional economic integration process, it is the relation between cities instead of the hierarchy system that forms the regional urban system. Therefore, the essential attribute of the regional urban system is the relational network. Approaches to assessment of regional economic integration are urgently needed because there is no uniform approach. The primary reason is that it is difficult to collect relational data based on traditional approaches. Big relational data provides a new means to examine this issue. Regarding the process of regional economic integration as the flow of capital, cargo, information, and mobility, big relational data are collected to characterize these indicators. Based on the database of company registration information from China's State Administration for Industry and Commerce, the firm-based network is analyzed to characterize the flow of capital and cargo. Based on the database of the Baidu Search Index, the information-based network is analyzed to characterize the flow of information. Based on the railway database, the transportation-based network is analyzed to characterize the flow of mobility. These three city networks are then compared and combined using multivariate analysis and interlocking network approaches. Then, to measure the degree of regional economic integration of the main cities, the regional economic integration index is determined. Finally, suggestions are proposed to study the regional issues in the use of big relational data. How to evaluate the process of regional economic integration is a question that needs to

T. Li (✉) • Y. Zhang
Urban Development Institute of Fudan University, School of Social Development and Public Policy, Fudan University, Shanghai, China
e-mail: little_li@163.com

J. Miao
Department of Economics, Stony Brook University, New York, USA

Z. Shen, M. Li (eds.), *Big Data Support of Urban Planning and Management*, Advances in Geographic Information Science, DOI 10.1007/978-3-319-51929-6_5

be viewed from multiple perspectives. Big relational data shows good prospects in this study area and needs to be focused on in the future.

Keywords Regional economic integration • Interlocking network • Yangtze River Delta

5.1 Introduction

Regional economic integration has always been a hot topic in the study of economic geography and urban development. Conventional research on regional economic integration is affected by the theory of trade and industry. Scholars usually apply the traditional methods such as the gravity model, the boundary effects model, the general equilibrium model, the input-output model, and so on to assess regional economic integration based on attribute data, which include GDP, population, fiscal expenditure, and price (e.g., Barro et al. 1992; Armstrong 1996; Yu and Huang 1998; Naughton 1999; Sun (2003), Zhao and Chen (2006), Han (2012), Li and Luo (2007), Hu (2011), Wang (2009, 2011), Zhao and Cai 2009; Liu and Hu 2011).

In the context of informatization and globalization, the network society begins to rise. Friedmann (1986) noted that cities played roles as the centers of the flows of capital, cargo, information, and mobility in the regional and global system. Castells (2006) proposed the dual attributes of space. It is the space of flows rather than the space of spaces that shapes the regional urban system. The space of flows refers to the flow of capital, cargo, information, and mobility, whereas the space of spaces refers to the cities that play the roles of hubs or nodes in the city network. The space of flows is the relation of cities in the network rather than the hierarchy that defines the global and regional urban network (Sassen 1991; Taylor 2004). This argument provides a new perspective to study regional problems. The methodology of analyzing regional issues has begun to transfer from the traditional attribute method to the new relational method. This change is major progress in the field of regional studies.

Economic geographers conventionally study regional problems based on attribute data which reflects the attribute of the cities themselves. This appears more appropriate for the analysis of an individual city and may not be reliable to discern the relation between cities (Beaverstock and Boardwell 2000). It is necessary to choose appropriate relational data to describe the interaction between cities. However, it is difficult to obtain access to the relational data as a result of the limitations of the traditional means of data collection. Smith and Timberlake (1995) listed 12 types of relational data, which they called "the wish list" due to the unavailability of these data. Since the late 1990s, certain scholars have attempted to use many types of relational data to study regional and urban issues and have made progress, in which the most representative are the studies sponsored by the Globalization and World Cities (GaWC) Research Network at Loughborough University in London (e.g., Taylor 1997, 2004; Beaverstock et al. 2000.

In terms of regional economic integration, domestic and foreign scholars have conducted tentative explorations on this issue. Specifically, these scholars (Naughton 1999; Kumur 1994; Lu and Li 2009; Zhou and Xia 2010) apply "flow data" such as volume of trade or capital, labor flow, or passenger capacity to assess regional economic integration. Through the results of these analyses, we found that the "flow data" does not technically meet the meaning of "relational data." "Relational data" in a specific region should be expressed as a data matrix composed of "directed and multiple-valued" data. The matrix is based on the horizontal and vertical axis measured by the relational data across the cities in this region. Thus, the "flow data" in the studies noted above are primarily attributed data and could not be expressed as a data matrix. Thus, the data could not be used for in-depth analysis of the relation between cities.

Therefore, within the considerable literature, there is a deficiency of lacking analyses of regional economic integration based on relational data and methods so that we cannot get insight into the regional economic integration from the perspective of city network. To rectify this deficiency, we intend to measure regional economic integration on the basis of "relational data." With the development of information and communications technology (ICT), research based on relational data has begun to increase. New types of data and methods have begun to emerge to provide new perspectives to study regional problems (e.g., Alderson and Beckfield 2007; Tang and Zhao 2010; Tang and Li 2014a; Xiong et al. 2013). These studies provide new data and methods for us to measure regional economic integration and make in-depth analysis.

This chapter contributes to literature by measuring regional economic integration based on the relational data and methods. This complements the related literature dominated by the studies based on the attribute data and provides a new perspective to get insight into the regional economic integration. This chapter is divided into three parts. The next section puts forward the research approach, analytical methods, and data sources, following with a section that uses the Yangtze River Delta as an example to assess its regional economic integration. The last section discusses the conclusions.

5.2 Theoretical Framework, Methods, and Data

5.2.1 *Review on the Relational Data*

Before applying relational data to assess analysis on regional economic integration, it is essential to provide a deep understanding of the concept of regional economic integration and then to find the appropriate indicators to characterize this concept. "Appropriate indicator" means the data which could represent the concept of regional economic integration well and could be easily obtained from open channels to aid subsequent research.

Table 5.1 Summary of relational data

Levels of space of flows	Relational data	Representative studies
1st level—infrastructure	Flow of traffic	Smith and Timberlake (2002), Derudder et al. (2005), Taylor et al. (2007), Van Nuffel et al. (2010), Luo et al. (2011)
	Flow of telecommunication	Moss and Townsend (2000), Townsend (2001), Rutherford et al. (2004), Hall and Pain (2006)
2nd level—the nod and hub	Firm branches of advanced producer services	Beaverstock et al. (1999, 2000, Taylor (2001), Taylor et al. (2001, 2002), Derudder et al. (2005), Derudder and Taylor (2005), Derudder and Witlox (2005), Derudder (2006), Taylor (2006), Pain (2008), Taylor and Aranya (2008), Taylor et al. (2010), Tang and Zhao (2010), Zhao and Tang (2010), Yin et al. (2011)
	Firm branches of all industrial sectors	Alderson et al. (2004, 2007), Wall (2009), Tang and Zhao (2010), Zhang (2010), Lv (2012), Tang and Li (2014a, b), Zhu et al. (2014)
	Flow of news and information	Taylor (1997), Tang and Zhao (2009a, b), Xiong et al. (2013, 2014), Liu and Zhang (2014)
	Flow of business	Hall and Pain (2006)
	Commodity chains	Rossi et al. (2007)
3rd level—elite space	Spatial distribution and flow of elite labor force	Beaverstock et al. (2000, 2004), Beaverstock (2007)

Source: Author summaries

The concept of regional economic integration was first put forward by Dutch economist Tinbergen (1954) who noted that the purpose of regional economic integration was to remove human-induced impediments that hindered economic development in a region. Thereafter, production factors such as capital, cargo, information, and mobility could flow freely to realize collaboration in this region. Balassa (1967) argued the regional economic integration was not only a process but also a type of state. Integration referred to the exchange of all types of production factors that were not blocked by governmental restriction. Meng (2001) also provided a similar perspective. Therefore, the concept of regional economic integration can be summarized as the free flow of all types of production factors in a certain region. Thus, finding the suitable data that represent the intercity flow of the production factors becomes the starting point for measuring regional economic integration. According to research conducted by Beaverstock et al. (2000), relational data, which represent intercity connections, can be divided into three categories: infrastructure, node and hub, and elite space. Furthermore, the researchers connected these three categories to the three levels of the space of flows proposed by Castells (Table 5.1). The first type of data, which refers to passenger capacity and telecommunication connections, features flow of traffic and telecommunication between cities, corresponding to the infrastructure networks between cities. The second type of data, which corresponds to the nodes or hubs of networks, is current mainstream data for the analysis of city networks. These data consist of firm branches, flows of news and information, and commodity chains. The third type of data, which corresponds to elite space, means spatial distribution and flow of the elite labor force in the city network.

In the first type of data, the flow of traffic reflects the mobility of people between cities. This data's advantage is that a large amount of data is relatively easy to obtain because the official websites are open to public. However, the defect is that it is difficult to distinguish business people, commuters, and tourists. Therefore, these data are commonly used as an auxiliary analysis. The flow of telecommunication reflects the connections of telecommunication information transferred between cities. However, it is difficult to popularize this flow's application due to the difficulty of accessing these data because the telecommunication companies are keeping them for commercial use for themselves. There are obvious disadvantages to the third type of data, including the limited sample size and the challenging role of elite space. According to representativeness and availability, the second type of data is currently the most popular relational data source for research. The flow of news and information can indirectly reflect the informational linkage between cities. This flow has been gradually focused on in recent studies because of its advantages of large sample size and timeliness. Flow of business information and commodity chain data are not widely employed due to small sample sizes and slow feedback on questionnaires.

It is notable that the most extensive type of relational data is firm data. Sassen (1991) and Taylor (2004) argued that the firm is the "agent" of regional economic activity so that the location strategy of firms defines the interlocking network between cities. In a direct manner (of the establishment of branches) and in an indirect manner (of interfirm trade), firms can integrate all types of production factors in the region and establish economic ties between cities, forming the regional city network. Consequently, the interlocking network established by all firms contributes to promote the regional economic integration process. It is difficult to obtain trade data between firms because of business secrets. In contrast, it is easier to gain data on firm branches. Therefore, the data on firm branches becomes a good data source on behalf of the flow of capital, cargo, information, and mobility between cities. With the development of ICT, the search and retrieval technology of big data can be implemented to obtain useful information. This development has made relational analysis based on the firm branches data a widely used method. In conclusion, this data has four advantages. First, the data is representative because it is the firm that interlocks the cities into a network (Taylor 2004). Second, the data is easy to be obtained from the open information released by the Chinese Industrial and Commercial Bureau. Third, the data is based on a large amount of dataset of 8.75 million firms in 2010. The large quantity ensures the data can reflect the macroscopic phenomena. Fourth, through specific data statistical method, the location information of headquarters and branches of firms could be collected which is further used to analyze the city network.

From these perspectives, this chapter screens all the relational data and selects the data on firm as the most representative and dominant. At the same time, this research also chooses data on the flow of information and traffic, which are less representative but easy to obtain, as complementary data to assess regional economic integration.

5.2.2 Data Sources

Although many scholars (e.g., Beaverstock et al. 1999; Taylor 2004) use the data on advanced producer service (APS) firms to analyze the city network, this study selects data on firm branches covering all industrial sectors to analyze regional economic integration. APS firms usually establish branches in the cities that are at a higher level in the global urban system; therefore, the data of APS firms is more suitable for the analysis of the global and national city network. However, for the cities in most developing countries, particularly in the Asian-Pacific region, the regional network is primarily bolstered by manufacturing. Thus, the data on firm branches that cover all industrial sectors is more convincing. The data source of this chapter is retrieved from the database of the firms registered with the State Administration for Industry and Commerce in 2010. This database contains various types of information, including firm name, registered address, industrial category, zip code, and administrative code. First, by writing SQL programs through Access and Excel, the firm database is translated into standard items such as name, city, etc. Then, in the manner of a keyword search (such as *company*company, *company *office, etc.), the branch information of cross regional firms with their headquarters and branches in the Yangtze River Delta is collected. Finally, the matrix data of the firm network is developed by means of a manual search to supplement the information that the program cannot cover and match.

In terms of the flow of information, drawing lessons from the achievements made by relevant scholars (Xiong et al. 2013, 2014; Liu 2010), this chapter analyzes the flow of information between cities through the "Baidu Index." Since June 2013, the number of Chinese Internet users has grown to 591 million, of which the search engine users are 470 million, which represents 79.5%. Search engines are the primary means for Internet users to obtain access to information, and Baidu, which is the first choice for Chinese users among the many search engines, has a utilization rate that has remained at more than 70% in recent years. Therefore, search engines based on the Baidu Index can make it convenient to obtain the data on flow of information between cities. The specific procedure is to log into the Baidu Index page, query the Baidu Index by using city names as keywords, and then determine the distribution of cities where users search for that keyword. Finally, the data matrix of information network between cities is formulated.

In terms of the flow of traffic, according to the study by Luo et al. (2011, 2015), the "Nanjing-Shanghai-Hangzhou-Ningbo" high-speed railway corridor has become the main mode of transportation in the Yangtze River Delta. The data regarding the number of personnel who board and disembark from the trains is difficult to obtain because the railway administration do not announce these data publicly. However, the number of high-speed railway divisions could be chosen to indirectly characterize the mobility of personnel in the region. Although the railway divisions do not exactly match the mobility of people due to the problem of attendance, it is possible to obtain this open data through the railway official website; this means that this method could be applied widely. Thus, the high-speed railway divisions' data are chosen to characterize the flow of traffic indirectly. Specifically, we visit the high-speed railway's official website and select information regarding the departure and arrival stations in the region. Then, the data matrix of the traffic network between cities is found.

5.2.3 Connectivity and Relational Data

Using the data on firm, flow of information, and traffic, this chapter puts forward a methodology to assess regional economic integration by applying network analysis and multivariate analysis.

First, network connectivity between cities (CBC) is calculated based on a data matrix of the three types of relational data to make further analysis. Then, the data are standardized, and a weight is assigned. According to the research conducted by relevant scholars (Tang and Zhao 2010), the three indicators are weighted as 0.6, 0.2, and 0.2. Finally, the regional economic integration index (REII) is generated. The regional economic integration process is characterized by these three indicators which are, respectively, the flow of commodities and capital, the flow of information, and the flow of people. Among these three indicators, the data on firm branches is the core indicator to characterize regional economic integration. Therefore, this indicator is accorded more weight.

According to the research of Alderson et al. (2004), Wall (2009), and Tang et al. (2014a, b), the formula for CBC is as follows. T_{ij} is the number of firm branches in city j with its headquarters in city i, while T_{ji} is the number of firm branches in city i with its headquarters in city j. V_{ij}, the network connectivity between city i and j (CBC), is defined as:

$$V_{ij} = T_{ij} + T_{ji} \tag{5.1}$$

The n*n matrix is found to depict CBC in the city network based on V_{ij} in the region. The aggregated connectivity N_i of city i (CNC) is defined as the sum of the connectivity with other cities (j = 1 … n) in this region:

$$N_i = \sum_{n}^{j=1} V_{ij} \tag{5.2}$$

To facilitate comparative analysis, the maximum of the CNCs in the region is defined as 100, and the CNCs of other cities are standardized by the percentage of maximum value.

According to the related literature (Xiong et al. 2013; Luo 2011), CBCs and CNCs based on the data on information and traffic could also be calculated by the same method as formulas (5.1) and (5.2); the meaning of the formula is defined below.

As information connectivity, T_{ij} refers to the volume of searches for specific keyword j among Internet users in the city i. Similarly, T_{ji} is the search volume of searches for specific keyword i in city j. The information connectivity between city i and j is equivalent to the sum of T_{ij} and T_{ji} (formula 5.1), and the information CNC of each city is equal to the sum of all information CBCs with other cities (formula 5.2).

For traffic connectivity, T_{ij} denotes the number of railway divisions that depart from city i to city j. At the same time, T_{ji} is the number of railway divisions that are going in the opposite direction. The traffic connectivity between city i and j equals the sum of the two (formula 5.1), and the CNC of each city is equal to the sum of all CBCs with other cities (formula 5.2).

After calculating the three CNCs, the REII of each city is computed as follows:

$$n_i = 0.6^* N_{i1} + 0.2^* N_{i2} + 0.2^* N_{i3} \tag{5.3}$$

$$N_{iz} = 100^* \lg n_i / \lg n_{\text{imax}} \tag{5.4}$$

where

N_{i1}: firm CNC of city i

N_{i2}: information CNC of city i

N_{i3}: traffic CNC of city i

n_i: weighted value of the three indicators of city i

N_{iz}: REII of city i

N_{imax}: the maximum of the weighted values among all cities in the region

5.3 Empirical Analysis

5.3.1 *Comparison of the City Networks Based on the Three Types of Relational Data*

Based on the data matrix of firm branches, flow of information, and flow of traffic, the CNCs of all cities are calculated. By assigning the CNC of Shanghai as 100, which is the maximum value in the region, all the CNCs of cities are standardized. The result shows that the CNCs in the region feature a gradient structure (see Tables 5.2, 5.3, and 5.4 and Fig. 5.1). Comparing the city networks formed by the three relational data sources, four features are notable. First, the CNCs in the three types of city networks all present a gradient structure, and this pattern conforms to the obvious "core-periphery" feature in the region. The CNCs of cities along the

Table 5.2 CNCs of the firm network in the Yangtze River Delta

Level	City(CNC)
High(1st)	Shanghai(100)
Upper-middle(2nd)	Hangzhou(53), Nanjing(51), Suzhou(39)
Middle(3rd)	Ningbo(29), Wuxi(27)
Low-middle(4th)	Nantong(14), Wenzhou(13), Changzhou(12), Jiaxing(12), Shaoxing(12), Hefei(11), Taizhou of Zhejiang Province(10), Jinhua(9), Yangzhou(9), Taizhou of Jiangsu Province(8), Zhenjiang(7), Huzhou(7), Yancheng(7)
Low(5th)	Xuzhou(5), Huai'an(5), Lianyungang(4), Suqian(3), Zhoushan(3), Wuhu(2), Quzhou(2), Lishui(2), Maanshan(2), Huainan(1), Chuzhou(1)

Table 5.3 CNCs of the information network in the Yangtze River Delta

Level	The name of the city(CNC)
High(1st)	Shanghai(100)
Upper-middle(2nd)	Nanjing(68), Suzhou(68), Hangzhou(67)
Middle(3rd)	Ningbo(50), Wuxi(48), Wenzhou(48), Hefei(46), Changzhou(44)
Low-middle(4th)	Xuzhou(39), Yangzhou(38), Jiaxing(38), Taizhou of Zhejiang Province(38), Shaoxing(36), Nantong(36), Yancheng(36), Lianyungang(35), Jinhua(35)
Low(5th)	Wuhu(34), Taizhou of Jiangsu Province(34), Zhenjiang(33), Huai'an(32), Huzhou(32), Suqian(30), Zhoushan(28), Chuzhou(28), Maanshan(27), Quzhou(26), Lishui(26), Huainan(24)

Table 5.4 CNCs of the traffic network in the Yangtze River Delta

Level	The name of the city(CNC)
High(1st)	Shanghai(100)
Upper-middle(2nd)	Nanjing(95), Suzhou(71), Wuxi(68), Hangzhou(67), Changzhou(65)
Middle(3rd)	Zhenjiang(54), Xuzhou(41), Jiaxing(37), Ningbo(32), Hefei(32), Shaoxing(29)
Low-middle(4th)	Wenzhou(27), Jinhua(26), Taizhou of Zhejiang Province(20), Chuzhou(17), Quzhou(17), Huzhou(13), Wuhu(10), Huainan(9)
Low(5th)	Maanshan(5), Lishui(4), Taizhou of Jiangsu Province(4), Nantong(3), Yangzhou(3), Yancheng(2), Huai'an(2), Lianyungang(0), Suqian(0), Zhoushan(0)

"Nanjing-Shanghai-Hangzhou-Ningbo" railway corridor are higher, whereas the cities in Anhui, northern Jiangsu, and southern Zhejiang are relatively low. Second, among these three types of city networks, Shanghai always holds the first position in the regional city network. This indicates that Shanghai is definitely the core city in the region. The primacy ratio of Shanghai in the firm network and information network is 1.9 and 1.5, respectively. This means that these two types of networks belong to a typical single-center system. In the traffic network, the primacy ratio of Shanghai is only 1.1. This phenomenon embodies the double-center characteristics of a traffic network, and it is closely connected with the crucial status of Nanjing in the transportation junction. Third, Nanjing, Hangzhou, and Suzhou hold the

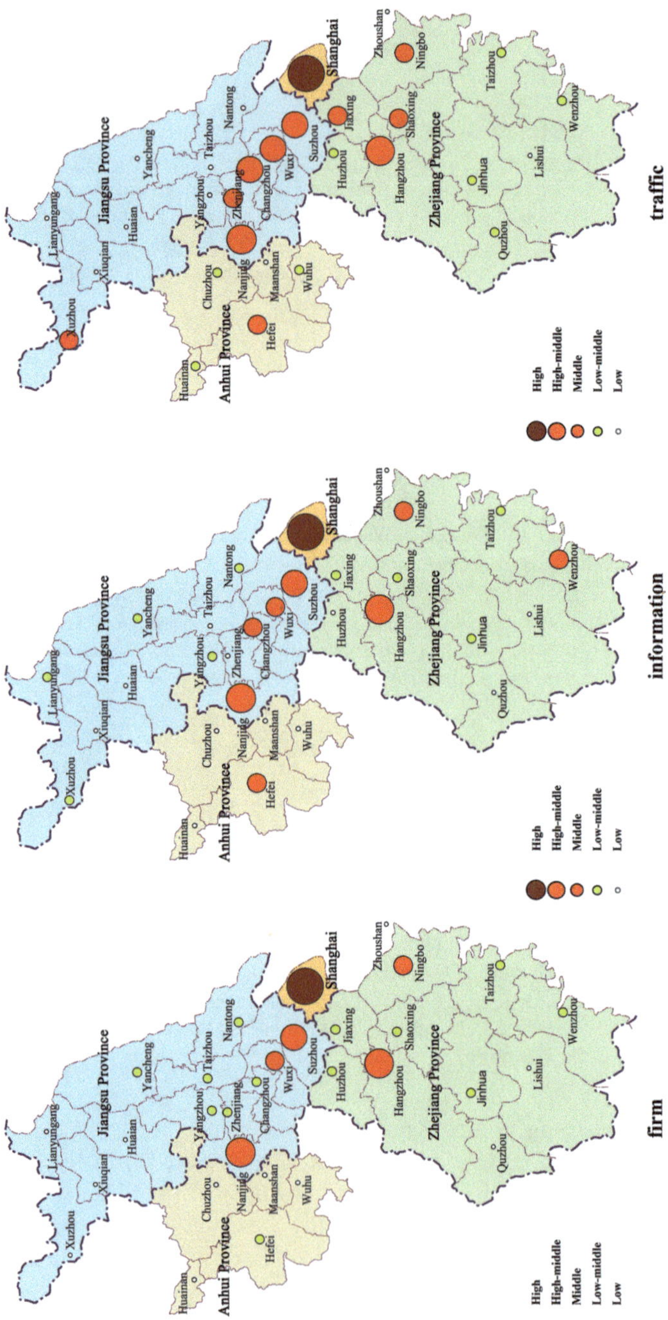

Fig. 5.1 CNCs of firm, information, and traffic networks in the Yangtze River Delta

Table 5.5 Regional economic integration index (REII) in the Yangtze River Delta

REII	City(REII)
100	Shanghai(100)
90–99	Nanjing(93), Hangzhou(92), Suzhou(91)
80–89	Wuxi(86), Ningbo(84), Changzhou(83)
70–79	Wenzhou(79), Hefei(79), Jiaxing(79), Zhenjiang(79), Shaoxing(78), Xuzhou(78), Taizhou of Zhejiang Province(76), Jinhua(76), Nantong(74), Yangzhou(73), Huzhou(72), Taizhou of Jiangsu Province(71), Yancheng(71), Wuhu(70)
60–69	Huai'an(69), Lianyungang(69), Quzhou(68), Chuzhou(68), Xiuqian(66), Maanshan(65), Zhoushan(65), Lishui(64), Huainan(64)

positions of second-level cities in the three city networks. This finding means that these cities are the subcenter cities in the region. Fourth, the cities with higher CNCs are all concentrated in the "Nanjing-Shanghai-Hangzhou-Ningbo" railway corridor not only in the traffic network but also in the firm network and in the information network. This finding means the spatial distribution of CNCs is closely associated with the regional high-speed railway corridor.

5.3.2 Regional Economic Integration Index in the Yangtze River Delta Region

Based on the three types of city networks, the REII of each city is calculated in the region (Table 5.5, Fig. 5.2). This index shows an obvious gradient pattern. The highest index is Shanghai (the value is 100). Nanjing (92), Hangzhou (93), and Suzhou (91) are at the second level. Wuxi (86), Ningbo (84), and Changzhou (83) are at the third level. The fourth-level cities include Wenzhou (79), Hefei (79), and another 14 cities. The fifth-level cities include Huai'an (69), Lianyungang (69), and another nine cities. It is found that the cities with higher REIIs concentrate in the "Nanjing-Shanghai-Hangzhou-Ningbo" railway corridor. It is also evident that the index of each city reduces, in turn, from south to north and from east to west in Jiangsu province, and the index decreases in Zhejiang province from north to south and from east to west. It is notable that the cities along the railway and harbors (such as Shanghai and Ningbo) generally have a high index. It is of great significance for the development of major transport infrastructure to promote the CBCs and accelerate the process of regional economic integration. In addition, the spatial distribution of REII is very similar to the regional urban system structure. Both show the "core-periphery" spatial pattern, which is in accordance with the results of previous scholars. This result means that the cities at a higher level in the region have a more significant impact on the regional city network and play a more pivotal role in promoting the flow of commodities, capital, information, and personnel.

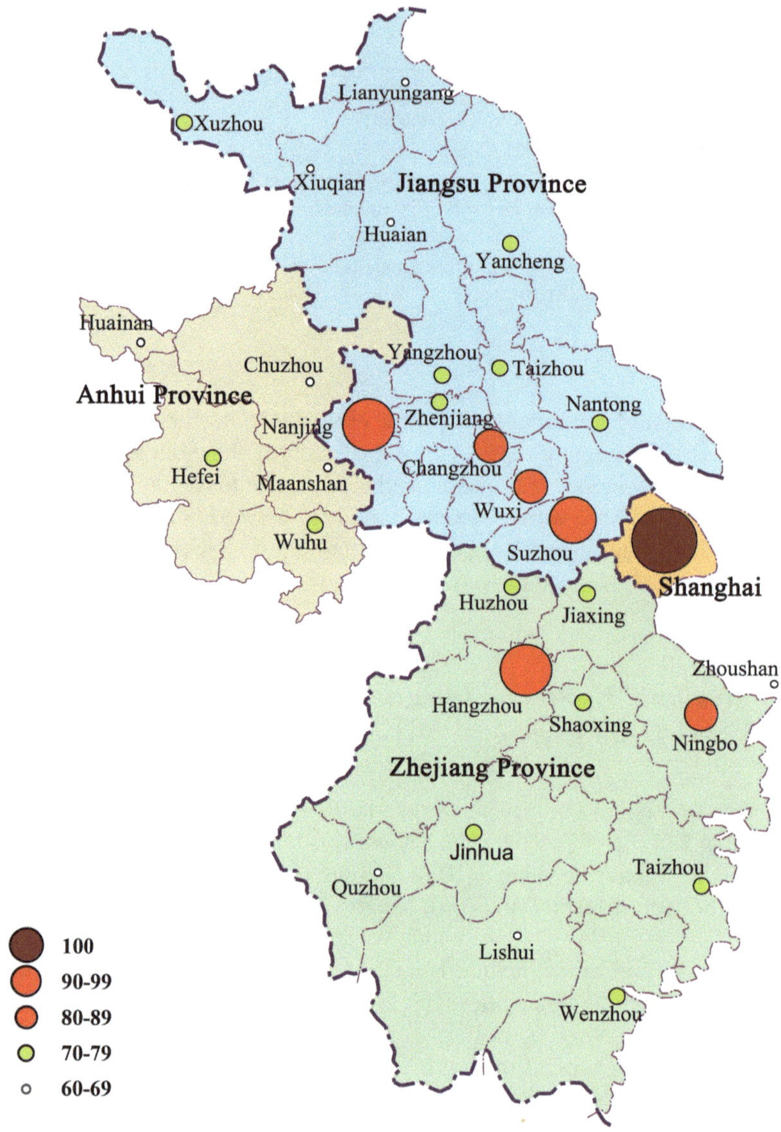

Fig. 5.2 Regional economic integration index (REII) in the Yangtze River Delta

We can observe each city's main interlocking cities through the network data matrix (Fig. 5.3). From the perspective of regional economic integration, each city maintains strong mobility of capital, information, personnel, and commodities with its most closely associated city in the region. The pattern of the city network made by those cities is the manifestation of regional economic integration. Each row in Table 5.6 lists the top three cities that are strongly linked to each city. The results show three features. First, Shanghai is the first or second interlocking city

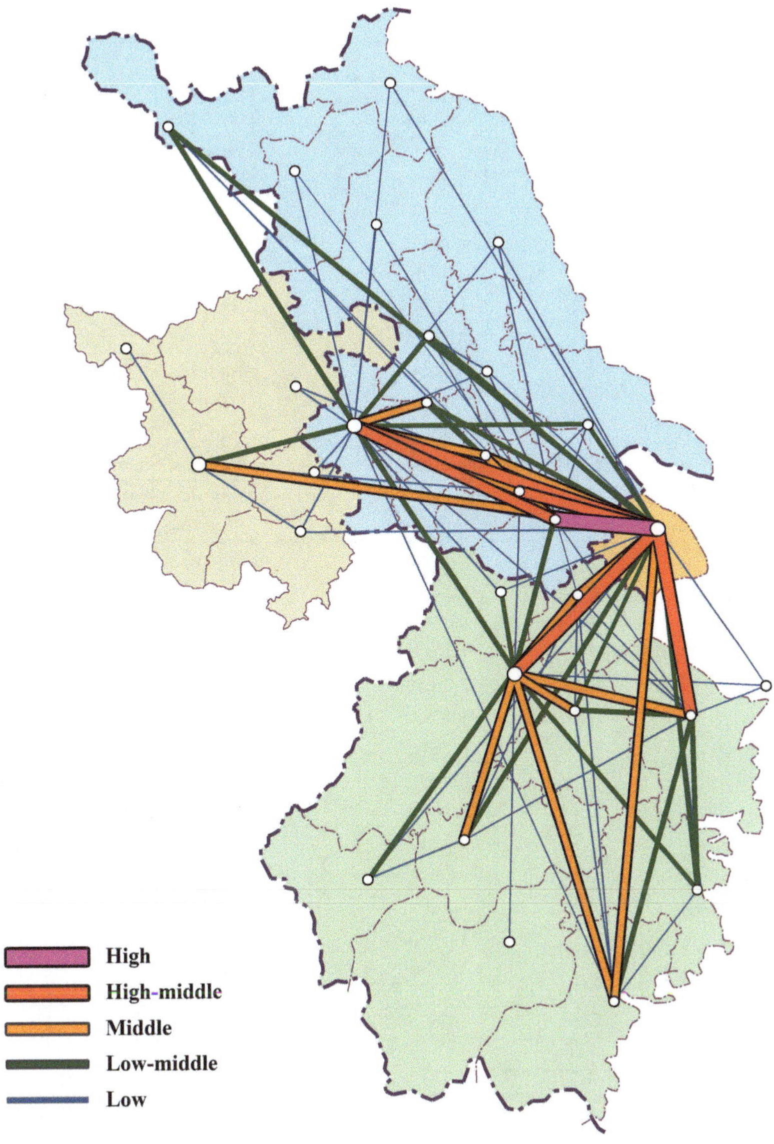

Fig. 5.3 Pattern diagram of interlocking network in the Yangtze River Delta

of most cities. This indicates that the influence of Shanghai has affected the majority of cities. Thus, Shanghai is definitely the primary city in the region. Second, southern Jiangsu and northern Zhejiang cities have more close CBCs with Shanghai than their provincial capital cities, Nanjing and Hangzhou, whereas northern Jiangsu and southern Zhejiang cities are in the opposite situation. This finding explains why southern Jiangsu and the northern Zhejiang cities have a high degree of economic integration. Third, most cities in Jiangsu, Zhejiang, and Anhui

Table 5.6 The main interlocking cities of each city in the Yangtze River Delta

Municipality/province	City	Main interlocking cities
Shanghai	Shanghai	Suzhou, Hangzhou, Nanjing
Jiangsu Province	Nanjing	Shanghai, Suzhou, Wuxi
	Wuxi	Shanghai, Nanjing, Suzhou
	Changzhou	Shanghai, Nanjing, Wuxi
	Suzhou	Shanghai, Nanjing, Wuxi
	Nantong	Shanghai, Nanjing, Suzhou
	Yangzhou	Nanjing, Shanghai, Taizhou
	Zhenjiang	Nanjing, Shanghai, Suzhou
	Taizhou	Nanjing, Shanghai, Yangzhou
	Xuzhou	Nanjing, Shanghai, Suzhou
	Yancheng	Nanjing, Shanghai, Suzhou
	Lianyungang	Nanjing, Shanghai, Suzhou
	Huai'an	Nanjing, Shanghai, Suzhou
	Suqian	Nanjing, Shanghai, Suzhou
Zhejiang Province	Hangzhou	Shanghai, Ningbo, Jiaxing
	Ningbo	Shanghai, Hangzhou, Shaoxing
	Jiaxing	Shanghai, Hangzhou, Ningbo
	Huzhou	Hangzhou, Shanghai, Nanjing
	Shaoxing	Hangzhou, Shanghai, Ningbo
	Zhoushan	Ningbo, Hangzhou, Shanghai
	Taizhou	Hangzhou, Shanghai, Ningbo
	Wenzhou	Hangzhou, Shanghai, Ningbo
	Jinhua	Hangzhou, Shanghai, Jiaxing
	Quzhou	Hangzhou, Shanghai, Jiaxing
	Lishui	Hangzhou, Wenzhou, Shanghai
Anhui Province	Hefei	Shanghai, Nanjing, Wuhu
	Wuhu	Hefei, Shanghai, Nanjing
	Maanshan	Nanjing, Shanghai, Hefei
	Chuzhou	Nanjing, Shanghai, Hefei
	Huainan	Hefei, Shanghai, Nanjing

provinces have fewer trans-provincial contacts, and the major connected cities of these are in their own provinces. This shows that the administrative regional economy still plays a very obvious role; in addition, the administrative divisions do have a blocking effect on the flow of production factors in the region.

5.4 Discussion and Conclusion

Regional economic integration can be viewed as the flow of capital, cargo, information, and mobility between cities in a region. On the basis of the big relational data of firm branches, information, and traffic, this chapter uses the interlocking network and multivariate analysis methods to assess the regional economic integration index (REII) among 30 cities in the Yangtze River Delta. The regional firm network and the information network feature only one single center, whereas the traffic network is characterized by double centers. The spatial distribution of the regional economic integration index shows the features of a "core-periphery" pattern. The cities with a higher regional economic integration index are concentrated in the "Nanjing-Shanghai-Hangzhou-Ningbo" corridor; this phenomenon has an obvious relevance with the regional major transport infrastructure such as high-speed railway and harbors.

From the relational perspective, this study substantiates the conclusion made by previous scholars (Tang and Zhao 2010) that the urban spatial system in Yangtze River Delta region presents "core-periphery" characteristics. At the same time, it shows that the development of major transport infrastructure is conducive to promoting the flow of production factors between cities. Most noteworthy is the phenomenon that in the regional traffic network, the cities along the traffic corridor always have high degrees of relevance, which shows that it is of great significance for the traffic flow on the regional railway to promote regional economic integration. In the regional economic integration process, the "space of flows," which is constituted by the interlocking network of capital, cargo, information, and mobility, breaks through the regional restriction to a certain extent. On the one hand, the city's development is no longer limited to the dependence on the nearby city, and it could seek development opportunities through cross regional contact. On the other hand, the administrative economy remains obvious, and administrative divisions continue to hinder the flow of production factors in the region. It is imperative for us to accelerate the progress of regional economic integration. These findings have explicit policy significance and provide a new theoretical basis for coordinated development in the region.

The question regarding the assessment of regional economic integration is intricate and requires cautious answers. The introduction of big relational data offers us a new perspective to understand the regional economic integration process. Previous studies on this issue have often been based on attribute data and thus ignore the core characteristics of regional economic integration, namely, the flow of all types of production factors between cities. Research based on big relational data reflects the shift of the cognitive paradigm on regional issues from "rank-hierarchy" to "relationship-network" in academia. Furthermore, the empirical conclusions regarding the assessment of regional economic integration in the Yangtze River Delta have enriched our understanding of relational data, and this provided us a brand new perspective and a novel means to analyze and describe the regional economic integration process. With the new era's background and theoretical system, only through analysis and observation from many different angles, combined with a variety of methods, can we have a more in-depth understanding of regional economic integration? This will also be a pivotal direction for future research.

Notes

1. This study uses the Yangtze River Delta as the research scope. "The main functional area plan," which was published by the National Development and Reform Commission of China in 2010, defined two provinces and one city (Shanghai, Jiangsu and Zhejiang provinces) as "the Yangtze River Delta." In the operational level of regional coordination, the "Economic Coordination Committee of the Yangtze Delta Cities" is the core organization. Currently, in addition to covering the two provinces and one city, more cities are added in to the region including Hefei, Wuhu, Maanshan, Chuzhou, and Huainan, and the regional scale has expanded to 30 cities as "three provinces and one city." Therefore, we use the 30 cities as the research scope of space.
2. The data of Baidu Index from 2011 to present is available in the Internet. Therefore, this article selects 2011 data as the criterion.
3. This study began in 2014; therefore, we choose the data of railway divisions at the end of 2014 as the data source.

Fund This work is funded by National Science Foundation of China (51708123, 51508108) and China Postdoctoral Science Foundation (2016M590320).

References

Alderson, A. S., & Beckfield, J. (2004). Power and Position in the World City System. *American Journal of Sociology, 109*(4), 811–851.

Alderson, A. S., & Beckfield, J. (2007). Globalization and the world city system: Preliminary results from a longitudinal data set. In P. J. Taylor et al. (Eds.), *Cities in globalization: Practices, policies and theories* (pp. 21–36). London: Routledge.

Armstrong, H. W. (1996). An appraisal of the evidence from cross-sectional analysis of the regional growth process within the European Union. In *Convergence and divergence among European regions* (pp. 40–64). London: Pion, 40–64.

Balassa, B. (1967). Trade creation and trade diversion in the European common market. *The Economic Journal, 77*, 1–21.

Barro, R., Sala, I., et al. (1992). Convergence. *Journal of Political Economy, 100*(2), 223–251.

Beaverstock, J. V., Smith, R. G., & Taylor, P. J. (1999). A roster of world cities. *Cities, 16*(6), 445–458.

Beaverstock, J. V., & Boardwell, J. T. (2000). Negotiating globalization, transnational corporations and global city financial centres in transient migration studies. *Applied Geography, 20*(3), 277–304.

Beaverstock, J. V., Smith, R. G., et al. (2000). World city network: A new metageography. *Annals of the Association of American Geographers, 90*(1), 123–134.

Beaverstock, J. V., Hubbard, P., & Short, J. R. (2004). Getting away with it? Exposing the geographies of the super-rich. *Geoforum, 35*(4), 401–407.

Beaverstock, J. V. (2007). World city networks 'from below': International mobility and inter-city relations in the global investment banking industry. In P. J. Taylor (Ed.), *Cities in globalization: Practices, policies and theories* (pp. 52–71). London: Routledge.

Castells, M. (2006). *The rise of network society*. Beijing: Social Sciences Academic Press.

Derudder, B. (2006). On conceptual confusion in empirical analyses of a transnational urban network. *Urban Studies, 43*(11), 2027–2046.

Derudder, B., & Taylor, P. J. (2005). The cliquishness of world cities. *Global Networks, 5*(1), 71–91.

Derudder, B., & Witlox, F. (2005). An appraisal of the use of airline data in assessing the world city network: A research note on data. *Urban Studies, 42*(13), 2371–2388.
Friedmann, J. (1986). The world city hypothesis. *Development and Change, 17*(1), 69–83.
Hall P, Pain K (2006) The polycentric metropolis: learning from mega-city regions in Europe. London: Earthscan.
Han, J. (2012). *Research of the development of regional economic integration in the Yangtze River Delta*. Shanghai: East China Normal University.
Hu, B. (2011). *Urban cluster in the Yangtze River delta-motivation of networked organization and governance mode*. Shanghai: Shanghai University of finance and Economics Press.
Hu, Y., & Zhang, J. X. (2010). Economic space evolution of Yangtze River Delta city group based on county scale. *Economic Geography, 30*(7), 1112–1117.
Kumur, A. (1994). Economic reform and the internal division of labor in China: Production, trade and marketing. In D. Goodman & G. Segal (Eds.), *China deconstructs: Politics, trade and regionalism* (pp. 99–130). London: Routledge.
Li, R. L., & Luo, H. S. (2007). Regional economic integration: The connotation, effect and realization. *Inquiry to Economic Issues, 1*, 52–57.
Liu, Z. B. (2010). *Regional economic integration in the Yangtze River Delta region*. Beijing: China Renmin University Press.
Liu, S. L., & Hu, A. G. (2011). Transportation infrastructure and regional economic integration in China. *Economic Research Journal, 3*, 72–82.
Liu, X. L., & Zhang, X. P. (2014). Exploratory Analysis of Urban Network Based on Baidu Index and 3s Framework- a case study of Shandong Province. In *Proceedings of 2014 Annual Meeting of China's Urban Planning* (pp. 1–9).
Lu, X. D., & Li, R. L. (2009). Regional integration, FDI and international production location: A footloose capital model. *China Economic Quarterly, 8*(4), 1475–1496.
Luo, Z. D. (2011). Analysis of the polycentric structure of Yangtze River Delta based on passenger traffic flow. *Urban Planning Forum, 2*, 16–23.
Luo, Z., Zhu, C., & Xue, W. (2015). The analysis on spatial structure of Yangtze River Delta based on passenger flow of high-speed railway. *Shanghai Urban Planning Review, 4*, 74–80.
Lv, K. J. (2012). *Theory and empirical research of complex urban network and industry network*. Beijing: China Architecture & Building Press.
Meng, Q. M. (2001). Concept and mechanism of regional economic integration. *Research On Development, 2*, 47–49.
Moss, M. L., & Townsend, A. M. (2000). The internet backbone and the American metropolis. *The Information Society, 16*, 35–47.
Naughton, B. (1999). *How much can regional integration do to unify China's markets?* Paper on conference for research on economic development and policy research. Stanford: Stanford University.
Nuffel, N. V., Saey, P., et al. (2010). Measuring hierarchical differentiation: Connectivity and dominance in the European urban network. *Transportation Planning and Technology, 33*(4), 343–366.
Pain, K. (2008). Examining 'core-periphery' relationships in a global city-region: The case of London and south east England. *Regional Studies, 42*(8), 1161–1172.
Rossi, E. C., Beaverstock, J. V., et al. (2007). Transaction links through cities: 'Decision cities' and 'service cities' in outsourcing by leading Brazilian firms. *Geoforum, 38*(4), 628–642.
Rutherford, J., Gillespie, A., et al. (2004). The territoriality of pan-European telecommunications backbone networks. *Journal of Urban Technology, 11*(3), 1–34.
Sassen, S. (1991). *The global city: New York, London, Tokyo*. Princeton: Princeton University Press.
Smith, D. A., & Timberlake, M. (1995). *Conceptualizing and mapping the structure of the world systems city system. Urban Studies, 32*, 287–302.
Smith, D. A., & Timberlake, M. (2002). Hierarchies of dominance among world cities: A network approach. In S. Sassen (Ed.), *Global networks, linked cities* (pp. 117–141). London: Routledge.

Sun, D. B. (2003). A probe into the dynamic mechanism of regional economic integration in China. *International Economics and Trade Research, 19*(6), 71–74.
Tang, Z. L., & Li.T. (2014a). A comparative analysis of urban system in the Yangtze Delta region and the middle Yangtze region: An approach of firm-based interlocking network. *Urban Planning Forum, 2*, 24–31.
Tang, Z. L., & Li.T. (2014b). A comparative analysis of Urban Systems in the Beijing-Tianjin-Hebei region, the Yangtze River Delta region and the Pearl River Delta region: An approach of firm-based interlocking network. *Shanghai Urban Planning Review, 6*, 37–45.
Tang, Z. L., & Zhao, M. (2009a). Local network research of the Yangtze River Delta based on economic news. *GaWC Research Bulletin, 312*.
Tang, Z. L., & Zhao, M. X. (2009b). Time and spatial transformation of economic globalization in the Yangtze River Delta region. *Urban Planning Forum, 1*, 38–45.
Tang, Z. L., & Zhao, M. X. (2010). Economic globalization and transformation of urban system in the Yangtze River Delta region: Interlocking network and value-added hierarchy. *Urban Planning Forum, 1*, 29–34.
Taylor, P. z. (1997). Hierarchical tendencies amongst world cities: A global research proposal. *Cities, 14*(6), 323–332.
Taylor, P. J. (2001). Urban hinterworlds: geographies of corporate service provision under conditions of contemporary globalization [J]. *Geography, 86*(1), 51–60.
Taylor, P. J. (2004). *World city network: A global urban analysis*. London/New York: Routledge.
Taylor, P. J. (2006). Parallel paths to understanding global inter-city relations. *American Journal of Sociology, 112*(3), 881–894.
Taylor, P. J., & Aranya, R. (2008). A global 'urban roller coaster'? Connectivity changes in the world city network, 2000-04. *Regional Studies, 42*(1), 1–16.
Taylor, P. J., Catalano, G., et al. (2002). Exploratory analysis of the world city network. *Urban Studies, 39*(13), 2377–2394.
Taylor, P. J., Derudder, B., et al. (2007). Comparing airline passenger destinations with global service connectivities: A worldwide empirical study of 214 cities. *Urban Geography, 28*(3), 232–248.
Taylor, P.J., Ni, P., et al. (2010). Measuring the world city network: New results and developments. *GaWC Research Bulletin, 300*.
Tinbergen, J. (1954). *International economic integration*. Amsterdam: Elsevier.
Townsend, A. M. (2001). Network cities and the global structure of the internet. *American Behavioral Scientist, 44*, 1697–1716.
Wall, R. S. (2009). *Netscape: Cities and global corporate networks*. Rotterdam: Haveka.
Wang, D. Z. (2009). Source, development and defect of the regional economic integration. *Commercial Research, 2*, 18–21.
Wang, H. X. (2011). Quantitative research on factors flows, agglomeration and urban interactive development: A case study of the Yangtze Delta area cities. *Shanghai Journal of Economics, 12*, 45–55.
Xiong, L. F., Zhen, F., et al. (2013). The research of the Yangtze River Delta Core Area's City network characteristics based on Baidu index. *Economic Geography, 33*(7), 67–73.
Xiong, L. F., Zhen, F., et al. (2014). Characteristics of the City network in the three major economic zones of China: A study based on Baidu information flow. *Tropical Geography, 34*(1), 34–43.
Yin, J., Zhen, F., et al. (2011). China's City network pattern: An empirical analysis based on financial Enterprise layout. *Economic Geography, 21*(5), 754–759.
Yu, W., & Huang, J. K. (1998). The reform of Rice market from the perspective of the degree of integration. *Economic Research Journal, 3*, 50–57.
Zhang, C. (2010). *The structure evolution of the integrity distribution network in China: Theory building and empirical research*. Beijing: People's Publishing House.
Zhao, Y. L., & Cai, G. W. (2009). Border effect of market potential and inter-external market integration. *Economic Research Journal, 7*, 119–130.
Zhao, W., & Chen, Y. (2006). The Review and recent development of regional economic integration. *Business Economics and Administration, 6*, 58–62.

Zhao, M. X., & Tang, Z. L. (2010). A network-based measurement for urban hinterland area in Yangtze River Delta. *Economic Geography, 30*(3), 371–376.

Zhou, L. Q., & Xia, L. K. (2010). The measurement and comparison of regional economic integration: Evidence from the Beijing-Tianjin-Hebei region, the Yangtze River Delta and the Pearl River Delta. *Jianghai Academic Journal, 4*, 81–87.

Zhu, C. S., Wang, D., et al. (2014). Centrality and power: A method of analyzing city network spatial structure. *Urban Planning Forum, 4*, 24–30.

Chapter 6
The Recognition of CAZ in Shanghai Based on Evaluated POI

Liu Liu and Zhuqing Liu

Abstract The upgrade of urban core from central business district (CBD) to Central Activities Zone (CAZ) becomes a trend that many cities in China follows, yet it turns out to be ambiguous during the specification of its boundary. This project proposes a relatively clear definition of CAZ through a comparison with CBD and then innovates a method to identify the potential CAZ in a city based on evaluated point of interest (POI) dataset. Taking Shanghai as an example, we first go through the background of CAZ, and decompose its concept, we propose a data process to calibrate the degree of functional mixture in aspects of retail, catering, office, leisure, culture, and recreation. Through the compilation of rasterized maps, we find the top fourteen dynamic places in downtown Shanghai and deliver three scenarios for the discussion of CAZ boundary. Moreover, this replicable method has already been applied in many planning practical analysis and documents.

Keywords CAZ • POI • Functional mixture • Land use

6.1 Introduction

Central Activities Zone, briefly named as CAZ, emphasizes supporting multiple functions, improving quality of open spaces, and keeping 24-h dynamic places. As the economic center in mainland China, Shanghai is upgrading its central business district (CBD) to CAZ in its latest comprehensive plan. However, the geographical determination of such a district has faced a wide variety of opinions. In planning practice, no criterion has been confirmed for delimiting its boundary, yet clarifying this is meaningful not simply out of a research purpose. The opportunity of receiving investment and policy support may change massively once an area designated with a title

L. Liu (✉)
CitoryTech Limited, HongKong, China
e-mail: liuliu@citorytech.com

Z. Liu
China Academy of Urban Planning and Design Shanghai Branch, Shanghai, China

Z. Shen, M. Li (eds.), *Big Data Support of Urban Planning and Management*,
Advances in Geographic Information Science, DOI 10.1007/978-3-319-51929-6_6

such as CAZ. Given these political considerations, we focus on the core definition of multi-functional diversity and propose a tentative method to recognize the boundary of the most dynamic districts in Shanghai based on evaluated point of interest (POI). The result of our research partially explained what is CAZ and how its boundary could be calculated. This method has been first applied since 2014 in the spatial improvement planning project as a part of the 13th Five-Year Plan in Jingan District. Later on, this method has been applied to different planning projects in other cities.

The term “central business district” (CBD), which comes from the concentric zone model, demonstrates the commercial and business center of a city and follows the bid rent curve (Burgess 1925). Usually with the most accessible location, city centers supports expensive land price, thus only attracting industries who expected higher rates for better locations. As a result, the development of CBD in North America has created a compacted urban core consists of financial companies, headquarters, and luxury stores.

The prosperity of downtown has been destroyed by a nationwide trend of suburbanization. Accompanied with the decentralization of employment, its decline is explained by two reasons: one is called natural evolution theory, which related to the public transportation innovation, and the other is considering fiscal and social problems of central cities (Mieszkowski and Mills 1993). No matter which reason or their combination with mutual interaction, the suburbanization during the middle of the twentieth century has put an end to the development of CBD in North America (Liston 1968). Some researchers believed that the decline may have actually happened since the 1920s based on population data (Edmonston 1975) and even in the 1900s for large older cities such as Boston, Chicago, and Philadelphia (Bogue 1953).

When it comes to the context of mainland China, many municipal governments believe CBD as a representation of the central nucleus motivating the development of a city. The movement of constructing CBD has been popular during past decades. In 2003, over 30 cities were attempting for their own well-developed city center, which is not but easily confused with CBD (Weixin 2003). Such proliferation driven by GDP growth pressure in China uniformed those districts with office towers in city centers. Industries including banking, securities, trading, insurance, consulting, and other financial companies are clustered in those planned districts. Financial activity covers the most part of the life here.

In contrast to the original theory, the concept CBD in China later on tended to be abused in plenty of planning projects. This abbreviation is simply like a drug and almost every commercial development has invited it as their promotion tool. Nevertheless, in this paper our research narrow down its scope only to the central district led by “business” instead of public commercial activities, which differs from the vast development almost everywhere in China.

Interestingly, as the second largest economic entity in the world, China avoided from ethnic conflicts and high unemployment rate. The country nowadays has faced another series of problems while running its localized CBD. Shanghai has played an explanatory role for this circumstance, with its earliest CBD Lujiazui. Started ever since the 1990s, Lujiazui has always been symbolizing its reform of mainland China and stands for the most dynamic economic center. With convenient public transportation access, yet its highly controlled proportion of commercial space brought limited choice of lunch to white collars. Without the support of recreational and leisure facilities, this place turns less attractive and fails to keep its population while off work, which also raises traffic burden since everyone escapes here after work.

The physical environment of Lujiazui has also been designed less friendly. Streets seem to be truncated, because there are no on-street stores to hold pedestrian. Traffic burden has no chance to be mitigated since no one after work would be willing to stay in this boring place.

On the flip side, places with a mixture of retailers, restaurants, and recreational facilities turn out to attract more visitors and office workers; companies shift from pure CBD areas towards those areas. For instance, in Manhattan, one can easily find almost everything he needs in Time Square: food, alcohol, dressing, leisure, entertainment, art, sports, and also work. Very likely, this can partially explain why a bundle of financial companies have moved their headquarters from downtown to midtown.

The comparison above indicates the importance of fostering urban vitality during planning process, and it also has been widely accepted by planners. Distinguished from CBD, the word CAZ has been invented by the Greater London Authority in its plan in 2004. For the first time, world city puts priority on "activities" other than "business" in its official planning document. Built on this concept, our research puts forward a method to recognize the most dynamic places via evaluated POI dataset. To some extent, these places stand for the power that motivate city lives and boost urban dynamics.

6.2 The Development Of CAZ

Standing for "Central Activities Zone," CAZ has been proposed by the Greater London Plan since 2004. Other similar terms like "Central Activities District" or "Activity Center" have also been mentioned as alternative names for this concept. Since then, the concept of CAZ has been widely adopted in many metropolitan planning documents and policy making. However, the CAZ boundary is still yet to be finalized as it was illustratively shown in London Plan Consultation 2009.

For decades, many attempts have been applied to the delimitation of CBD. Murphy and Vance defined a list of central business and delimitated CBD based on a threshold of 1.0 in CBHI and 50% in CBII (Murphy and Vance 1954). Derived from the Murphy-Vance technique, others have improved indicators such as CBI.

Later Bowden adopted an improved indicator CBI to detect the sensitive changes of downtown during time (Bowden 1971). Other methods based on market analysis and land value book, (Proudfoot 1937), the functional approach.

6.2.1 CAZ Concepts

For strategic planning, since 2011, there has been a Central London subregion comprising the boroughs of Camden, Islington, Kensington and Chelsea, Lambeth, Southwark, Westminster, and the City of London. From 2004 to 2008, the London

Plan included a subregion called Central London comprising Camden, Islington, Kensington and Chelsea, Lambeth, Southwark, Wandsworth, and Westminster (Mayor of London 2004). It had a 2001 population of 1,525,000. The subregion was replaced in 2008 with a new structure which amalgamated inner and outer boroughs together. This was altered in 2011 when a new Central London subregion was created, now including the City of London and excluding Wandsworth. However, districts at the outer edge of this subregion such as Streatham, Dulwich, and Holloway are not generally considered as Central London (Greater London Authority 2011).

In general, the boundary of the CAZ area defined in the Greater London Plan embraces the City of London, most of Westminster and the inner parts of Camden, Islington, Hackney, Tower Hamlets, Southwark, Lambeth, and Kensington and Chelsea. It is described as a unique cluster of vitally important activities including central government offices, headquarters, and embassies, the largest concentration of London's financial and business services sector and the offices of trade, professional bodies, institutions, associations, communications, publishing, advertising, and the media (Mayor of London 2008).

6.2.2 CAZ Features

Different from the concept of CBD, CAZ specifically emphasizes a mixture of a series of functions. The CAZ area in London covers its geographic, economic, and administrative core. It brings together the largest concentration of London's financial and globally oriented business services. Almost a third of all London jobs are based there, and together with Canary Wharf, it has historically experienced the highest rate of growth in London.

A variety of function mixtures give the CAZ a unique character and feel across its hugely varied quarters and neighborhoods. As the seat of national government, it includes parliament, the headquarters of central government, and the range of organizations and associations linked with the legislative and administrative processes. It is also a cultural center, providing the base of theaters, concert halls, and other facilities of national and international significance, as well as the base for a range of cultural industries of retail centers, from the internationally important West End and Knightsbridge to more local centers primarily meeting the needs of residents. It is also home for 275,000 Londoners, providing a large amount of housing to meet local and citywide needs. Finally, it embraces much of what is recognized across the world as iconic London—the sweep of the inner Royal Parks and the Thames combined with a mixture of unrivaled and sometimes ancient heritage and more modern architecture (Mayor of London 2011) (Fig. 6.1).

To sum up, CAZ supports many different activities, including retail, recreation, culture, etc. As well as being an economic hub (CBD), the CAZ is a place where many people not only work but also live. A certain amount of housing within CAZ supports a low-carbon lifestyle and reduces the traveling time between home and

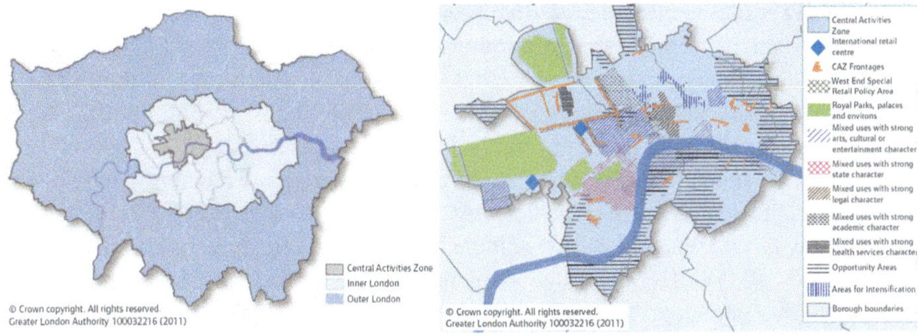

Fig. 6.1 Outer London, Inner London and Central Activities Zone (The London Plan, Spatial Development Strategy for Greater London, July 2011), and the CAZ Diagram (The London Plan, Spatial Development Strategy for Greater London, July 2011)

office. The highly mixed functionality with in a condensed space entitles the CAZ a unique character and feel across its diverse quarters and neighborhoods, which the city is committed to protecting and enhancing.

6.3 Dataset Preparation

Since the original method applied in planning project was built on dataset in 2014, and related data, models, and conclusions are all included in the document, we update our dataset and take a deeper look into the result for this paper.

As the discussion above, CAZ represents a multifunctional zone within a city. Usually a range of land use will be included, such as franchise stores, recreational facilities, theaters, museums, etc. As the keyword "activities" emphasizes, to qualify a typical CAZ depends on whether a place can support diverse events and popularity.

Taking the first place of GDP rank all these years, Shanghai is considering the concept of CAZ instead of CBD. Confirming the current qualified places is important before designating the district which will support all types of activities in the future. To recognize the boundary of CAZ, we suggest an approach for the detection of functionally mixed places via evaluated POI, which is crawled from the website of "Dianping." It is a Chinese alternative to Yelp, and its data is collected from the public through mobile clients.

6.3.1 *Evaluated POI*

Updated to August in 2016, we have captured a total number of 386,628 POI records within the outer ring of downtown Shanghai. Apart of geometry characters, the dataset includes other attributes such as shopID, score, category, and average price.

The shopID is the primary key to distinguish different records. The score for every single POI is a 10-point-based comprehensive assessment collected and calculated from all the users. The category is the default classifier which tags all the POI with many types based on their function: there are over 480 types, including office work, market, florist, furniture, and even nail salon.

The spatial distribution of the POI turns out to be sparser when reaching outwards from the core. Within the outer ring, Pudong holds over 80,000 POI due to its enormously large area, followed by Jingan, Xuhui, and Huangpu, whose number of POI exceeds 40,000. Huangpu becomes the most condensed district with a density of nearly 1500 per square kilometer, followed by Jingan, Hongkou, Changning, and Xuhui (Fig. 6.2).

Taken the district area into consideration, Huangpu ranks the top one in POI density among the 11 districts, together with about 1490 POI per square kilometers, which is almost twice as that of Jingan district. The result of the top two matches the impression of local Shanghainese, as Huangpu, Luwan, and Jingan have always been the most dynamic urban core districts during the past decades. With the administrative adjustment, today's Huangpu stands for a combination of itself and Luwan, while Jingan stands for a merge with Zhabei, which is a less developed district and very likely dilutes the POI density in Jingan. In spite of having the most POI, Pudong still suffers the lowest density of 215 POI per square kilometer due to its vast area (Fig. 6.3).

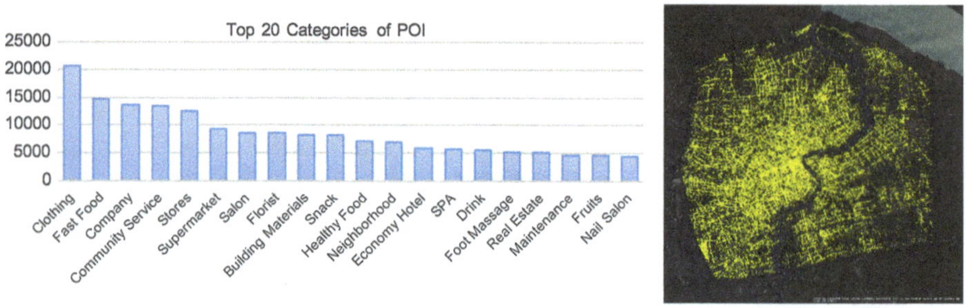

Fig. 6.2 Top 20 categories of POI (*left*) and overall map within downtown Shanghai (*right*)

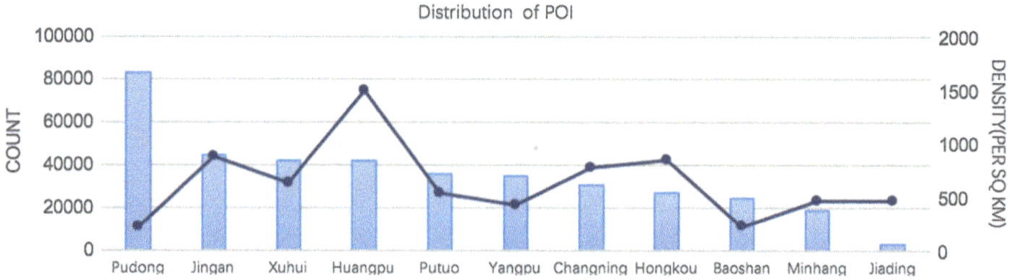

Fig. 6.3 Distribution of POI in all administrative districts in Shanghai

6.3.2 *Category Reclassification*

The attribute of category has over precisely classified the dataset, as there are a total number of 484 distinctive items. Yet according to the analysis of London's CAZ, those categories need regrouping to general urban functions. In our research, we treat the attribute "category" as subcategory and regroup them into six general categories: "retail," "catering," "office," "recreation," "leisure," and "culture." This categorization has also taken the diversity of CAZ into consideration. The relative portion of each category is mostly similar over districts. Districts such as Xuhui and Jingan have a relative higher ratio of office POI. When it comes to retail, Huangpu edges out its competitors with a ratio over 40% (Fig. 6.4).

The definition of CAZ put emphasis on the activities, which means detecting the existence of people is simply finding not only the place with specific functions but also dynamics brought by visitors. The raw dataset contains variable dimensions of subjective evaluation on POI, and they are suitable for a calibration of the power of attracting customers.

To plot the evaluated POI map of downtown, we create kernel density map in ArcMap to describe the level of aggregation for all the six general categories. In addition, score has been selected as the population parameter for the heat map, with a search radius of 300 meters. The stretch method is neutral break, which brings a clear illustration after calculation (Fig. 6.5).

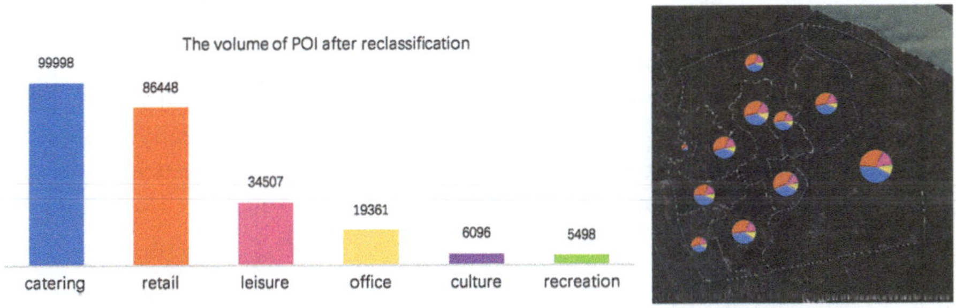

Fig. 6.4 The histogram of different functional POI after re-categorization

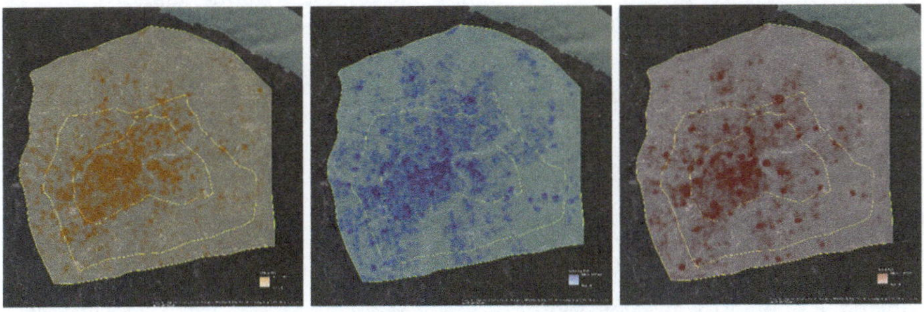

Fig. 6.5 Kernel density map of office (*left*), catering (*middle*), and retail (*right*)

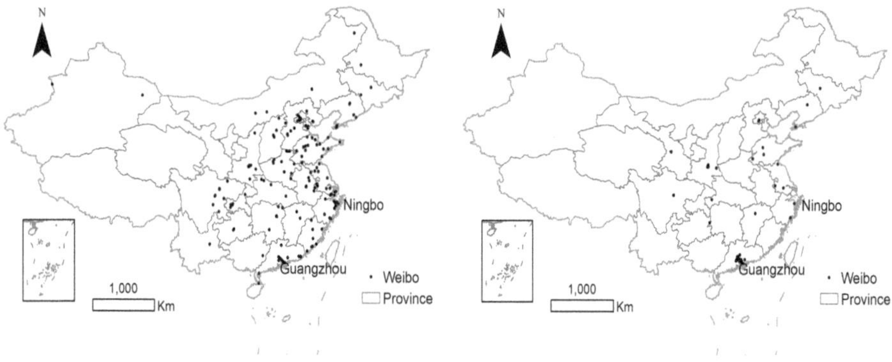

Fig. 6.6 Kernel density map of culture (*left*), leisure (*middle*), and recreation (*right*)

The distribution of office map is highly condensed near Nanjing Road, which continuously takes the largest area in the map. Other districts like Lujiazui and Xujiahui have also been quite large. Catering map spreads a lot, many sub dot-like places jumps out other than the core district, and quite often they stand for a large shopping center, such as Wanda Plaza, Bailian Youyicheng, and Changtai Plaza. Other linear places like Xianxia Road food street has also been shown on the map. Maps for retail display traditional commercial clusters such as People's Square, Riyueguang Center, Jingan Temple, etc. (Fig. 6.6).

As for culture map, the major condensed places are clustered in the southwest area within inner ring. A relatively large continuous "culture district" is located around the boundary among Jingan, Huangpu, and Xuhui districts. The heat map of recreation displayed more polarized than others. It is because most of the recreational facilities such as Tom's World, KTV, and billiard room usually go together with shopping mall or even urban complex, while places for leisure such sauna, spa, and beauty center are less necessary to be geographically bundled.

6.4 Calibration of Mixture

Two dimensions reflect the degree of functional mixture: volume and diversity. To achieve the comprehensive value, we aggregate the normalized heat maps of the six general categories. To measure the level of mixture use, we count the number of layers where its pixel value exceeds the threshold. Beyond that, a deeper look towards the top 14 highlighted places is added after the measurement.

6.4.1 Value Aggregation Map

Given the popularity and density for different POI categories vary, it becomes necessary to normalize the rasterized heat maps before aggregation. For every general category, in this case, we map the pixel value with a score from 0 to 100 using the

Fig. 6.7 Value aggregation map of Shanghai (*left*), the magnified core area (*top right*)

tool "Rescale by Function" in ArcMap, which also means we suppose equal weight among all the six aspects, and theoretically speaking, the available range of pixel value in those aggregated maps should cover 0 to 600.

Compiling those normalized maps, we visualize the heat map in three dimensions via ArcScene. The extrusion of height in the heat map reflects the degree of complied functional intensity. Many places are highlighted, including a large area of the downtown center within the city's inner ring. The peak areas are mainly distributed in Puxi, which is located to the west of Huangpu River, while in Pudong only places near the Lujiazui CBD illustrate a high functional intensity. The highest peak, with a value of 402, is located around Yaohan Center (Fig. 6.7).

6.4.2 Function Intersection Map

To find those places with most diverse functions, we use raster calculator in ArcMap to discard all the pixels under the threshold of ten for each heat map, leaving only the hottest places for each category on the map. These processed heat maps are overlaid together to create a complied map reflecting the degree of functional diversity in geographic distribution. As a result, places covered by more layers of high density of heatmap reflect a much more diversed CAZ. In this case, we rank the degree of mixture with level one to level six, when places of level one are covered by heat map from only one category, and places of level six are covered by heat maps from all the six categories.

After the compilation we vectorize the boundary for each level. Given the purpose of detecting diverse places, we focus on mix level of five and six, which equal to districts highly clustered by at least five different functions.

Interestingly, famous tourism spots such as the Bund and Lujiazui are not among the most comprehensively dynamic places. Comparing to their peers, these areas tend to be less versatile and attractive. Some spots such as Jingan Temple and Nanjing

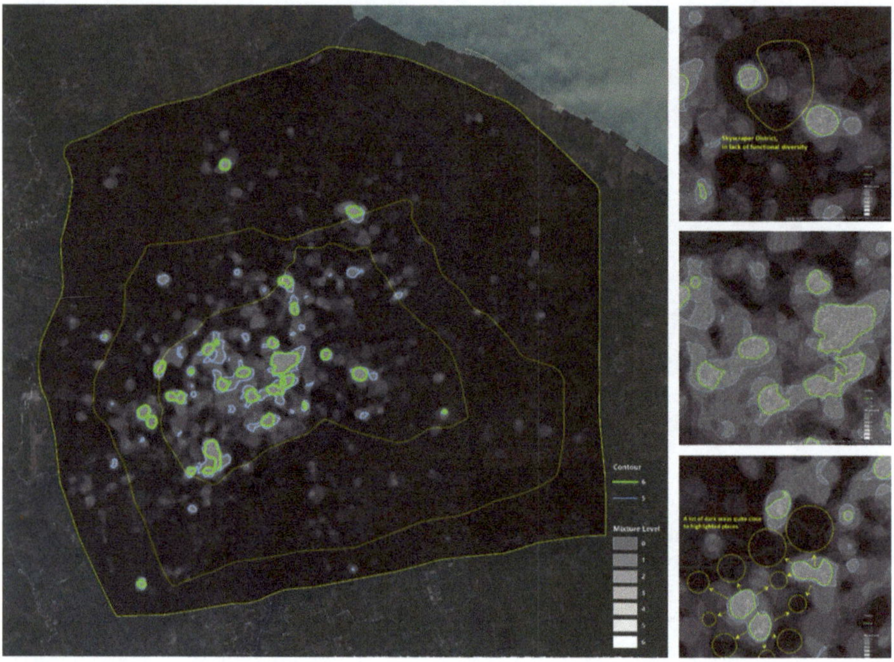

Fig. 6.8 Function intersection map of Shanghai (*left*), and magnified three places: Lujiazui (*top*), Nanjing Road (*middle*), Zhongshan Park (*bottom*)

Road pedestrian zone, are both instilled with scenic views and multifunctional places. The dark area between Zhengda Plaza and Yaohan Center on the magnified map locates the most of office towers in Lujiazui, yet its mix level is just around two, which means only two functions can be highlighted there.

The radiant coverage also differs between the downtown center and peripheral places relying on one gigantic commercial complex. For instance, the mixture level decreases dramatically around Parkson Hongqiao and Zhongshan Park. On the map where greyscale reflects how many different functions are overlaid, the juxtaposition of black places and white places demonstrates that streets surrounding those functionally concentrated super blocks holds much less stores.

Compared to the monofunctionality of Lujiazui, and superblock in places such as Zhongshan Park, the downtown area demonstrates a relatively homogeneous picture: the lighted areas are mostly connected, surrounded by a mildly low-density area (Fig. 6.8).

6.4.3 Top Dynamic Places

With both aggregation and intersection maps, it becomes possible to figure out places with large overall value or with multiple functions. After combining them together, we found the top 14 places which are enriched by popularity and

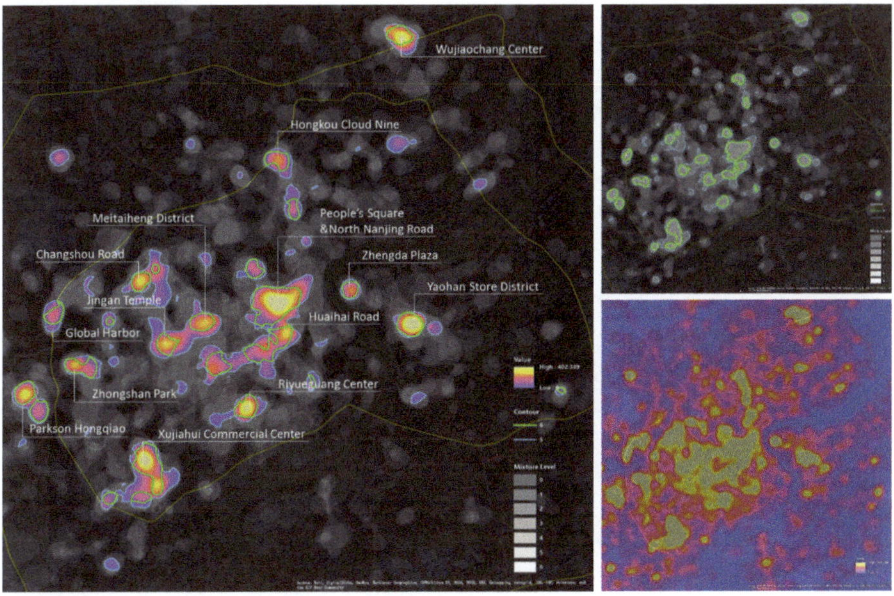

Fig. 6.9 The heat map of compiled functions (*left*) and the contour map of mixture degree (*right*) and the top 14 mixed-use districts within middle ring of Shanghai

diversity: (1) People's Square and Nanjing Road, (2) Xujiahui Commercial Center, (3) Huaihai Road, (4) Yaohan Center, (5) Zhongshan Park, (6) Riyueguang Center, (7) Parkson Hongqiao, (8) Wujiaochang Center, (9) Mei Tai Heng Plazas, (10) Hongkou Cloud Nine, (11) Jingan Temple, (12) Changshou Road, (13) Global Harbor, and (14) Zhengda Plaza (Fig. 6.9).

We plotted the average value of each category separately with spider map. Findings are generalized as below:

1. Places supported by a super commercial complex such as Global Harbor, Zhongshan Park, or Hongkou Cloud Nine have smaller hexagons than others. Although as pure business case they have achieved great success when introducing large amount of experience economy, their power is locked in one single but huge block. Such introversive places possess short radiant coverage and limited resources in all aspects.
2. There is no all-round candidate throughout the top places. Obtaining one of the largest hexagons, Yaohan Center shows its weakness in retail and culture. Scoring highly in culture, Riyueguang Center does not have enough votes from office. As described in the intersection map, there exist mismatches among highlighted places from different categories. For the purpose of defining CAZ, extending and uniting different hot spots would benefit each other.
3. Comparing to the other functions, the score of culture is weak in all the places. The one who has the highest score of culture among them is Riyueguang, which is perhaps because of its closeness to Tianzifang. It is an elegant cultural park

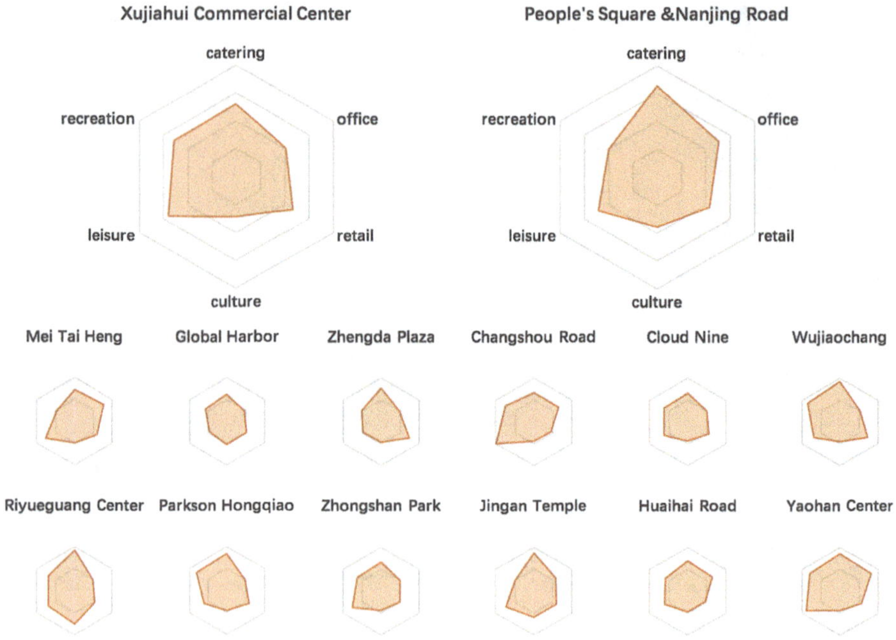

Fig. 6.10 The radar map of the 14 top dynamic places

transformed from Shanghai's old Linong housing, and many artist studios have been located there. Unfortunately, this is the only case among all the places we find. Culture places are usually not close to other functions (Fig. 6.10).

6.5 Conclusion

As for the original target of this project, we demonstrate three possible scenarios for defining the CAZ boundary in Shanghai. Meanwhile, the exploration of districts that breeds urban vitality is far from to stop. Many attempts with different parameters or inference need further testing.

6.5.1 Suggested CAZ Boundary

The discussion about defining CAZ boundary has always been contentious in local planning authority. Targeted for being zoned, different government districts located in downtown have their own versions of CAZ. The inner ring, middle ring, and even the outer ring have all been suggested to be the boundary.

The result of our project is simply based on the past data collected from website, yet defining the boundary faces an expected future. It is true that powerful shifts may happen in China, where policy-driven development is quite often. Given these uncertainties, we suggest three scenarios for possible CAZ boundary.

Scenario A
In current situation, the most dynamic area with coherent geographic proximity sits in the center of West Shanghai. From the pedestrian parts of Nanjing Road in the east, until to the Jingan Temple in the west, including the Huaihai Road paralleled in the south, enormous recreational facilities, department stores, and commercial centers are closely located, thus producing the prosperity of activities. This scenario provides a relatively clear and continuous boundary, which reflects the most diverse and popular urban areas in Shanghai, and can be viewed as the current CAZ in general.

Scenario B
This scenario merges the identity of Shanghai-Lujiazui and the Bund into CAZ. It is true that the commercial diversity there is not as high as its neighbors. However, inviting these two landmarks does not only mean covering scenic spots but also a balance between West and East Shanghai. Besides, such a belt-like zone is roughly consistent with the development axis decided many years ago from experts and local government. This scenario can be understood as a near-future CAZ once those gaps of either less popularity or less diversity are filled.

Scenario C
Taking an ambitious step, this scenario extends the boundary from the previous two to swallow the nearest dynamic places. While keeping the zone itself still geographically continuous, this version of CAZ covers a fairly large area. One can easily lose focus and walk into some "dull places" when he pays a visit to this zone. Dozens of dark places influence the realization of Scenario C, and the scale of it is too broad comparing to the one in London. This scenario can be suggested as a long-term goal based on its huge coverage (Fig. 6.11).

Both Scenario A and B are applicable plans, but the latter one seems to be more favored. Plan A defines a pure district with top popularity where diverse activities are highly clustered. Plan B keeps a balance on the development for both West and East Shanghai, though the completion of that zone requires efforts in overcoming dark places.

6.5.2 Research Extension

Many of the steps in this project have alternatives, and tuning parameters can possibly lead to different maps. For instance, choosing the six general categories relies on a certain degree of our subjective judgment, which omits community services, education support, and traffic conditions. Settings like the search radius for the kernel density map or the threshold for the intersection map can also be different.

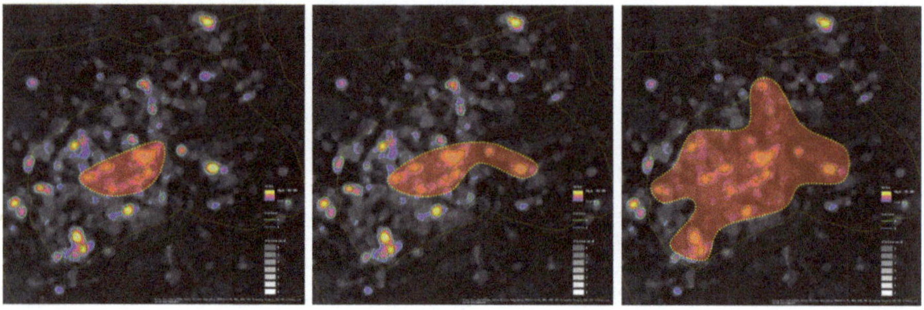

Fig. 6.11 Scenario A (*left*), Scenario B (*middle*), and Scenario C (*right*)

To sum up, this project provides a scope in disclosing the most dynamic district for a city via evaluated datasets. Findings such as the dark area in Lujiazui and characters generalized for superblocks are reasonable and instructive for future plan. Besides, the attempt in drawing out the CAZ boundary with different scenarios connects data research to practical policy making process.

References

Bogue, D. J. (1953). *Population growth in standard metropolitan areas, 1900–1950: With an explanatory analysis of urbanized areas*. Housing and Home Finance Agency, Office of the Administrator, Division of Housing Research.

Bowden, M. J. (1971). Downtown through time: Delimitation, expansion, and internal growth. *Economic Geography, 47*(2), 121–135.

Burgess, E. W. (1925). *The growth of the city: An introduction to a research project.* 收入 *The Trend of Population (页 85–97).* American Sociological Society.

Edmonston, B. (1975). *Population distribution in American cities*. Lexington Books.

Greater London Authority. (2011). *London's places*. London: The London Plan.

Liston, R. A. (1968). *Downtown: Our challenging urban problems (卷 2139).* Dell Publishing Company.

Mayor of London. (2004). *The London Plan: Chapter 5*. Greater London Authority.

Mayor of London. (2008). *The London Plan*. Greater London Authority.

Mayor of London. (2011). *The London Plan*. Greater London Authority.

Mieszkowski, P., & Mills, E. S. (1993). The causes of metropolitan suburbanization. *The Journal of Economic Perspectives, 7*(3), 135–147.

Murphy, R. E., & Vance, J. E., Jr. (1954). A comparative study of nine central business districts. *Economic Geography, 30*(4), 301–336.

Proudfoot, M. J. (1937). The outlying business centers of Chicago. *The Journal of Land & Public Utility Economics, 13*(1), 57–70.

Weixin, C. (2003). Investigation on CBD development of Chinese cities [J]. *City Planning Review, 12*, 18.

Chapter 7
The Fear of Ebola: A Tale of Two Cities in China

Xinyue Ye, Shengwen Li, Xining Yang, Jay Lee, and Ling Wu

Abstract Emerging social issues have often led to rumors breeding and propagation in social media in China. Public health-related rumors will harm social stability, and such noise negatively affects the quality of disease outbreak detection and prediction. In this chapter, we use the diffusion of Ebola rumors in social media networks as a case study. The topic of rumors is identified based on latent Dirichlet allocation method, and the diffusion process is explored using the space-time methods. By comparing Ebola rumors in the two cities, the chapter explores the relationship between the spread of rumors, user factors, and contents. The results show that: (1) rumors have a self-verification process; (2) rumors have strong aggregation characteristics, and similar rumors in different regions at the same period of time will lead to a synergistic effect; (3) non-authenticated users are more inclined to believe the rumors, while the official users play a major role in stopping rumors as they pay more attention to the fact; (4) the spread and elimination of rumors largely depend on the users who have more followers and friends; and (5) the topics of rumors are closely related to the local event.

Keywords Ebola • Rumor • LDA • Social media • China

X. Ye
Department of Geography, Kent Sate University, Kent, OH, USA
e-mail: xye5@kent.edu

S. Li
Department of Information Engineering, China University of Geosciences, Wuhan, China
e-mail: lsw4000@gmail.com

X. Yang (✉)
Department of Geography and Geology, Eastern Michigan University, Ypsilanti, MI, USA
e-mail: westoxgeorge@gmail.com

J. Lee
College of Environment and Planning, Henan University, Kaifeng, Henan, China

Department of Geography, Kent Sate University, Kent, OH, USA
e-mail: jlee@kent.edu

L. Wu (✉)
Department of Sociology, Kent State University, Kent, OH, USA
e-mail: lwu18@kent.edu

Z. Shen, M. Li (eds.), *Big Data Support of Urban Planning and Management*, Advances in Geographic Information Science, DOI 10.1007/978-3-319-51929-6_7

Abbreviations

ESDA Exploratory spatial data analysis
LDA Latent Dirichlet allocation

7.1 Introduction

In August 2014, the worst Ebola outbreak on record conceded by the World Health Organization had become an international public health emergent event. Communication channels during times of emerging social crises and natural disasters play an important role before, during, and after such events, and social media have become important channels for communications with the advancement of Internet communication technology. However, emerging social issues have often led to rumor breeding and propagation in social media in China. Public health-related rumors will harm social stability, and such noise negatively affects the quality of disease outbreak detection and prediction based on social media. To study the diffusion of social media rumor towards such social events, in this chapter we present a case study of the diffusion of Ebola rumors in social media networks in China along with the outbreak period.

This chapter contributes to the literatures as an empirical study to apply social media and volunteered geographic data to understand human behavior and social phenomena. Recently a new source of geographic information has become available through user-generated efforts, enabled by technologies loosely regarded as Web 2.0 (Goodchild 2007). Geographic information can now be found in the contents of wikis, blogs, tweets, photos, and many other forms of user-generated content. Goodchild defines this new source of data as volunteered geographic information (VGI), which emphasizes the role of voluntary efforts in producing geographic information, to differentiate with the conventional geographic information production process. Sui (2008) further pointed out that VGI has transformed the conventional way in how geographic information is created and used so it represents a "wikification" of GIS, which is not simply confined to new ways of data productions but also includes the development of GIS software, hardware, organizations, and people. Already we have witnessed the popularity of VGI has provided many research opportunities for geographers and GIScientists in the digital age. Many studies found out VGI to be a valuable source because of its huge potential to engage citizens and to be a significant, timely, and cost-effective source for geographers' understanding of the earth (Yang et al. 2016). Our study resonates some other recent successful cases using VGI to help disaster relief, such as the provision of useful location-based information to the victims after the 7.0 earthquake in Haiti in 2010 and the Great East Japan Earthquake and Tsunami in 2011 (Zook et al. 2010; Yap et al. 2012).

7.2 Literature Review

7.2.1 *Social Media and Public Health*

Social media produce large volumes of real-time information every day. Messages via social media have supported research and analysis in a wide variety of fields, such as crisis analysis (MacEachren et al. 2011), detecting opinion events (Maynard et al. 2014), assessment of disaster damages (Sakaki et al. 2013), strategic environmental assessment (Floris and Zoppi 2015), and air pollution (Kay et al. 2014), among others. Scholars have also explored how social media information conveyed issues in public health (Hay et al. 2013). For example, Windener (Widener and Li 2014) used sentiment analysis to explore the prevalence of healthy and unhealthy food in the USA. King et al. (2013) analyzed the role of social media in informing, debating and influencing opinions in a specific area of health policy.

People began to realize that we have an extremely poor knowledge of how the vast majority of infectious diseases spread across spatial scales (Hay et al. 2013). The process of understanding starts with records of disease occurrences estimated from the literature (Rogers et al. 2006), reports of disease distribution on the web (Brownstein et al. 2008), and the trends suggested by GenBank (Benson et al. 2012). Together, these could be used to define a definitive extent of the disease (Brady et al. 2012) and to populate a database of occurrence locations (i.e., points) where the disease had been reported. For example, Messina et al. (2014) manually built a comprehensive global database of confirmed human dengue infection in March 2014, consisting of 8309 geo-positioned occurrence. However, it is labor-intensive to keep track of the current trend of infectious diseases and distribution. It should be noted that social media messages are self-reported volunteered information, unlike those from official or government outlets with delayed announcements. Tweets can be used to track and predict the emergence and spread of an epidemic (Achrekar et al. 2011; Velardi et al. 2014). The spatial pattern of flu risk was dynamically mapped across various spatial and temporal scales by FluMapper (Padmanabhan et al. 2014).

7.2.2 *Rumor Theory and Social Media*

Unlike lies, rumors are invalidated news that may include speculation and uncertainty of certain events (Buckner 1965). Rumors can be thought as a message or a communication between people about certain events or certain issues of concern that may or may not correspond to the truth (Mullen 1972). Rumors not only communicate messages but also emerge a form of interactions among people in a group of individuals when they face the threat of a crisis (Indrawan-Santiago et al. 2014; Peterson and Gist 1951; Rosnow 1988). Social media platforms, such as Twitter and microblog, sometimes spread wrong information or rumor because they are citizen-based non-professional media. However, some information with reliable sources can minimize rumor propagation, suppressing the level of anxiety in the virtual community (Indrawan-Santiago et al. 2014; Oh et al. 2010; Okada et al. 2014; Spiro et al. 2012; Starbird et al. 2014). The social media messages containing rumors are different from those spreading news (Mendoza et al. 2010). Liao and Shi (2013) also reported the spreading of rumor during a nation-scale scandal using microblog.

7.2.3 The Outbreak of Ebola

To detect the disease outbreaks, a process was developed by the World Health Organization to action smartly in the big data context (Grein et al. 2000; Samaan et al. 2005). However, it seemed that this did not work well in West Africa because of the prevailing poverty in the region (Chan 2014). In September 2014, the US Center for Disease Control (CDC) predicted that 1.4 million people in Sierra Leone and Liberia would have succumbed to the disease if the epidemic continued its swift spread (Epstein 2014). Ebola has been frequently highlighted by newspaper, TV, and social media, after the virus has spread into many countries and has taken a lot of lives in 2014. More than 10 million tweets mentioned the word "Ebola" on Twitter within 20 days from 15 September 2014 in 170 countries (Luckerson 2014).

Of course, social media has also become new tools for us to analyze and predict how infectious diseases develop and evolve spatially and temporally. Along with the advancement of Internet technology, the spread of messages through social media had become much accelerated, more specific, and more short-lived than before. With constantly emerging social issues to prompt new rumors, the Internet actually provided an incubator for rumors to grow. This is understandable when considering that social media users came from all walks of life and participations in social media were very much unstructured and unorganized. As a results, social media, although not intentionally, nurture and spread rumors (Jin et al. 2014). Due to the widespread fear of Ebola and the lack of timely and transparent reporting of Ebola cases to the general public, rumors about Ebola infection possessed the two critical elements of Ebola infection being an important issue and the ambiguity in all news and reports about it. As a result, Ebola infection was widely discussed on social media.

Rumors about public health often bring out discontent in a society. Rumors also add unwanted noise in the use of social media information for any analysis or data exploration, creating road blocks to studies of public health issues (Ruths and Pfeffer 2014). This chapter uses the social media message regarding the 2014 Ebola infection news discussed by users of social media in China to study how rumors spread via social media. Specifically, this chapter addresses the following questions:

- What is the rumor diffusion process of Ebola in China (Ningbo and Guangzhou)?
- What are different topics regarding Ebola from social media in China (Ningbo and Guangzhou)?

- Who are those influential individuals or organizations involved in discussing the Ebola?
- What are the diffusion patterns of fear of Ebola?

This chapter is organized as the following. Section 2 reviews the research related social media, public health, and rumors. The paper discusses the method used to collect social media information and methods for unsupervised content classification, analysis, and the trends detection in Sect. 3. The following section summarizes the analysis and findings on the unique characteristics of how rumors spread in social media. Finally, we suggest future directions for additional research on this topic.

7.3 Method

In this article, we are interested in the outbreak of microblogging spatial patterns and variation-related rumors. We collected geotagged-Sina microblogging data, classified by topics using latent Dirichlet allocation (LDA) topic model (Blei et al. 2003) and analyzed by exploratory spatial data analysis.

7.3.1 Data Collection

Sina Weibo is the largest social network in China. By December 2013, the number of Weibo monthly active users reached 129.1 million, and daily active users reached 61.4 million. The posts in Sina Weibo reached the number of more than 2.8 billion in December, 2013. Sina Weibo provides developers with APIs (application programming interface) to facilitate users to extend application in Weibo. A limited sample of Sina Weibo can be collected through the official API. However, this will pose significant limitations on the detail of microblogs and the amount of data. Because web pages are unstructured, most crawlers are based on the simple definitions of the scope of web pages, which are subject to efficiency analysis (Plachouras et al. 2014). We then use a hybrid method, by integrating the API with a web crawler developed by our own to harvest the experimental data used in this study. In mid-October 2014, a rumor was spread on the Internet (Sohu 2014) about China's first Ebola-infected patient who was found in Ningbo, a coastal city in southeast China. This rumor also claimed the mortality rate of Ebola disease to be more than 90%. Almost in the same time, similar rumors about Ebola emerged in Guangzhou, another coastal city situated at the Pearl River delta (Xinhuanet 2014). Given the similarity of the characteristic of these two cities, we take the rumors in Ningbo and Guangzhou as our research objects. Considering the rumors may have some pre-warning stage, we began collecting data from the first day of October, 2014. To capture a full cycle of rumor diffusion process, we set the duration of our data collection as 60 days.

Previous study indicated that original posts on social media tend to have higher and more significant correlations than its retweets (Aslam et al. 2014); we thus solely collected original posts on Sina Weibo. After data cleaning and formatting, we obtained 1163 original microblogs about "Ningbo Ebola" and 1944 about "Guangzhou Ebola" in October and November 2014.

7.3.2 LDA

Text in each microblog contains the sentiment of user on Ebola rumors in different aspects. We use LDA model to classify the microblog content in order to explore topics which the rumors involved. LDA is an unsupervised machine learning techniques, used to identify the underlying topics from large-scale document collection and corpus, which has been already widely used for information extraction from tweets (Mehrotra et al. 2013).

LDA (Blei et al. 2003) is a widely used topic model based on the probabilistic theory. In this model, documents are composed of a series of random latent topics represented by the distribution of words. It has been used to automatically generate expertise profiles of online community (Liu et al. 2014), to quantify the distribution of probes between subcellular locations (Coelho et al. 2010), to mine opinion (Pang and Lee 2008), and to assign the annotation of large satellite images (Lienou et al. 2010).

First, we extract the microblog content from our processed dataset. After that the common stop words (such as "Ebola," because too often in the text sample, the keyword appears frequently) was removed. Then we consider a microblog as a document, using Gibbs sampling as the realization of LDA to carry topic extraction. There are two probabilistic parameters, α and β, describing the topic distribution. α represents the probabilities of topic distributions of per-document, and β is the probabilities of word distributions of per-topic.

Recent years have witnessed significant interest in exploratory spatial data analysis (ESDA) techniques and application, such as crime (Grubesic and Mack 2008), regional economics (Ye and Carroll 2011; Ye and Rey 2013), transportation (Whalen et al. 2013), environment (Xie et al. 2012, 2013), and public health (Milinovich et al. 2014). ESDA should be considered as a descriptive step before suggesting dynamic factors to explain the spatial patterns under study and before estimating and testing more sophisticated regression models (Anselin 2004). ESDA can help detect the spatiotemporal complexity towards forming novel research questions.

Through deriving related Weibo information through LDA, we conduct ESDA to explore spatial, temporal, and user distribution of Ebola rumors.

7.4 Analysis and Interpretation

Ningbo is a seaport city in Zhejiang province in East China. Guangzhou is the largest city in South China and is the capital of Guangdong Province. In mid-October 2014, a rumor was spread on the Internet (Sohu 2014) saying China's first

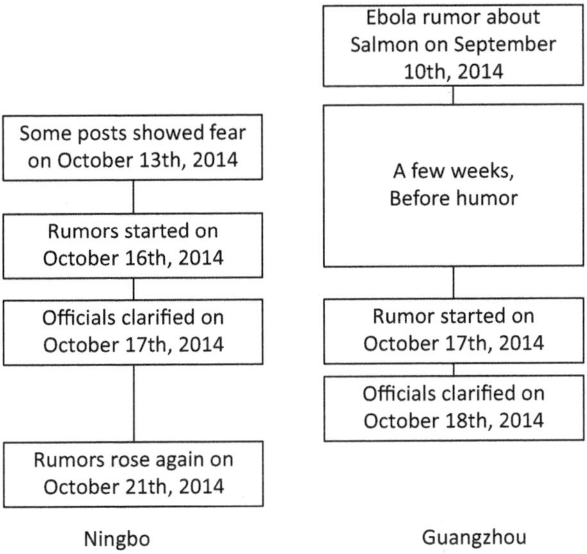

Fig. 7.1 Rumor diffusion process of Ebola in Ningbo and Guangzhou

Ebola-infected patient was found in Ningbo. This rumor also claimed the mortality rate of Ebola disease to be more than 90%. Almost in the same time, similar rumors about Ebola emerged in Guangzhou (Xinhuanet 2014) (Fig. 7.1). The fear of infectious disease such as dengue and SARS was the root for rumor to disperse among the mass media. It is also the case for Ebola. We notice that the first microblog that mentioned Ebola was posted in 13 October 2014. "Now the Ebola virus is very scary. There are a lot of Africans in Ningbo as well. Are they carrying the Ebola here?"

Our analysis shows that the rumor officially started on the night of 16 October 2014. A microblog post claimed that: "Is Ebola in China now? I am flying Taipei from Ningbo and the flight was held back for sanitation check. Someone on the plane appears to have Ebola symptom." Out of the attention and fear of Ebola, this microblog was reposted 181 times in a short while.

The former rumor microblog drew the attention from official department in Ningbo city. In the midnight of 17 October 2014, an official Weibo account named "Ningbo Health" was created. "There is a passenger from Nigeria who travelled from Taiwan to Ningbo on October 16th. He was diagnosed with certain fever symptoms. He is still under observation in a hospital." This microblog was then reposted by more than 2000 other Sina Weibo users.

Shortly after this official microblog, another official channel "Ningbo Announcement" posted a similar microblog and reposted the previous one from "Ningbo Health."

Eight hours after its initial announcement, "Ningbo Health" posted a follow-up microblog to clarify the Ebola rumor. "Update: the foreigner from Nigeria observed in the hospital who was detected slight fever yesterday has resumed to normal temperature now." Three hundred seventy-nine Sina Weibo users forwarded this post.

"Ningbo Announcement" also reposted the update microblog from "Ningbo Health" to clear the rumor. It restated that there is no clarified Ebola case in Ningbo city and advised people not to disperse and make up rumor on this event. On 21 October 2014, another similar rumor was dispersed in Sina Weibo: "the first case about Ebola was discovered in Ningbo. This disease has a mortality rate of more than 90%."

Rumor about Ebola was found in Guangzhou city 1 month before its emergence in Ningbo. On 10 September 2014, a post was spread on Sina Weibo. "By eating Salmon you will get infected with Ebola." Ever since the salmon rumor, there are occasional microblogs claiming that Ebola has emerged in Guangzhou. The Ebola rumors about Guangzhou first spread from WeChat (MicroMsg), a popular instant message tools, on the night, 17 October 2014. In a few hours later, some people sent a Weibo to request for clarifying. Similar to the official channel in Ningbo, an official channel "Sina Guangdong" got involved in the incident and stepped out to clarify this rumor. "According to Health Department of Guangdong Province, there is no Ebola case in Guangdong province so far." This clarification is against a previous rumor on Sina Weibo claiming that "a case of Ebola was identified in Canton fair. This patient is from Nigeria and he was sent to hospital for isolation." On 19 October, "Sina Guangdong" clarified against the previous rumor again.

As a transitional society, China is sensitive to many pressing issues, such as economic inequality, housing, food safety, and environmental pollution. Through forwarding and commenting negative news, people express their complaints to the virtual space, which easily lead to a variety of rumors.

7.4.1 Topics Regarding Ebola from Social Media

The content of microblog is often the description of certain event as well as a channel to disclose user's sentiment. In order to capture a comprehensive knowledge on what has been discussing on Sina Weibo, we used topic modeling method to dig out the topics among these Ebola-relevant microblogs. We treated each microblog as a document and classified them in different topics.

To extract the topics in Ningbo, we first delete the redundant microblogs as well as remove our thematic keyword "Ebola" and system keyword "Weibo." After the data preprocess of 1163 microblogs, we have 903 microblogs to run the LDA using the parameters (alpha = 2.5, beta = 0.01, and 1000 iteration) (Blei et al. 2003; Porteous et al. 2008). After generalization we have four topics (Table 7.1). The first category of topic captures the source of the rumor. It contains keywords such as "Nigeria," "men," "hospital," "fever," and "immigration." This also suggested that people know where Ebola might have come from or where Ebola infection was the most severe. Our second category of topic shows people fear towards Ebola. For example, keywords such as "intensive," "worry," "China," "prevention," and "aware" are related to the high populated area which is regarded more vulnerable to infectious disease. It also suggests that people were more concerned about their immediate

Table 7.1 Topics in Ningbo

Topic 1		Topic 2		Topic 3		Topic 4	
Word	Ratio	Word	Ratio	Word	Ratio	Word	Ratio
Nigeria	0.103	China	0.144	Eliminate	0.053	Virus	0.261
Men	0.085	Prevention	0.144	Already	0.051	Spread	0.026
Mild	0.071	Population	0.131	No	0.043	Attention	0.026
Hospital	0.066	Intensive	0.128	Epidemic	0.039	US	0.021
Fever	0.059	Aware	0.121	China	0.033	Clarify	0.020
Immigration	0.047	Worry	0.110	Scare	0.032	Mutate	0.017
Taiwan	0.036	Epidemic	0.045	Do not	0.029	Up to date	0.017
Examine	0.035	SARS	0.040	Beijing	0.025	Death	0.013
Afternoon	0.034	Afraid	0.036	Yesterday	0.023	Confirm	0.013
Transit	0.033	Possible	0.020	Nowadays	0.022	Prevention	0.013
Now	0.033	Sarsaparilla	0.007	Bye-bye	0.022	Tea leaf	0.012
Health bureau	0.028	Care	0.006	Confirm	0.020	False alarm	0.011
Report	0.021	Life	0.005	Rumor	0.019	Infectious	0.010
Epidemic	0.020	Fragile	0.004	Before	0.017	Overall	0.010
West Africa	0.020	Hurry	0.003	Center	0.014	Export	0.010
At present	0.019	Deliver	0.002	True	0.014	Occur	0.010
Infect	0.018	Container	0.001	Recently	0.014	Student	0.009
Prevention	0.018	Tencent.com	0.001	Foreigner	0.013	Said	0.009
Carry out	0.017	Sina.com	0.001	Risk	0.013	Global	0.009
Planning commission	0.016	A brand of cloth	0.001	Surprise	0.013	Decline	0.009

environment and their own well-being and people were conscious about what to do to prevent contracting the Ebola virus. In the third category of topic, we notice there are some keywords that show the negativity, such as "eliminate," "already," "no," "epidemic," "scare," etc. This gives a negative answer to whether Ebola has been emerging in China. The fourth category presents a global attention to Ebola. This includes keywords "virus," "spread," "attention," "US," and "clarify". The global spread of Ebola disease strengthens the fear of Ebola at regional and local areas, leveraging the root for rumor dispersion.

After the data preprocess, we have 1503 microblogs to run the LDA using the same parameters as Ningbo. Five topics are identified in Guangzhou Weibo messages (Table 7.2).

The first topic reflects the masses' fear towards Ebola. The keywords "black," "China," and "affected" show that due to the fact of large number African people living in Guangzhou, people tend to relate it to Ebola as it was outbreak in West Africa.

We notice the process of rumor clarification in the second topic. The keyword "news" shows a traditional channel to clear rumor, and "share" is how microblog got spread. "First case," "clarify," "official," and "rumor" show message from official source is crucial in clarification a rumor.

Table 7.2 Topics in Guangzhou

Topic 1		Topic 2		Topic 3		Topic 4		Topic 5	
Word	Ratio	Word	Ratio	Word	Ratio	Word	Ratio	Word	Ratio
Black	0.148	News	0.144	Virus	0.1	The China Import and Export Fair	0.111	Salmon	0.085
China	0.054	Share	0.067	Detect	0.037	Dengue	0.102	Friends	0.054
Affected	0.027	First case	0.039	Disease	0.033	Epidemic	0.061	Norway	0.044
Look	0.025	Clarify	0.038	Related	0.021	Strong	0.023	Safe	0.036
Hope	0.025	Official	0.036	Company	0.021	Enterprise	0.018	Network	0.025
Nowadays	0.024	Rumor	0.035	Asia	0.019	Influence	0.018	Post	0.024
Problem	0.022	NetEase.com	0.032	Local	0.019	Exhibiting	0.015	Eat	0.024
Cannot	0.018	Fear	0.022	Death	0.018	Today	0.014	Recently	0.024
True	0.018	Expert	0.022	Field	0.018	Opening	0.014	Repost	0.023
Control	0.018	Message	0.019	Infect	0.018	Yesterday	0.014	Sales	0.02
Illegal	0.017	Sina.com	0.013	Gene	0.017	Data	0.012	Halved	0.017
Should	0.016	Cause	0.013	A company	0.016	Pay attention	0.011	Rumor	0.014
Seen	0.016	Client	0.012	Protection	0.015	Decline	0.01	Never	0.014
African	0.015	Public	0.012	Dose	0.015	Current fair	0.01	Seafood	0.013
Worry	0.015	Up to date	0.01	Locate	0.013	Client	0.009	Like	0.013
Ill	0.014	ifeng.com	0.009	Success	0.012	Foreigner	0.009	Really	0.013
Lives	0.01	Tonight	0.009	In China	0.012	Attend	0.009	Reporter	0.012
Predict	0.009	Appear	0.009	Produce	0.011	Buyer	0.008	Import	0.011
Come in	0.009	Channel	0.008	Respirator	0.011	Product	0.008	Relief	0.011
Terrific	0.008	TV program	0.008	Market	0.01	Fair	0.006	Declare	0.011

Table 7.3 The post numbers of users in Ningbo

Posts	Users
1	1089
2	31
3	2
4	3
6	1

The third topic shows how virus was detected and prevented. "Virus," "detect," "disease," and "related suggestion" indicate that detection and prevention towards virus are still the key process of public health. Topic 4 relates to China Import and Export Fair which is held every spring and autumn in Guangzhou. This event is a famous and top fair for overseas trade, attracting many international buyers and sellers. This represents general public's concern of Ebola on Import and Export Fair.

Topic 5 describes the issue of salmon. Salmon eaters in China have been frightened by titles of the popular articles in the social media like Weibo and WeChat as "China has stopped the import of a whole Norwegian salmon," "People are likely to get Ebola from eating salmon," and "Salmon carry with them the Ebola virus" in September 2014.

By comparing topics from Ebola rumors in Guangzhou and Ningbo, we notice two similar topics: fear towards Ebola and the clarification of rumor. Fear towards Ebola is the foundation of rumor, while the clarification of rumor terminates the rumor dispersion process.

There are different perspectives of two rumor events. First of all, the source of rumor is different. Rumor in Ningbo started from a scam microblog while rumor in Guangzhou was initially communicated in WeChat. Secondly, due to the fact that a large number of Africans are living in Guangzhou, there is delusion from some people that Ebola is close to them in Guangzhou. That explains people in Guangzhou pay more attention to the topic on detection and prevention of infectious disease. In addition, in the same period, an import and export fair was held in Guangzhou, and this brings in several businessmen from African. Again, people in Guangzhou show great speculation on whether the fair is related to Ebola. Last but not least, a rumor about eating salmon will lead to Ebola was pushing the rumor in Guangzhou as well. In a word, Guangzhou being a city in an economically prosperous region has more opportunity for being exposed to Western culture/influences. In Ningbo, the Ebola rumor is an accident breakout wherein a passenger from Nigeria was sent to a hospital in airport.

We summarize the post number of each Weibo user, and the result is presented in Table 7.3. As we can see, most users (1089) contributed just one post during the rumor process. The maximum number of posts a user publishes is six. All users who contribute more than four posts are individual type of users. This indicates that in the process of special event, official channel tends to be cautious.

Statistics on user post number in Guangzhou shows similar pattern except that there are more people who publish large number of microblogs (Table 7.4). This is due to the fact that Guangzhou has large population than Ningbo and also the mass media in Guangzhou is more active.

Table 7.4 The post numbers of users in Guangzhou

Posts	Users
1	1426
2	121
3	26
4	9
5	9
6	4
7	2
8	1
11	2

Table 7.5 Top five reposted users in Ningbo

# of reposts	User	Type	City	Followers
7930	Jiangning Police Online	Official	Nanjing	834,668
1318	Top news	Official	Beijing	37,060,313
1043	CCTV News	Official	Beijing	25,825,867
959	Ningbo Health	Official	Ningbo	16,832
809	News in Realtime	Official	USA	107,728

Table 7.6 Top five reposted users in Ningbo

# of reposts	User		Type	City	Followers
1366	Sina Guangdong	Official	Guangzhou	1,227,876	1366
1029	People.com.cn	Official	Beijing	26,197,418	1029
908	huanqiu.com	Official	US	5,138,602	908
791	CCTV News	Official	Beijing	25,825,867	791
711	caijing.com.cn	Official	Beijing	11,287,244	711

Study of reposts shows that a total of 233 posts have been forwarded or reposted. The top five users that receive reposts are official channels from the local government. This indicates the trustiness from local citizens on their government (Tables 7.5 and 7.6).

Similar to Ningbo, the top five users that have been reposted are official users. This again shows the authority of official account on Sina Weibo.

7.4.1.1 Users Asked for Clarification

In Ningbo, five users asked for verification from seven official accounts: two police accounts, two news accounts, two other government accounts, and one Weibo company account. Because the rumor in Guangzhou did not originate in Weibo, not many verification requests were identified in Weibo. The main verification is on the report from Zhujiang News.

7.4.1.2 Users Spread the Rumor

In Ningbo, 455 Weibo messages were rumor. Four hundred forty-five users (68 certified users) were involved. Thirty-one Weibo messages and 31 users (11 certified users) spread the rumor in Guangzhou. In both cities, most common users (uncertified users) form the main user group in spreading the rumor.

7.4.1.3 Users Helping the Official Source Confirm the Accurate Information

In Ningbo, 233 Weibo messages confirmed the accurate information. Among 229 users involved, there were 59 official accounts, 46 certified accounts, and 124 uncertified accounts. In Guangzhou, 194 Weibo messages confirmed the accurate information. Among 180 users involved, there were 52 official accounts, 94 certified accounts, and 34 uncertified accounts. In both cities, common users (uncertified users) are minority on helping the official source confirm the accurate information.

7.4.1.4 User and Levels of Postings

To explore the relationship between user characteristics and message repost, we present the scatter plot involving followers and repost (Fig. 7.2). Ningbo and Guangzhou have similar patterns: more followers, more reposts.

At the same time, we examine the relationship between number of friends and number of message reposts. Both cities witnessed the similar patterns again. As (Rapoport and Rebhun 1952) argued, a higher degree of connection in friendship will facilitate the spread of rumor (Fig. 7.3).

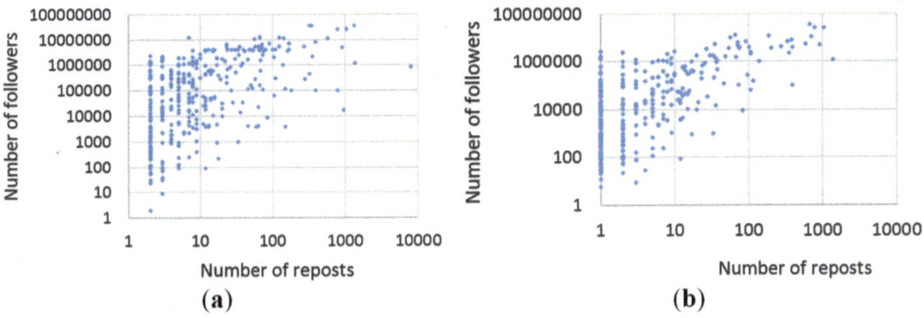

Fig. 7.2 Repost and followers (**a**) in Ningbo, (**b**) in Guangzhou

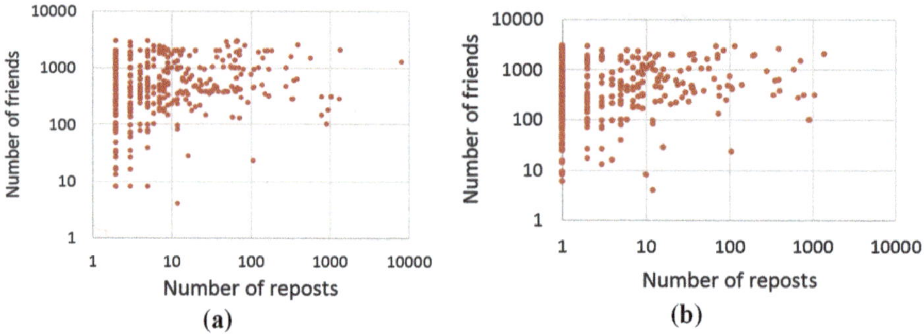

Fig. 7.3 Reposts and friends (**a**) in Ningbo, (**b**) in Guangzhou

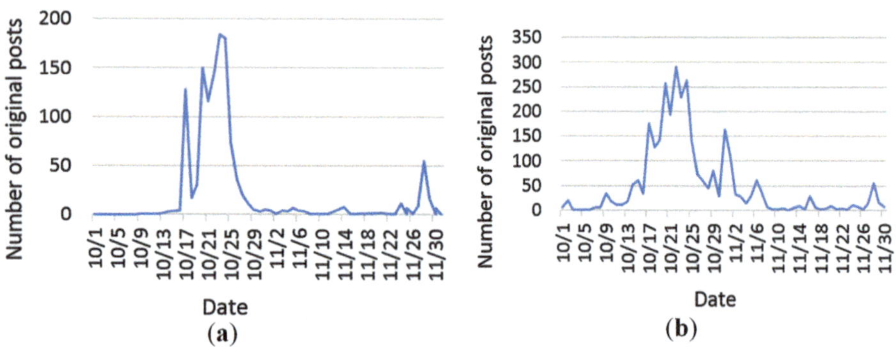

Fig. 7.4. Trend of two months (**a**) in Ningbo, (**b**) in Guangzhou

7.4.2 Diffusion Patterns of Fear of Ebola in Space and Time

7.4.2.1 Trend in Time

1. By days

By plotting the number of microblogs related to Ebola by days, we can track the temporal trend of the evolvement of rumor process. Figure 7.4. (a) presents that from the rumor time 16 October, the number of microblogs started to surge, and there are two peaks in Ningbo's rumor. One indicates the fear to Ebola and another one corresponds to the prevalence of rumor.

Figure 7.4.b shows that there is one peak in terms of the number of microblog posts in Guangzhou by each date. This is majorly due to the "salmon" rumor related to Ebola. Another find is that the temporal peak of both cities appears to be in the same period, indicating that the same topic of rumor correlates in time

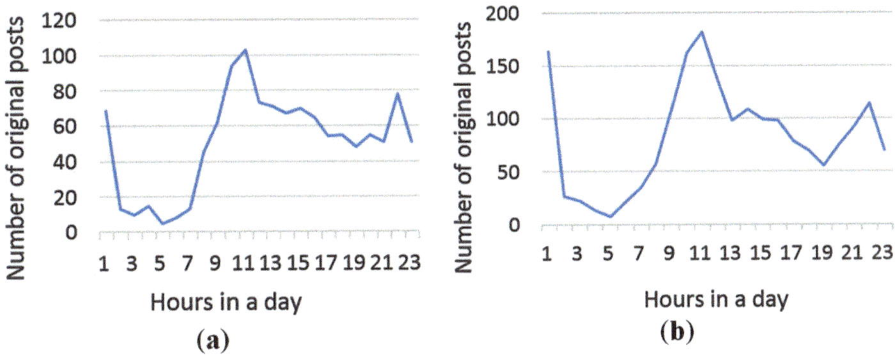

Fig. 7.5 Trend of a day (**a**) in Ningbo, (**b**) in Guangzhou

2. By hours

By plotting the time of each Ebola microblog of two cities in 24 h every day, we can see the temporal trend of Guangzhou and Ningbo corresponds well to the masses' daily activities (Fig. 7.5).

7.4.2.2 Trend in Space

The source of Ebola rumor in Ningbo originated from the Nigerian passenger who traveled to Ningbo international airport while the rumor in Guangzhou came from the local WeChat community with no further source. We then use geographic information system to map each microblog according to its location in order to explore the trend of rumor in space.

1. Distribution by locations

Figure 7.6 shows the geographic distribution of related microblog messages in China. As we can see, rumors about Ningbo's Ebola rumor were scattered beyond Zhejiang province. However, users who got involved in Guangzhou's Ebola rumor were mostly from Guangdong Province.

In order to further test the spatial cluster, we use Average Nearest Neighbor (Wong and Lee 2005) to test the degree of spatial cluster of points. The results show that the score of nearest neighbor ratios of users' location on Ningbo rumor is 0.2932 while the score of Guangzhou users is 0.4526. These numbers are less than the expect value 1, which indicates both point patterns appear to be cluster in space. By standardizing the two scores using Z-score, with Ningbo's Z-score being −10.1524 while Z-score in Guangzhou being −18.6868, both cities' NNR results are statistically significant.

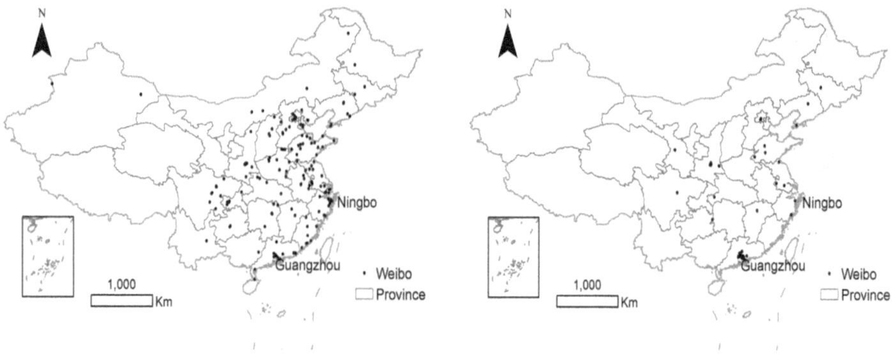

Fig. 7.6 The locations of Weibo related to (**a**) Ningbo, (**b**) Guangzhou

2. Distribution by profile

Users' profile can be used to identify a coarse spatial location (Hecht et al. 2011). In this study, we classify user location by province. In the Ningbo case, Zhejiang (where Ningbo is located) has the most users, followed by Beijing and Guangdong which also witnessed quite a few users. Regarding the Guangzhou case (where Guangdong is located), most users are identified in Guangdong, followed by Zhejiang. The rumors in these two places seem to be locally associated and also related to each other due to the same theme.

7.5 Discussion and Conclusion

During emergencies such as natural disasters or disease outbreak, it is often of critical importance to assess impact and response of people and the surrounding environment as an indicator for better evacuation planning and relief operations. Communication channels during times of emerging social crises and natural disasters are critical before, during, and after such events. Our study shows that social media messages are rich in content, which are ideal to capture and reflect a multitude of aspects of attitude, behaviors, and reactions of social media users to a specific topic or event. We thus argue that there is a pressing need to understand social media use during natural disasters and emergent social event from the perspective of social media users and affected citizens. This is also reflected in other study indicating that during a disaster, affected citizens are usually the most valuable sources who are capable of providing information before the public channel and agency. Therefore, it has been widely acknowledged that Humanitarian Assistance and Disaster Relief (HADR) responders can gain valuable insights and situational awareness by monitoring social media and volunteered geographic information related to the emergent event (that study also examined a set of factors that can explain those uses, such as time of the tweets, the location and characteristics of the users, previous patterns of social media use, and the degree to which the crisis has directly or indirectly affected the users). Therefore, these messages can be used to monitor and track geopolitical and disaster events, support emergency response and coordination, and serve as a measure of public interest or concern about events (Huang and Xiao 2015).

In conclusion, this study is an exploratory research conducted for the 2014 Ebola season that uses social media as a possible method for identifying trends in fear of Ebola incidence. The specific Ebola rates and trends are not included in this paper because we want to address the rumors spread in social media and not the detection and prediction of the spread of influenza. There is no Ebola patient found in China in the incidence. In terms of temporal dimension, across the two cities that Weibo messages were collected from, Weibo rates and rumor rates both peaked around 23 November 2014, and both peaked on 11 o'clock in 1 day. Though Weibo is not the birthplace of the original rumor, it facilitates its spread because most Weibo users are anonymous. Driven by fear, the rumor spread demonstrates a fast outbreak within very short time period. LDA method illustrates that rumors relate to real events associated with Ebola, which represents the collective efforts in the context of emergency (Peterson and Gist 1951; Rosnow 1988). In addition, rumors have localized spatial pattern. The advantages to using social media to survey rumor incidence is that it not only would quicken response time for official, but also it would give a various aspects, which will help us to find the sources and reason, about the rumors. When a rumor spreads, it will spread to social network in a few hours or in seconds even. The topics discussed in Weibo about Ebola rumor show us the source of rumor, the event, and focus of people.

Although the attributes of social media messages, such as user name, text of post, time, location, as well as the number of followers and being followed, can be collected by crawler and API, age, gender, and race are not available in most social media platforms. Hence, it is difficult to detect who spread rumors from demographic information. According to a report in 2013, about 53% of users are under 23 years old, who might be significantly affected by rumors (Bai 2014). In the growing social media platform, millions of people create, share, and exchange information and thoughts. Further research can be investigated in other rumors with various spatiotemporal features. Social media platform can be enhanced through the development of an early warning system predicting the outbreaks of rumor diffusion, which can help prevent the spread of humors.

Acknowledgments This material is based upon the work supported by the National Science Foundation under Grant No. 1416509, with the project titled "Spatiotemporal Modeling of Human Dynamics Across Social Media and Social Networks." Any opinions, findings, and conclusions or recommendations expressed in this material are those of the author and do not necessarily reflect the views of the National Science Foundation.

References

Achrekar, H., Gandhe, A., Lazarus, R., Yu, S. H., & Liu, B. (2011, April). Predicting flu trends using twitter data. In Computer Communications Workshops (INFOCOM WKSHPS), 2011 IEEE Conference on (pp. 702–707). IEEE.

Anselin, L. (2004). Exploring spatial data with GeoDaTM: A workbook. *Urbana, 51*, 61801.

Aslam, A. A., Tsou, M., Spitzberg, H. B., An, L., Gawron, M. J., Gupta, K. D., Peddecord, M. K., Nagel, C. A., Allen, C., Yang, J., & Lindsay, S. (2014). The reliability of tweets as a supplementary method of seasonal influenza surveillance. *Journal of medical Internet research, 16*, e250.

Bai, S., Yuan, S., Hao, B., & Zhu, T. (2014). Predicting personality traits of microblog users. Web Intelligence and Agent Systems: An International Journal, 12(3), 249–265.

Benson, D. A., Cavanaugh, M., Clark, K., Karsch-Mizrachi, I., Lipman, D. J., Ostell, J., & Sayers, E. W. (2012). GenBank. *Nucleic Acids Research, 41*, s1195.

Blei, D. M., Ng, A. Y., & Jordan, M. I. (2003). Latent dirichlet allocation. *The Journal of Machine Learning Research, 3*, 993–1022.

Brady, O. J., Gething, P. W., Bhatt, S., Messina, J. P., Brownstein, J. S., Hoen, A. G., Moyes, C. L., Farlow, A. W., Scott, T. W., & Hay, S. I. (2012). Refining the global spatial limits of dengue virus transmission by evidence-based consensus. *PLoS Neglected Tropical Diseases, 6*, e1760.

Brownstein, J. S., Freifeld, C. C., Reis, B. Y., & Mandl, K. D. (2008). Surveillance sans Frontieres: Internet-based emerging infectious disease intelligence and the HealthMap project. *PLoS Medicine, 5*, e151.

Buckner, H. T. (1965). A theory of rumor transmission. *Public Opinion Quarterly, 29*, 54–70.

Chan, M. (2014). Ebola virus disease in West Africa--no early end to the outbreak. *The New England Journal of Medicine, 371*, 1183–1185.

Coelho, L. P., Peng, T., & Robert. (2010). Quantifying the distribution of probes between subcellular locations using unsupervised pattern unmixing. *Bioinformatics, 26*, i7–i12.

Epstein, H. (2014). Ebola in Liberia: An epidemic of rumors. *New York Review of Books, 61*, 91.

Floris, R., & Zoppi, C. (2015). Social media-related geographic information in the context of strategic environmental assessment of municipal Masterplans: A case study concerning Sardinia (Italy). *Future Internet, 7*, 276–293.

Goodchild, M. F. (2007). Citizens as sensors: The world of volunteered geography. *GeoJournal, 69*(4), 211–221.

Grein, T. W., Kamara, K. B., Rodier, G., Plant, A. J., Bovier, P., Ryan, M. J., Ohyama, T., & Heymann, D. L. (2000). Rumors of disease in the global village: Outbreak verification. *Emerging Infectious Diseases, 6*, 97–102.

Grubesic, T. H., & Mack, E. A. (2008). Spatio-temporal interaction of urban crime. *Journal of Quantitative Criminology, 24*, 285–306.

Hay, S. I., George, D. B., Moyes, C. L., & Brownstein, J. S. (2013). Big data opportunities for global infectious disease surveillance. *PLoS Medicine, 10*, e1001413.

Hecht, B., Hong, L., Suh, B., & Chi, E.H. (2011). Tweets from Justin Bieber's heart: The dynamics of the location field in user profiles. In *Proceedings of the SIGCHI Conference on Human Factors in Computing Systems* (pp. 237–246). Vancouver: ACM.

Huang, Q., & Xiao, Y. (2015). Geographic situational awareness: Mining tweets for disaster preparedness, emergency response, impact, and recovery. *ISPRS International Journal of Geo-Information, 4*(3), 1549–1568.

Indrawan-Santiago, M., Miyabe, M., Nadamoto, A., & Aramaki, E. (2014). How do rumors spread during a crisis? Analysis of rumor expansion and disaffirmation on twitter after 3.11 in Japan. *International Journal of Web Information Systems, 10*, 394–412.

Jin, F., Wang, W., Zhao, L., Dougherty, E., Cao, Y., Lu, C., & Ramakrishnan, N. (2014). Misinformation propagation in the age of twitter. *Computer, 47*, 90–94.

Kay, S., Zhao, B., & Sui, D. (2014). Can social media clear the air? A case study of the air pollution problem in Chinese cities. *The Professional Geographer, 67*, 1–13.

King, D., Ramirez-Cano, D., Greaves, F., Vlaev, I., Beales, S., & Darzi, A. (2013). Twitter and the health reforms in the English National Health Service. *Health Policy, 110*, 291–297.

Liao, Q., & Shi, L. (2013). She gets a sports car from our donation: Rumor transmission in a Chinese microblogging community. In *Proceedings of the 2013 Conference on Computer Supported Cooperative Work* (pp. 587–598). San Antonio: ACM.

Lienou, M., Maitre, H., & Datcu, M. (2010). Semantic annotation of satellite images using latent Dirichlet allocation. *Geoscience and Remote Sensing Letters, IEEE, 7*, 28–32.

Liu, X., Wang, G. A., Johri, A., Zhou, M., & Fan, W. (2014). Harnessing global expertise: A comparative study of expertise profiling methods for online communities. *Information Systems Frontiers, 16*, 715–727.

Luckerson, V. (2014, October 7). Watch how word of Ebola exploded in America. *Time*.

MacEachren, A.M., Robinson, A.C., Jaiswal, A., Pezanowski, S., Savelyev, A., Blanford, J., & Mitra, P. (2011). Geo-twitter analytics: Applications in crisis management. In *25th International Cartographic Conference* (pp. 3–8).

Maynard, D., Gossen, G., Funk, A., & Fisichella, M. (2014). Should I care about your opinion? Detection of opinion interestingness and dynamics in social media. *Future Internet, 6*, 457–481.

Mehrotra, R., Sanner, S., Buntine, W., & Xie, L. (2013). Improving LDA topic models for microblogs via tweet pooling and automatic labeling. In *Proceedings of the 36th International ACM SIGIR Conference on Research and Development in Information Retrieval* (pp. 889–892). Dublin: ACM.

Mendoza, M., Poblete, B., & Castillo, C. (2010). Twitter under crisis: Can we trust what we RT? In *Proceedings of the First Workshop on Social Media Analytics* (pp. 71–79). Washington DC: ACM.

Messina, J.P., Brady, O.J., Pigott, D.M., Brownstein, J.S., Hoen, A.G., & Hay, S.I. (2014). A global compendium of human dengue virus occurrence. EP.

Milinovich, G. J., Williams, G. M., Clements, A. C. A., & Hu, W. (2014). Internet-based surveillance systems for monitoring emerging infectious diseases. *The Lancet Infectious Diseases, 14*, 160–168.

Mullen, P. B. (1972). Modern legend and rumor theory. *Journal of the Folklore Institute*, 95–109.

Oh, O., Kwon, K. H., & Rao, H. R. (2010). An exploration of social media in extreme events: Rumor theory and twitter during the Haiti earthquake 2010. 1–14. *ICIS, 231*.

Okada, Y., Ikeda, K., Shinoda, K., Toriumi, F., Sakaki, T., Kazama, K., Numao, M., Noda, I., & Kurihara, S. (2014). SIR-extended information diffusion model of false rumor and its prevention strategy for twitter. *Journal Ref: Journal of Advanced Computational Intelligence and Intelligent Informatics, 18*, 598–607.

Padmanabhan, A., Wang, S., Cao, G., Hwang, M., Zhang, Z., Gao, Y., Soltani, K., & Liu, Y. (2014). FluMapper: A cyberGIS application for interactive analysis of massive location-based social media. *Concurrency and Computation: Practice and Experience, 26*, 2253–2265.

Pang, B., & Lee, L. (2008). Opinion mining and sentiment analysis. *Foundations and Trends in Information Retrieval, 2*, 1–135.

Peterson, W. A., & Gist, N. P. (1951). Rumor and public opinion. *American Journal of Sociology*, 159–167.

Plachouras, V., Carpentier, F., Faheem, M., Masanès, J., Risse, T., Senellart, P., Siehndel, P., & Stavrakas, Y. (2014). ARCOMEM crawling architecture. *Future Internet, 6*, 518–541.

Porteous, I., Newman, D., Ihler, A., Asuncion, A., Smyth, P. & Welling, M. (2008). Fast collapsed gibbs sampling for latent dirichlet allocation. In *Proceedings of the 14th ACM SIGKDD International Conference on Knowledge Discovery and Data Mining* (pp. 569–577). ACM.

Rapoport, A., & Rebhun, L. I. (1952). On the mathematical theory of rumor spread. *The Bulletin of Mathematical Biophysics, 14*, 375–383.

Rogers, D. J., Wilson, A. J., Hay, S. I., & Graham, A. J. (2006). The global distribution of yellow fever and dengue. *Advances in Parasitology, 62*, 181–220.

Rosnow, R. L. (1988). Rumor as communication: A contextualist approach. *Journal of Communication, 38*, 12–28.

Ruths, D., & Pfeffer, J. (2014). Social media for large studies of behavior. *Science, 346*, 1063–1064.
Sakaki, T., Okazaki, M., & Matsuo, Y. (2013). Tweet analysis for real-time event detection and earthquake reporting system development. *IEEE Transactions on Knowledge and Data Engineering, 25*, 919–931.
Samaan, G., Patel, M., Olowokure, B., Roces, M. C., & Oshitani, H. (2005). Rumor surveillance and avian influenza H5N1. *Emerging Infectious Diseases, 11*, 463–466.
Sohu. (2014). China's first Ebola-infected persons appeared Ningbo? Rumors!
Spiro, E.S., Fitzhugh, S., Sutton, J., Pierski, N., Greczek, M., & Butts, C.T. (2012). Rumoring during extreme events: a case study of deepwater horizon 2010. In *Proceedings of the 4th Annual ACM Web Science Conference* (pp. 275–283). Evanston: ACM.
Starbird, K., Maddock, J., Orand, M., Achterman, P., & Mason, R.M. (2014). Rumors, false flags, and digital vigilantes: Misinformation on twitter after the 2013 Boston marathon bombing. In *iConference 2014 Proceedings* (pp. 654–662). Boston.
Sui, D. Z. (2008). The wikification of GIS and its consequences: Or Angelina Jolie's new tattoo and the future of GIS. *Computers, Environment and Urban Systems, 32*(1), 1–5.
Velardi, P., Stilo, G., Tozzi, A. E., & Gesualdo, F. (2014). Twitter mining for fine-grained syndromic surveillance. *Artificial Intelligence in Medicine*, 153–163.
Whalen, K. E., Páez, A., & Carrasco, J. A. (2013). Mode choice of university students commuting to school and the role of active travel. *Journal of Transport Geography, 31*, 132–142.
Widener, M. J., & Li, W. (2014). Using geolocated twitter data to monitor the prevalence of healthy and unhealthy food references across the US. *Applied Geography, 54*, 189–197.
Wong, W., & Lee, J. (2005). *Statistical analysis of geographic information with ArcView GIS and ArcGIS*. Wiley: New Jersey.
Xie, H., Kung, C., & Zhao, Y. (2012). Spatial disparities of regional forest land change based on ESDA and GIS at the county level in Beijing-Tianjin-Hebei area. *Frontiers of Earth Science, 6*, 445–452.
Xie, H., Liu, Z., Wang, P., Liu, G., & Lu, F. (2013). Exploring the mechanisms of ecological land change based on the spatial autoregressive model: A case study of the Poyang lake eco-economic zone, China. *International Journal of Environmental Research and Public Health, 11*, 583–599.
Xinhuanet. (2014). Guangzhou now Ebola? Rumors! Guangdong Province Write appeared Ebola cases.
Yang, X., Ye, X., & Sui, D. Z. (2016). We know where you are: In space and place-enriching the geographical context through social media. *International Journal of Applied Geospatial Research (IJAGR), 7*(2), 61–75.
Yap, L. F., Bessho, M., Koshizuka, N., & Sakamura, K. (2012). User-generated content for location-based services: A review. In *Virtual communities, social networks and collaboration* (pp. 163–179). New York: Springer.
Ye, X., & Carroll, M. C. (2011). Exploratory space-time analysis of local economic development. *Applied Geography, 31*, 1049–1058.
Ye, X., & Rey, S. (2013). A framework for exploratory space-time analysis of economic data. *The Annals of Regional Science, 50*, 315–339.
Zook, M., Graham, M., Shelton, T., & Gorman, S. (2010). Volunteered geographic information and crowdsourcing disaster relief: A case study of the Haitian earthquake. *World Medical & Health Policy, 2*(2), 7–33.

Part II
POI for Exploring Urban Space Recognition

Chapter 8
Identifying and Evaluating Urban Centers for the Whole China Using Open Data

Yaotian Ma and Ying Long

Abstract The urban center is the core component of urban structure. Its identification and evaluation have long been a concern of the urban planning discipline. However, the central city areas (urban centers) have never been well delineated for the China city system, leading few urban studies on urban centers due to data unavailability. To address this gap and based on reviewing existing identification methods of the urban center, this chapter proposes a novel approach for identifying urban centers using increasingly ubiquitous open data points of interest (POIs) and evaluating the identified nationwide urban centers using various types of open data from four dimensions, respectively. These dimensions range from scale, morphology, function, to vitality aspects, thus providing opportunities for exploring the overall development characteristics of nationwide urban centers. We hope this chapter may shed light on future urban studies on urban centers of China.

Keywords Urban center • Open data • Point of interest • Road network • Urban form

8.1 Introduction

The urban center, as the core space in a city, is a product of a highly developed service industry economy and an important research direction of the urban planning discipline. After experiencing decades of rapid urbanization, China's urban centers have shown obvious differences from general urban areas in many aspects. Therefore, scientifically defining the scope of urban centers and comprehensively describing their development status in terms of scale, morphology, function, and vitality are of great significance and value. Despite the obvious importance, the urban centers have never been well delineated for the China city system. This is mainly because of the difficulty of collecting data on a national scope. However, despite the lack of studies on a wide scale, in the past few decades, domestic and

Y. Ma • Y. Long (✉)
School of Architecture, Tsinghua University, Beijing 10084, China
e-mail: ylong@tsinghua.edu.cn

Z. Shen, M. Li (eds.), *Big Data Support of Urban Planning and Management*,
Advances in Geographic Information Science, DOI 10.1007/978-3-319-51929-6_8

foreign scholars have conducted various research on the identification methods of urban centers aiming at a single or a couple of cities.

Foreign traditional center identification methods include the assumption boundary method, population density determinative method, and building density determinative method. The assumption boundary method is a method with high subjectivity and is based mainly on residents' psychological cognitions about the urban center, while the population density determinative and building density determinative methods are mainly defined by quantitative indicators. The population density determinative method takes the block as the basic unit and sets the threshold of block population density with the block exceeding this density being identified as a center block. Similarly, the building density determinative method also takes the block as the basic unit; however, in contrast with the population density determinative method, the building density determinative method judges whether the land belongs to the center area through setting the threshold of building density. These two methods are relatively objective compared with the assumption boundary method, but there still exists the possibility of confusion; for example, the population density determinative method may wrongly regard some densely populated areas in the surrounding areas as the urban center, while the building density determinative method only considers building density but neglects building function, which may lead to the possibility of placing some areas of high-rise residential blocks into the scope of the urban center.

With further understanding of urban centers and progress of data acquisition methods, recent center identification methods have made some improvements over conventional methods in recent years, but still remain the two basic threads, subjective cognition and objective quantification. The identification method based on subjective cognition is represented by the identification method of using user-generated content. After collecting massive amounts of user-uploaded Flickr photos, Hollenstein and other researchers then recognize the photo tags. Their next step is to calculate the location which has the highest density of Flickr photos and define it as the scope of the urban center in the users'subjective cognition (Hollenstein and Purves 2010). In addition, there are also a number of other researches such as the study by Montello in which subjects have been asked to indicate whether locations are found in a particular region and to delineate such regions on a map (Montello et al. 2003). In the identification method based on objective quantification, Speck believes that an urban center is the place where people pursue social, economic, political, and cultural life. Therefore, it can be defined from three perspectives: population, activities, and physical characteristics (Speck 2012). Based on Speck's research direction and many other related studies of New Urbanism (see especially Gupta and Terzano 2008), Malizia and other researchers further propose a more detailed urban center identification indicator system which includes nine indicators: compactness, connectivity, mixed-use (diversity) degree, etc. Furthermore, 65 representative urban centers in the United States have been defined and described using the indicators above (Malizia and Song 2015).

The study of urban center identification started relatively late in China, among which the most representative research is the "Public Service Facility Index Method" proposed by Junyan Yang. Yang and other researchers have drawn lessons from the

Murphy Index Method used for identifying the boundaries of urban central business districts (CBDs) in traditional studies and put forward their methods using the similar computing techniques. According to the investigation results of the typical urban status quo, they first determine the threshold of the public service facility index in urban centers and then place the urban region whose index is higher than the cutoff value into the scope of the urban center (Yang and Shi 2014).

The reviews above prove that there are two basic methods, subjective cognition and objective quantification, adopted in existing studies on identifying urban centers. Despite their strengths and weaknesses, both methods are limited by the enormous workload of basic data acquisition and processing, so it is difficult to apply for a large geographical area, e.g., a whole country with hundreds of cities. Consequently, the identification and further measurement of urban centers mainly take single city or some sample cities as research objects and substantially concentrate on the research of metropolitan centers.

Based on existing research results, this chapter focuses on innovating the research paradigm in this field with the aim of remedying excessively complex identification methods and the deficiencies of the wide-scale holistic approach currently in use. In this era of data explosion, our research has advanced from a traditional small data template to one employing big data. Big data and open data provide opportunities for innovation in research, so that we can conduct wide-scale, and even nationwide, analysis. Long and Shen (2015) have experimented with a number of valuable approaches in using big data to carry out nationwide urban studies, such as mapping built-up urban areas for all Chinese cities at the parcel/block level, simulating urban expansion at the parcel level for all cities in China, evaluating urban growth boundaries for 300 Chinese cities, and estimating population exposure to PM2.5. After years of practice, they further proposed the "big model" paradigm, the nature of which is the establishment of relatively detailed urban region analysis and simulation models of a large geographic area capable of taking both wide-scale scope and detailed research unit into account. The research has referenced the fundamental concept of big model in that it has extended the research scope of urban center identification and measurement to the whole of China while focusing the research scope to certain points, thus avoiding the limit of the single urban area study and simultaneously carrying out studies of cities nationwide, so as to explore the general development law of urban centers in China.

8.2 Data and Methods

8.2.1 Data

This study takes 287 Chinese cities at the prefecture level or above as the research object and mainly focuses on their urban centers. The basic data used include:

1. Points of interest (POIs) from all the 287 cities' urban centers gathered from the Baidu maps in 2014. The characteristics of this data are that it is huge in quantity, which is up to 419,052 in total and covers the whole country, and it is

detailed enough to identify urban function points. Its type includes enterprises, life services, leisure and entertainment, as well as many other urban functions. It not only meets the requirements of the big model paradigm but also directly reflects the spatial distribution of urban functions. As a result, it can be used as the basic data source to identify urban centers.
2. Detailed road networks of all the 287 cities' urban centers. The road networks mainly contain three kinds of urban roads which are arterial road, minor arterial road, and branches. And the total length of all types of roads is 6904 km.
3. Microblog check-in records for all the 287 cities' urban centers. By using intersect analysis function of ArcGIS, we finally got a total of 836,473 microblog check-in records which contain the specific location and check-in time in the urban centers above.
4. City boundaries of all the 287 cities including the city names and the administrative level.

8.2.2 Methods

8.2.2.1 Exploring New Methods Based on Big Data to Identify Urban Centers

This study is based on the defined urban center range of Nanjing in the existing research performed by Junyan Yang. We test different center identification methods based on POI dot density distribution and compare their degree of fit with the existing scope of urban centers. Ultimately, the most reasonable method of identifying urban centers is to select POI material in Chinese cities' center group of built-up areas from the 2014 data and then use ArcGIS classification method (Jenks) to conduct clustering analysis of the dot density of interest points. When the grouping number of POI dot density is eight and the search radius and output cell size are 1000 m and 20 units, respectively, the highest level of the POI dot density distribution area is most consistent with the scope of the urban center in existing research. Therefore, this method is used as the identification method of urban centers and applied to the promotion and calculation of other cities.

After applying the identification method above to the 287 aforementioned cities and carrying out calculations of the scope of each city's urban center on ArcGIS, we then regard the street block as the unit and carry out artificial detail processing on obtained tentative center ranges to further clarify the precise boundary of each urban center, combined with the boundaries of urban internal roads and river systems.

8.2.2.2 Evaluating Nationwide Urban Centers from Four Dimensions

On basis of identifying the scope of urban centers nationwide, we extracted road network data and microblog check-in data and other data from each urban center. We also used the proportion of the total urban area (%) occupied by the urban center to represent the center's scale, road network length in the unit area (km/km^2) to represent the center's morphology, POI dot density (each dot/km^2) inside the center to represent the center's function, and microblog check-in frequency density (person-time/km^2) to represent the vitality of urban center. We then described the general law of national urban center development by again utilizing ArcGIS and other software to further evaluate the scale, morphology, function, and vitality of urban centers throughout China.

During the evaluation of four dimensions, scale, morphology, function, and vitality, the methods used are as follows:

ArcGIS classification method (Jenks)
We can obtain the distribution patterns of scale, morphology, function, and vitality for urban centers mainly by using Jenks' analytical method of ArcGIS. The Jenks method can group the similar values of input data in the most appropriate way to minimize the differences within groups and maximize the differences between groups to ultimately form the appropriate data grouping.

Zipf rank-size rule
The hierarchical system patterns of scale, morphology, function, and vitality in urban centers are mainly studied by means of the Zipf rank-size rule. This rank-size rule investigates the scale distribution law of urban systems in one region from the relationship between urban scale and urban scale ranking, which is helpful in indicating the rank-size distribution of a research object and its differences. The formula of the Zipf rank-size rule is $P(r) = P_1 r^{-q}$, and simultaneously taking the logarithm on both sides, we can get:

$$\ln P(r) = \ln P_1 - q \ln(r)$$

In the formula, r is the serial number of cities, Pr represents the indices (scale, functional density, vitality density, and road network density) of the urban center whose serial number is r, P_1 denotes the indices (scale, functional density, vitality density, and road network density) of the primate city, and q is the dimensionality of Zipf. When $q < 1$, it indicates that each developmental level of urban centers nationwide is rather concentrated with small-scale differences and it has a greater number of cities with medium rank-size. When $q > 1$, it indicates that the distribution of each developmental level of national urban centers is dispersive and the scale difference is large.

Quadrifid graph analytical method
The development status of scale, morphology, function, and vitality in urban centers throughout China is mainly analyzed through the quadrifid graph model. Quadrifid graph models take scatter diagrams as a basis and are mainly used for evaluating one certain thing from two dimensions. It regards the average value of the two dimensions evaluated as an identification line to divide the scatter diagram into four regions and to conduct classification analysis of evaluated items. This analysis plots two of the variables of functional density, vitality density, and road network density of urban centers, which are respectively placed at the horizontal and

vertical axes of the coordinate system to create a scatter diagram. Next, according to the classification rules of a quadrifid graph, the average values of center road network density (7.35 km/km²), center function density (1080 pcs/km²), and center vitality density (1743 person-time/km²) are used to divide the scattered points representing the 287 urban centers into four parts and to further obtain the development state features of these urban centers.

Spatial autocorrelation analytical method

The spatial agglomeration features of scale, morphology, function, and vitality in nationwide urban centers are mainly analyzed through the spatial autocorrelation method. Spatial autocorrelation analysis includes two aspects: global spatial autocorrelation and local spatial autocorrelation. Global spatial autocorrelation mainly checks the overall trend of spatial correlation throughout the whole study area, which is measured by using Moran's I index, and the formula is as follows:

$$I = \frac{\sum_{n}^{i=1}\sum_{n}^{j=1} W_{ij}\left(x_i - \bar{x}\right)\left(x_j - \bar{x}\right)}{S^2 \sum_{n}^{i=1}\sum_{n}^{j=1} W_{ij}}$$

In the formula, I is Moran's I index, n is the number of research objects, x_i and x_j represent the attribute values of region units i and j, W_{ij} is spatial weight matrix, S^2 is the observation value variance, and x^{-} is the average value of observation. The value of I ranges from −1 to 1, and when $I > 0$, there is a positive spatial autocorrelation, which appears as an agglomeration pattern in space; the higher the value, the larger the agglomeration degree will be. When $I < 0$, the spatial autocorrelation is negative, and the smaller the value is, the stronger the negative correlation will be. If $I = 0$ then the space is not relevant.

The overall spatial autocorrelation can reflect the distribution pattern of the research object in regional space, but it has no access to the location of the research object's assembling area. Therefore, this study additionally adopts a local hot spot analysis tool (Getis-Ord General G) of ArcGIS to explore the cold-hot spots' agglomeration of the scale, morphology, function, and vitality in the urban center. Its formula is as follows:

$$G^* = \frac{\sum_{n}^{j=1} W_{ij} x_j - \bar{X} \sum_{n}^{j=1} W_{ij}}{S\sqrt{\left[n \sum_{n}^{j=1} W_{ij}^2 - \left(\sum_{n}^{j=1} W_{ij}\right)^2\right] / (n-1)}}$$

In the formula, x_j represents the attribute value of region j, S signifies the standard deviation of the regional observation value x, W_{ij} indicates the spatial weight between region i and region j, and n is the total region quantity.

8.3 Scale, Morphology, Function, and Vitality in Nationwide Urban Centers

8.3.1 The Distribution Pattern of Scale, Morphology, Function, and Vitality in Nationwide Urban Centers

It can be seen from the results of the Jenks analysis that the distribution patterns of scale, morphology, function, and vitality of nationwide city centers have shown a descending trend from east to west and south to north, but specific features vary. In the distribution pattern of scale in urban centers, the proportion that urban center areas occupy of the total urban area is generally high in Northern China and the Yangtze River Delta region. In other regions, especially in Northeast and Southwest China, the proportion that urban center area occupies of the total urban area is relatively small (Fig. 8.1). In the distribution pattern of morphology in urban centers, the coastal areas have a generally higher road network density than the inland areas. The city centers with lower road network density are mostly located in Northeast, Northwest, and Southwest China (Fig. 8.2). In the distribution pattern of function in urban centers, China's southeast coastal areas have a generally higher function density. Additionally, in the Yangtze River Delta region and Pearl River Delta region, especially the Chengdu-Chongqing area, there formed an obvious urban clustering area with high functional density centers. However, the advantages of function density in the Beijing-Tianjin-Hebei region are not prominent (Fig. 8.3). In the distribution pattern of vitality in urban centers, the Beijing-Tianjin-Hebei region, the Yangtze River Delta sector, and Chengdu-Chongqing area have formed urban clusters with high vitality density centers, while in the Pearl River Delta region, which has a fairly high functional density, the advantages of vitality density in urban centers are not prominent (Fig. 8.4).

Comprehensively speaking, the distribution patterns of scale, morphology, function, and vitality in nationwide urban centers all present obvious geographical differences. Two different basic geographic patterns have been formed between the three main urban agglomerations and inland areas of Northeast, Northwest, and Southwest China and between major prefecture-level cities and general prefecture-level cities in each province. Among them, the degree of similarity between the distribution patterns of function and morphology in nationwide urban centers is rather high as the southeast and southern coastal cities have more apparent density and morphology advantages. Nevertheless, in the distribution pattern of urban centers' scale and vitality, the main large scale and high-density concentration areas appear in the Beijing-Tianjin-Hebei region and Yangtze River Delta region, while the southern coastal cities do not present obvious advantages. In addition, except for the three traditional urban agglomerations, the Chengdu-Chongqing region forms clustering areas of high value in all of the distribution patterns above except the scale pattern, indicating that although the Chengdu-Chongqing area is still lagging behind the three traditional urban agglomerations in the aspect of comprehensive competitiveness,

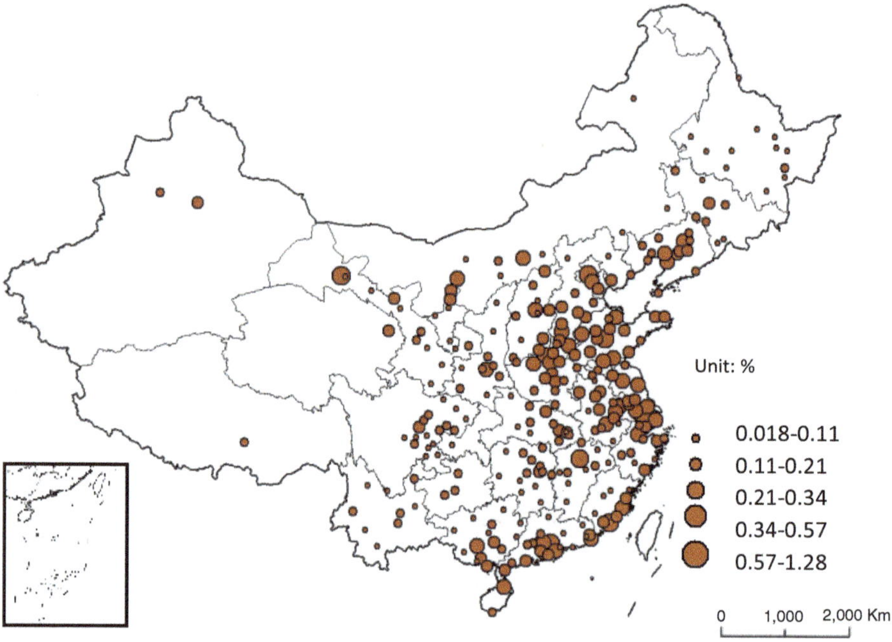

Fig. 8.1 Scale distribution pattern of nationwide urban centers

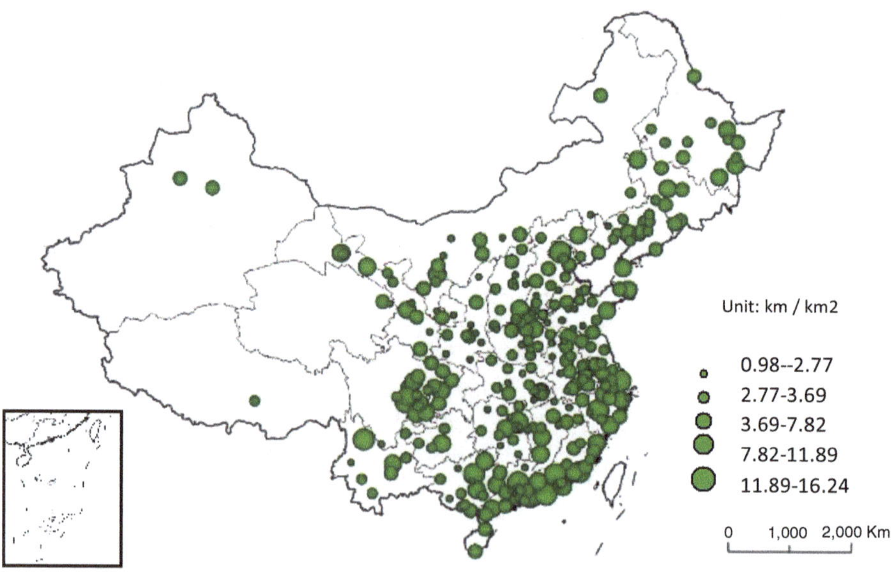

Fig. 8.2 Morphology distribution pattern of nationwide urban centers

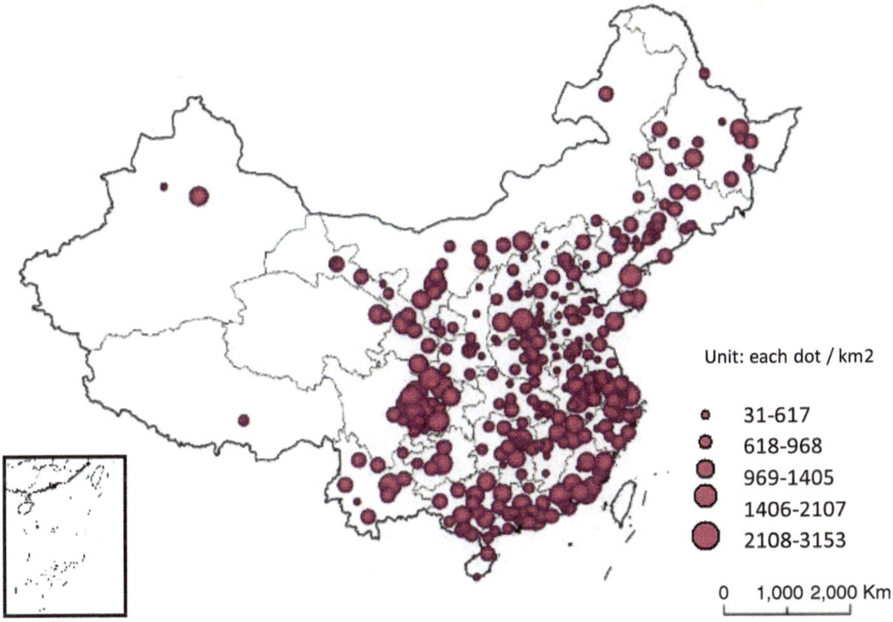

Fig. 8.3 Function distribution pattern of nationwide urban centers

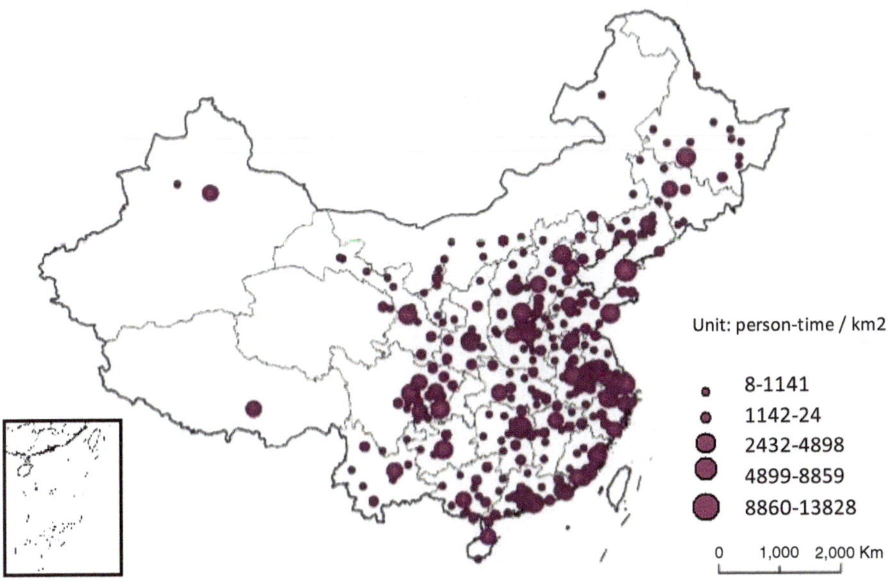

Fig. 8.4 Vitality distribution diagram of nationwide urban centers

the strength of each urban center in the Chengdu-Chongqing area should not be underestimated, especially when it comes to the density of function and validity in urban centers there.

8.3.2 *The Hierarchical System Pattern of Scale, Morphology, Function, and Vitality in Nationwide Urban Centers*

It can be seen from the *q* value calculated by the rank-size rule that the hierarchical system pattern of scale, morphology, function, and vitality varies a lot. In this system pattern of scale and vitality density, *q* values which are bigger than 0.5 and less than 1.0 belong to a weak equilibrium distribution (Figs. 8.5 and 8.8). Therefore, the scale and vitality density in nationwide urban centers all have large differences in distribution, which are embodied in the large differences between top-ranked cities and low-ranked cities in center scale and vitality density (same with primate and other cities). In the hierarchical system pattern of function and morphology, all *q* values are less than 0.5 and belong to the strong equilibrium distribution (Figs. 8.6 and 8.7). Therefore, the distribution of function and morphology in Chinese urban centers is relatively balanced. The specific manifestation is that the differences of function and morphology of urban centers among top-ranked and low-ranked cities are relatively small (Fig. 8.8).

Summing up the above analysis, we can infer that the hierarchical system pattern of function and morphology in urban centers in China is rather balanced. However, in the aspect of hierarchical system pattern of vitality and scale, the primate city enjoys a strong monopoly. That is to say, there exists a wide gap of vitality among urban centers, and although some urban centers also have relatively intensive functions, their vitality still lags behind those top-ranking centers. The differences among centers are not mainly reflected in function, morphology, and other infrastructure conditions but more in the aspect of scale and vitality.

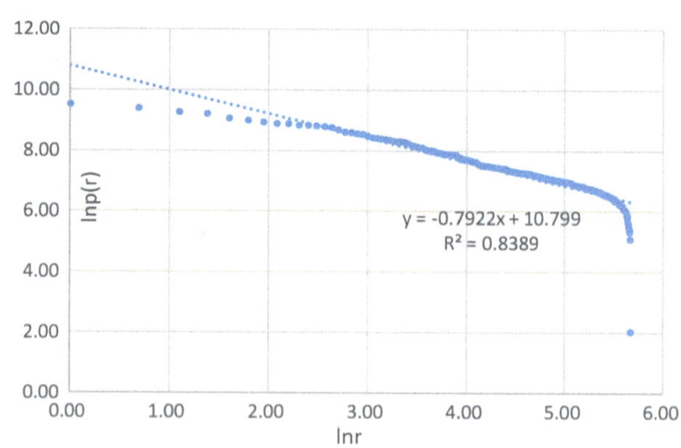

Fig. 8.5 ln-ln plot of the scale of nationwide urban centers

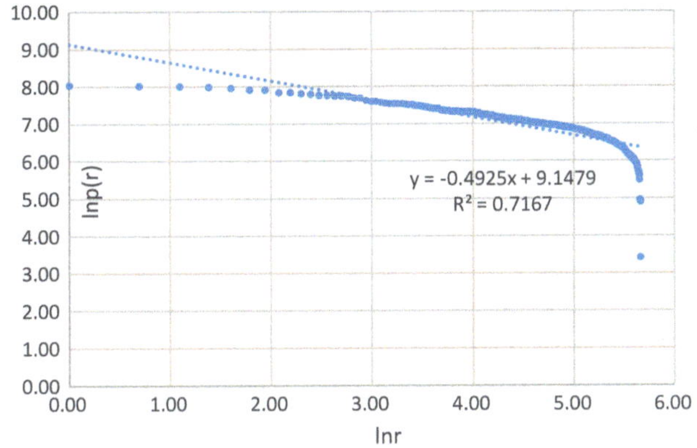

Fig. 8.6 ln-ln plot of the morphology of nationwide urban centers

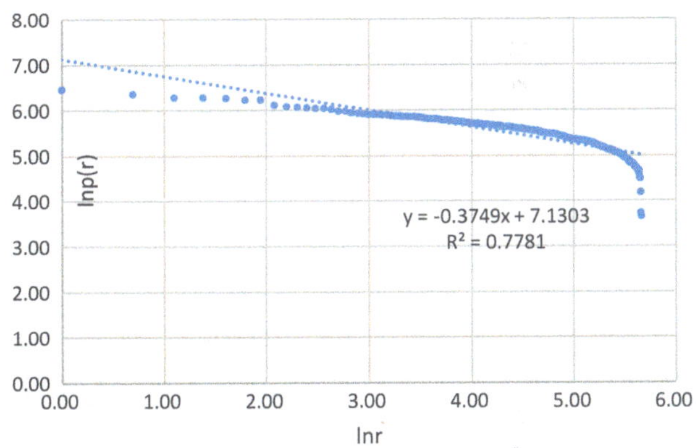

Fig. 8.7 ln-ln plot of the function of nationwide urban centers

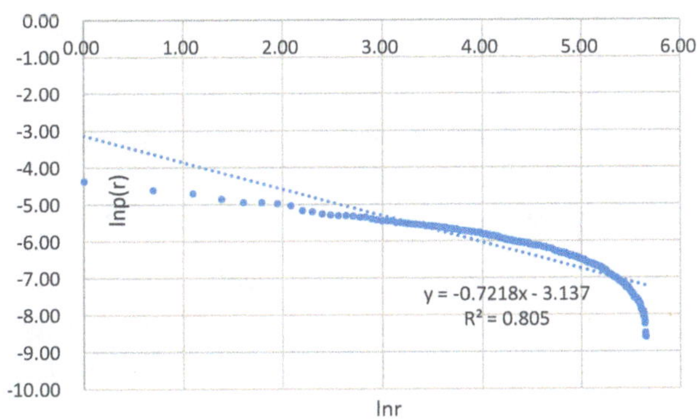

Fig. 8.8 ln-ln plot of the vitality of nationwide urban centers

Fig. 8.9 Quadrifid graph of functional density-road network density of nationwide urban centers

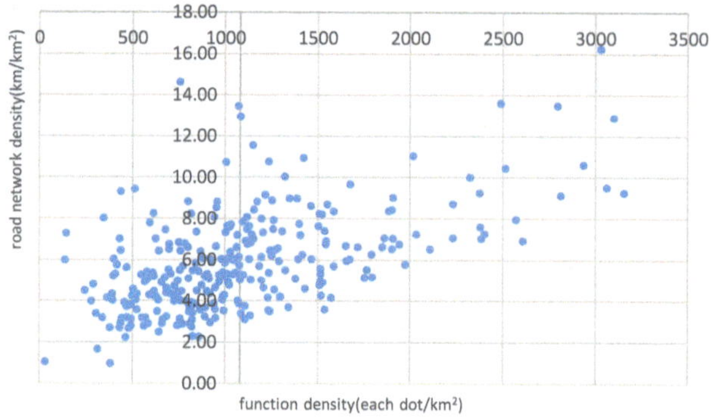

8.3.3 *The Development Status Features of Morphology, Function, and Vitality in Nationwide Urban Centers*

8.3.3.1 The Development Status of Function: Morphology in Nationwide Urban Centers

After conducting a quadrifid graph analysis of the function and morphology of nationwide urban centers, the average value of center functional density and road network density divides the coordinate system into four quadrants (Fig. 8.9): the first quadrant (high functional density-high road network density), the second quadrant (low functional density-high road network density), the third quadrant (low functional density-low road network density), and the fourth quadrant (high functional density-low road network density). The cities belonging to the quadrant of high functional density-high road network density include Shenzhen, Shanghai, and 81 other cities, mostly located in southeast coastal areas and accounting for 28% of total cities (Fig. 8.10). The cities belonging to high functional density-low road network density quadrant include Nanning, Quanzhou, and other 34 cities, which are dispersed throughout the inland provinces. The cities in the quadrant of low functional density-high road network density include Lijiang, Yibin, and other 45 cities, while the cities belonging to the quadrant of low functional density-low road network density include Ordos, Zhanjiang, and other 127 cities, accounting for the highest proportion (44%) among these four types of cities and are spatially concentrated in Central and Western China. As a whole, 287 urban centers in China are mainly of "high functional density-high road network density" and "low functional density-low road network density" status, with "low functional density-low road network density" leading the way. This indicates that the polarization phenomenon in urban centers' function and morphology is quite serious and there are still a large number of cities remaining in "double low" status.

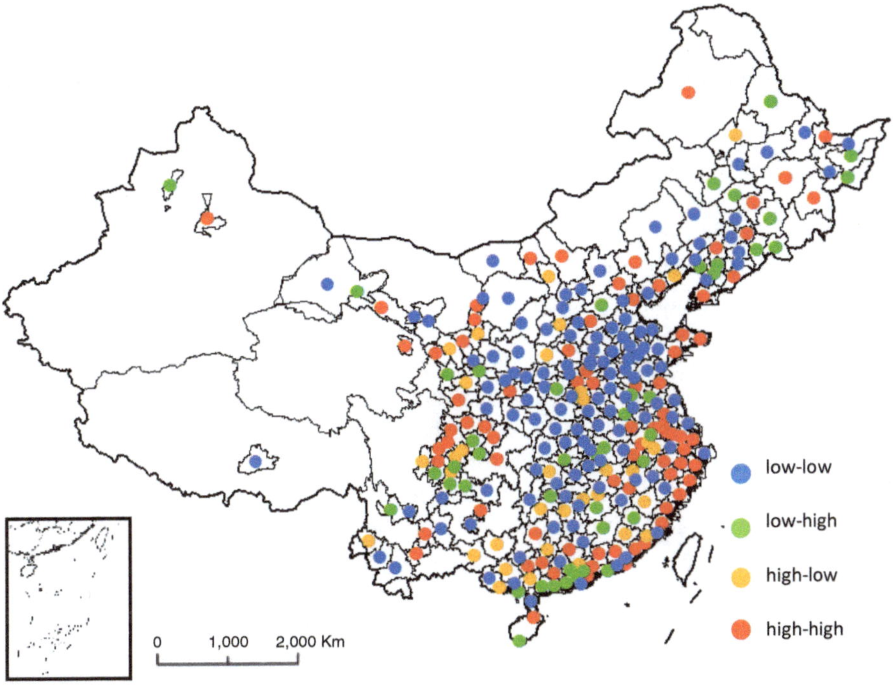

Fig. 8.10 Distribution of functional density-road network density of nationwide urban centers

Fig. 8.11 Quadrifid graph of functional density-vitality density of nationwide urban centers

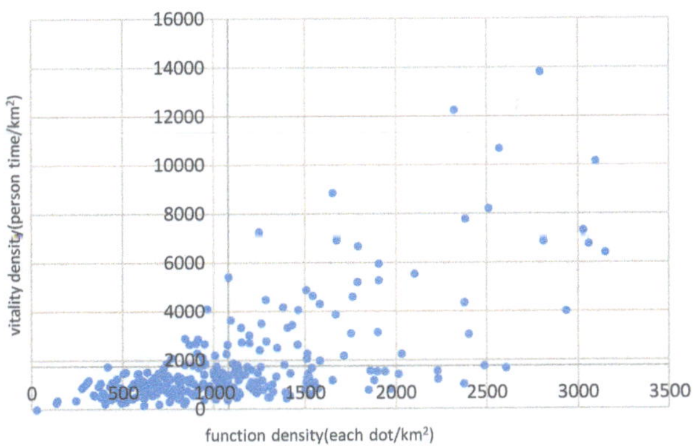

8.3.3.2 The Development Status of Function: Vitality in Nationwide Urban Centers

After conducting a quadrifid graph analysis of the function and vitality of nationwide urban centers, the average value of center functional density and vitality density divides the coordinate system into four quadrants (Fig. 8.11): the first quadrant (high functional density-high vitality density), the second quadrant (low functional density-high vitality density), the third quadrant (low functional density-low vitality density), and the fourth quadrant (high functional density-low vitality density). The cities belonging to the quadrant of high functional density-high vitality density encompass Shanghai, Dalian, and other 55 cities, accounting for 19% of total cities. Most first quadrant cities are located in the three major urban agglomerations including the Yangtze River Delta region, Beijing-Tianjin-Hebei region, and Chengdu-Chongqing area (Fig. 8.12). The cities belonging to the quadrant of high functional density-low vitality density cover Yinchuan, Hulun Buir, and other 60 cities. The cities belonging to the quadrant of low functional density-high vitality density comprise Lhasa, Lijiang, and other 19 cities, which are widely dispersed and mostly developed tourism cities. Fourth quadrant cities (low functional density-low vitality density) include Handan, Panjin, and other 153 cities, accounting for the highest proportion (53%) of these four types of cities, which are spatially concentrated in Central, Western, and Northeast China. As a whole, 287 prefecture-level and above urban centers in China are of two main types "high functional density-low vitality density" and "low functional density-low vitality density," the latter one being particularly dominant. This indicates that a large number of urban centers feature high functional density but lack vitality and there are still a large proportion of cities which remain in the "double low" status of vitality and function density.

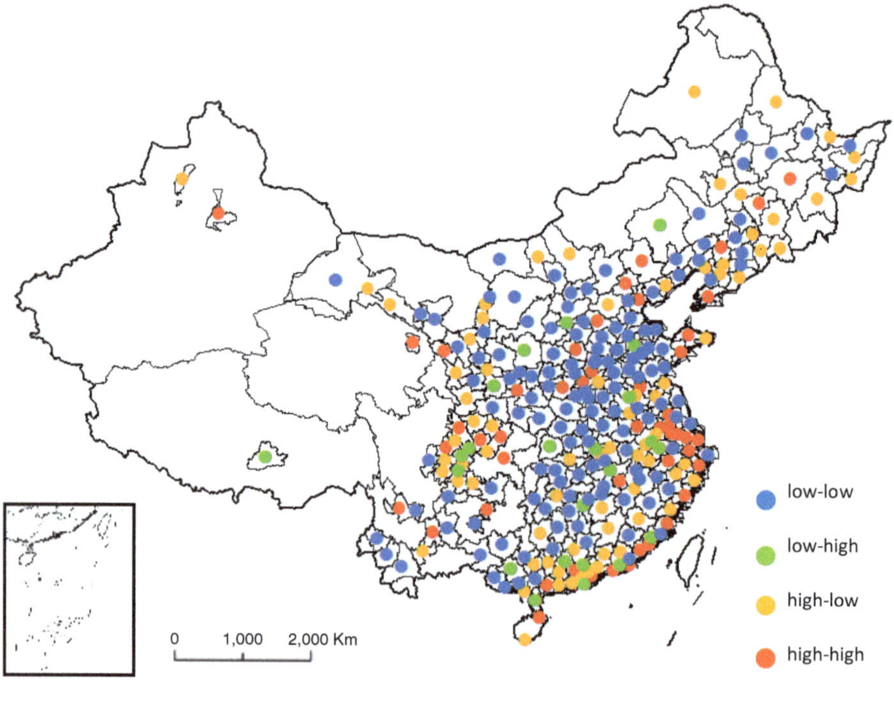

Fig. 8.12 Distribution of functional density-vitality density of nationwide urban centers

8.3.3.3 The Development Status of Morphology: Vitality in Nationwide Urban Centers

After conducting a quadrifid graph analysis of the morphology-vitality of nationwide urban centers, the average value of road network density and vitality density divides the coordinate system into four quadrants (Fig. 8.13): the first quadrant (high road network density-high vitality density), the second quadrant (low road network density-high vitality density), the third quadrant (low road network density-low vitality density), and the fourth quadrant (high road network density-low vitality density). The cities belonging to the quadrant of high road network density-high vitality density encompass Shanghai, Dalian, Nanjing, and other 74 cities, which are widely dispersed and account for 26% of total cities (Fig. 8.14). The cities belonging to the quadrant of high road network density-low vitality density include Leshan, Jixi, and other 75 cities. The cities belonging to the quadrant of low road network density-high vitality density contain Guangzhou, Taiyuan, and other 23 cities, including Guangzhou, Taiyuan, Nanning, Lhasa, Wuhan, and a plurality of provincial capitals. The cities belonging to the quadrant of low road network density-low vitality density cover Xinyang, Ya'an, and other 138 cities, accounting for the highest proportion (48%) of these four types of cities which are spatially concentrated in Central and Western China. As a whole, 287 prefecture-level and above urban centers in China are mainly "high road network density-low vitality density" and "low road network density-low vitality density," with the latter remaining dominant. This indicates that a large number of urban centers exhibit high road network density but still lack vitality and there are a large proportion of cities which are in the "double low" status of vitality and network density.

Based on the above analysis, we can find that a large number of urban centers present the "double low" development status of low road network density-low functional density, low functional density-low vitality density, and low road network density-low vitality density in China's urban centers. The cities of "double high" status account for a small proportion, and most of these are located along coastal areas or in the three major urban agglomerations. In addition to universal "double

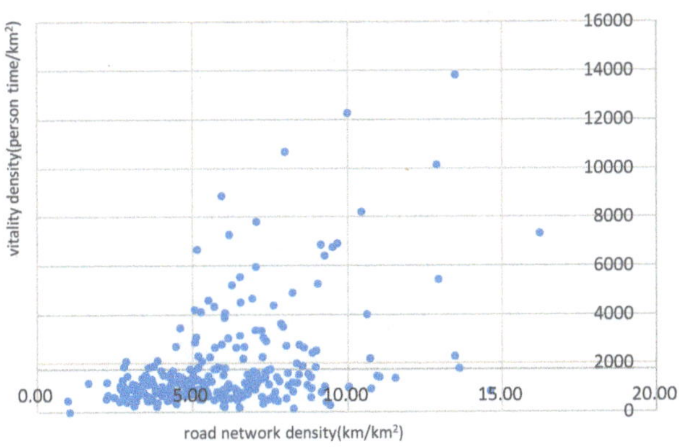

Fig. 8.13 Quadrifid graph of road network density-vitality density of nationwide urban centers

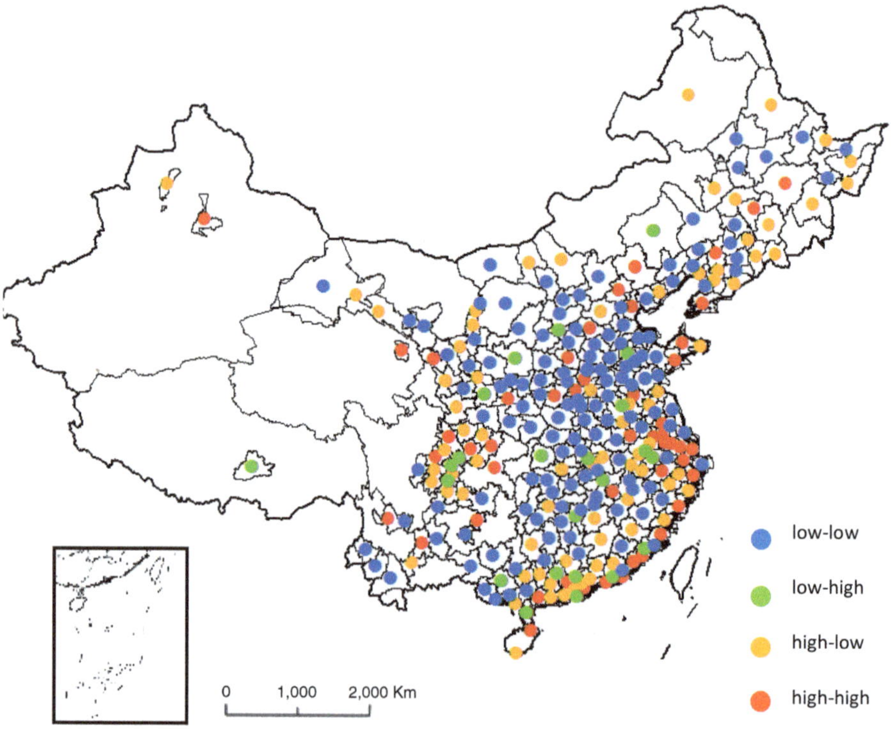

Fig. 8.14 Distribution of road network density-vitality density of nationwide urban centers

low" status, what is particularly noteworthy is that plenty of urban centers suffer from the serious problem of low vitality density. Although these cities possess high road network density or high functional density, their vitality density remains low, which does not seem to match the condition of high road network density and high functional density.

8.3.4 The Spatial Agglomeration Features of Scale, Morphology, Function, and Vitality in Nationwide Urban Centers

8.3.4.1 The Overall Spatial Agglomeration Features of Scale, Morphology, Function, and Vitality in Nationwide Urban Centers

To discuss the overall agglomeration pattern features of scale, morphology, function, and vitality in nationwide urban centers, this chapter calculates the Moran's I index (Table 8.1) through ArcGIS. From this table, we can see that all of the Moran's I index values are positive ($p \leq 0.01$), indicating that the scale, morphology,

Table 8.1 Moran's I index of nationwide urban centers

Type of index	Moran's I index	Z test value	Significance level
Center scale	0.26	14.56	0.01
Center road network density	0.094	5.33	0.01
Center functional density	0.21	11.79	0.01
Center vitality density	0.27	14.59	0.01

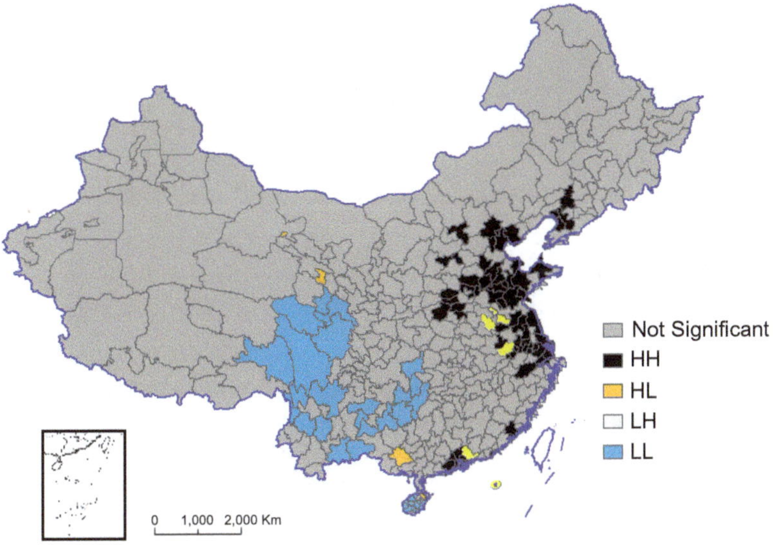

Fig. 8.15 Cold-hot agglomeration pattern of scale of nationwide urban centers

function, and vitality in nationwide urban centers are all positively correlated to space and the cities with high indicators tend to agglomerate in space. However, by comparing specific index values, it can be found that the cities with large center scale and high vitality density present more significant agglomerations in space.

8.3.4.2 The Partial Spatial Agglomeration Features of Scale, Morphology, Function, and Vitality in Nationwide Urban Centers

Through local spatial autocorrelation, we further identify the agglomeration pattern of cold-hot spots in nationwide urban centers. As for the partial spatial agglomeration features of urban centers' scale, there appears to be a high-value agglomeration area in the Beijing-Tianjin-Hebei region and North China, while there is a low-value agglomeration in the Chengdu-Chongqing region (Fig. 8.15).

As for the density of morphology in nationwide urban centers, there appears to be a high-value agglomeration area in the Yangtze River Delta and Pearl River Delta

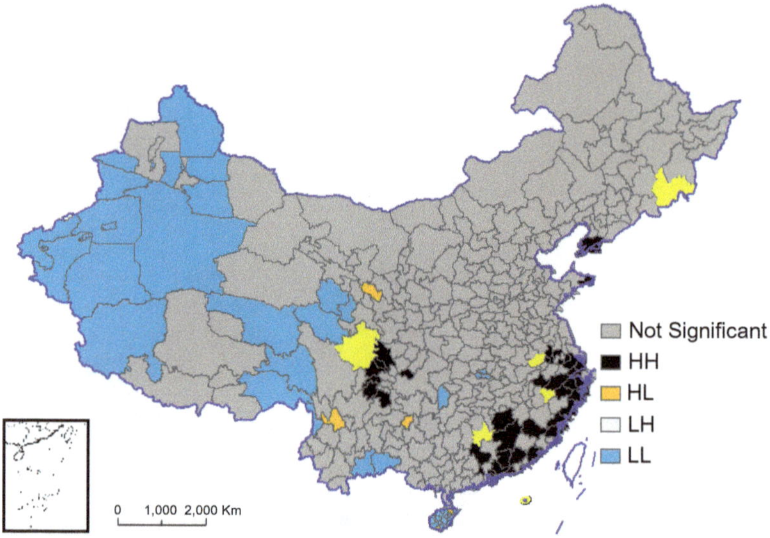

Fig. 8.16 Cold-hot agglomeration pattern of morphology of nationwide urban centers

regions, as well as the Chengdu-Chongqing area, while this phenomenon does not occur in the Beijing-Tianjin-Hebei urban agglomeration. The high-value areas in low value mainly appear in the provincial capital of Western China, while low-value clustering areas mainly appear in Western China outside the provincial capital (Fig. 8.16).

For the partial spatial agglomeration features of function in urban centers, there appears to be a high-value agglomeration area in the Yangtze River Delta and Pearl River Delta regions, as well as the Chengdu-Chongqing area, while this phenomenon does not occur in the Beijing-Tianjin-Hebei urban agglomeration. The high-value areas in low value mainly appear in the provincial capitals of Central and Southwest China, as well as some other regions. The low-value agglomeration areas occur in the outlying regions of Xinjiang and Tibet (Fig. 8.17).

When comparing the partial spatial agglomeration features of vitality in nationwide urban centers with the previous three features, the scope of agglomeration areas of both high and low value has been sharply reduced. The low-value areas in high value and high-value areas in low value have also been reduced. In addition, the high-value agglomeration regions mainly appear in the Yangtze River Delta region and the southeast coastal cities, while low-value agglomeration regions are not conspicuous (Fig. 8.18).

Taken together with global and local spatial agglomeration patterns, nationwide urban centers present obvious agglomerations with respect to scale distribution and functional distribution. High-value agglomeration areas of various indicators mostly appear in the three major urban agglomerations and southeast coastal areas, while low-value agglomeration areas are mainly concentrated in the western region. The high-value areas in low value mostly are located in the provincial capitals of each province.

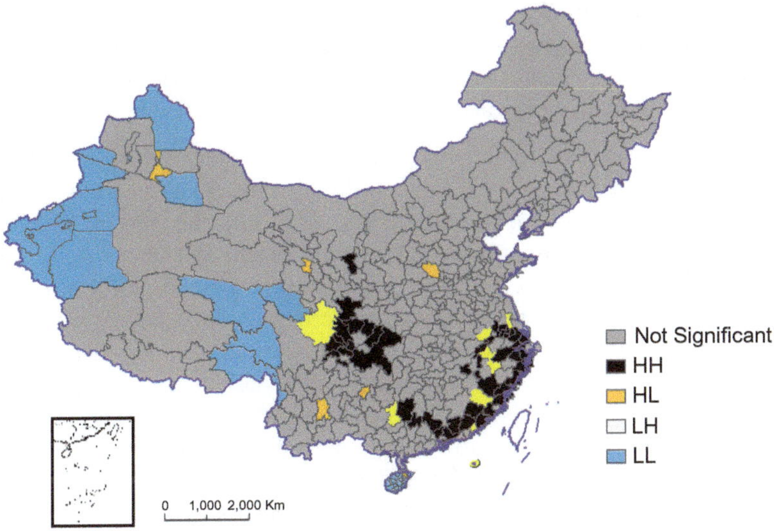

Fig. 8.17 Cold-hot agglomeration pattern of function of nationwide urban centers

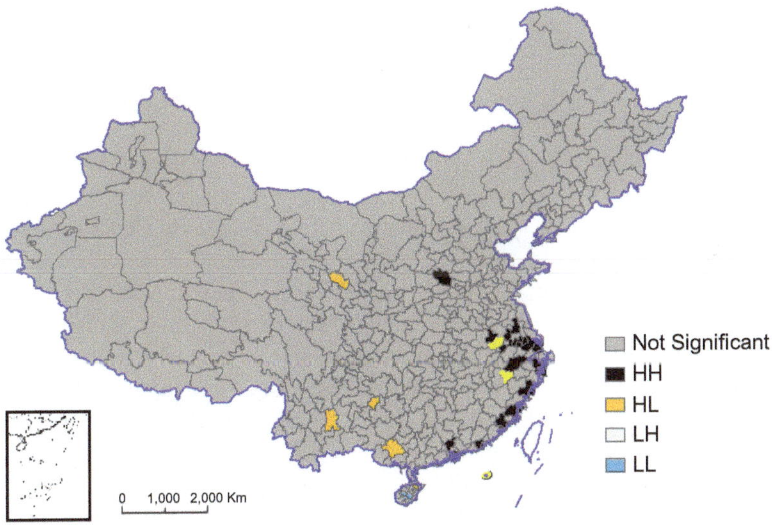

Fig. 8.18 Cold-hot agglomeration pattern of vitality of nationwide urban centers

8.4 Conclusion and Discussion

Considering being lacking of an urban center study for a whole country, this chapter proposes a novel approach for identifying urban centers using increasingly ubiquitous open data points of interest (POIs) and evaluating the identified nationwide urban centers using various types of open data from four dimensions, respectively. These dimensions range from scale, morphology, function, to vitality aspects, thus providing opportunities in an attempt to summarize the regularities and features of the development in nationwide urban centers. The research reveals that:

1. Regarding the distribution patterns of scale, morphology, function, and vitality, nationwide urban centers all present a declining trend from east to west and from south to north. In addition, two different basic geographic patterns have formed between the three main urban agglomerations and inland areas, as well as between those major prefecture-level cities and the general prefecture-level cities in each province. However, the specific distribution patterns of these four dimensions also differ from each other to some extent. Among them, the distribution patterns of scale and vitality in nationwide urban centers are rather similar, with the main high-value regions existing in the Yangtze River Delta region and Beijing-Tianjin-Hebei region. Additionally, the degree of similarity between the distribution patterns of function and morphology in nationwide urban centers is rather high, showing that the southeast and southern coastal cities have more apparent density advantages.
2. Regarding the hierarchical system patterns, the features of scale, morphology, function, and vitality in nationwide urban centers vary. In the hierarchical system pattern of scale and vitality in nationwide urban centers, there is a big difference between the primate city and other cities. In contrast, in the hierarchical system of function and morphology in nationwide urban centers, the differences between the primate city and other cities are relatively small, and the distribution is more balanced as well.
3. Regarding the features of the development status, a large proportion of urban centers present a "double low" development status such as "low functional density-low vitality density," and cities with a "double high" development status only account for a small portion. In addition to the common "double low" status, a multitude of urban centers also have the problem of low vitality. Although these cities possess the condition of high road network density or high functional density, the vitality density remains in low-value status, which does not match the condition of high road network density and high functional density.
4. Regarding the features of spatial agglomeration, there are obvious positive spatial correlations on scales, function, vitality, and morphology distribution which show that urban centers with a higher (or lower) index tend to agglomerate in space. Among them, cities with large (or small) center scale and high (or low) center vitality density present a more significant agglomeration in space.

Although this chapter identifies the nationwide urban centers innovatively and evaluates the development features of urban centers along four dimensions based on the identification results, the research can be extended via various avenues. On one hand, we will explore the influencing factors of the development features in nationwide urban centers in the close future. On the other hand, we will make a more in-depth description of the development features, thus gaining more knowledge on the formation of urban centers. Considering urban centers will play an even more crucial role in the future development of cities, it is of great importance to conduct these mentioned continuing studies of the urban centers from the quantitative perspective and the national scale.

References

Gupta, P. K., & Terzano, K. (2008). *Creating great town centers and urban villages [M]*. Washington, DC: Urban Land Institute.

Hollenstein, L., & Purves, R. (2010). Exploring place through user-generated content: Using Flickr tags to describe city cores [J]. *Journal of Spatial Information Science, 2010*(1), 21–48.

Long, Y., & Shen, Z. (2015). Big models: From Beijing to the whole China [M]. In *Geospatial analysis to support urban planning in Beijing* (pp. 255–272). Switzerland: Springer.

Malizia, E., & Song, Y. (2015). Does downtown office property perform better in live–work–play centers? [J]. *Journal of Urbanism: International Research on Place making and Urban Sustainability, 9*, 1–16.

Montello, D. R., Goodchild, M. F., Gottsegen, J., et al. (2003). Where's downtown? Behavioral methods for determining referents of vague spatial queries [J]. *Spatial Cognition & Computation, 3*(2–3), 185–204.

Speck, J. (2012). *Walkable city [J]*. New York: Farrar, Straus and Giroux.

Yang, J., & Shi, B. (2014). Research on the quantitative definition method of Urban Center boundaries [J]. *Journal of Human Settlements in West China, 29*(6), 17–21. (in Chinese with English abstract).

Chapter 9
Geographic Big Data's Applications in Retailing Business Market

Xin Chen, Fangcao Xu, Weili Wang, Yikang Du, and Miaoyi Li

Abstract Location has been considered as a determinant factor for retail location's success since the 1970s. Various researches have inspected different aspects in order to understand the market and select the best site for each store, from classifying the location to generating hierarchical trading areas. Previous researches done by human geographers may provide great insights into one specified site. However, their traditional methods are likely to be too expensive and difficult to expand to the whole city and even the whole nation. This chapter explores methods based on a large amount of data, which are currently being used as commercial solutions. By inspecting POI data and other social media data within pre-built grids, different types of business districts could be discovered, and site characteristics could be summarized. Two famous fast-food retailers have been explored in this study. Results indicate that more than 70% of stores can be explained by business district classifications. Also, results indicate that similar POI aggregations can be analyzed, from which 83% and 90% of stores' locations can be explained for two retailers separately. Besides, the distance relationship between POIs and stores can contribute to explain the location of retail stores.

Keywords Site selection • POI • Retail • Location • Spatial analyst • Big data

X. Chen (✉) • F. Xu • W. Wang • Y. Du
Beijing GISuni Information Co., Ltd, Beijing, China
e-mail: mapchenx@163.com

M. Li
Joint International FZUKU Lab SPSD, Fuzhou University, Fuzhou City, China

Z. Shen, M. Li (eds.), *Big Data Support of Urban Planning and Management*,
Advances in Geographic Information Science, DOI 10.1007/978-3-319-51929-6_9

9.1 Introduction

Retailing is a high-tech, global industry that offers customers with daily living goods and services to earn a profit. However, retailing business is not static and is constantly evolving and changing. Companies have to figure out how to make their own business more successful through researching the world of retailing, from interior factors like categories of retailers, their product types and multiple channels—stores, the Internet, and catalogs—they sell their products, to the exterior factors like the environment of their target market, consumer buying behaviors, and competitors in the same regions (Levy and Weitz 2013). For both small retail businesses and chain giants, a full broad overview of the market will provide background information needed to develop and implement an effective retail strategy (Yudelson 2009). At this point in the evolution of the Information Age, most retailing companies are now seeking for sorts of advanced methodologies as well as traditional questionnaire surveys or count measures to do market analysis and decision makings (Cressie and Wikle 2010). In recent years, the big data, including smart card data (SCD), social media data, and mobile phone records provided by location-based services (LBS) like Global System for Mobile communication (GSM), global positioning system (GPS), social networking services (SNS), and Wi-Fi, can be used to real time monitor the urban activities (Calabrese and Ratti 2006). Then chain retailers could use geographic information science (GIS) software, combining computer-generated mapping (Hernandez 2007) with key locational data to graphically delineate extensive demographic and business characteristics and analyze the intensity and temporal distribution of people mobility patterns (Ratti et al. 2006), thus helping do market analysis (Rachel 2007; Duggal 2008).

This study will introduce some business solutions developed by Location Intelligence department of GISuni and its own commercial business software—GeoQ—to provide location analytics and consulting services for part of the world's top 500 enterprises and some famous institutions with on-demand analysis, geo-visualization maps, and ready reports.

9.1.1 Site Selection

Finding the right location is definitely one of the most influential paths to success in a customer's store choice decision for any retailer (Mover 1972). Industry research suggests that location explains up to 70% of the variance in people's supermarket choice decisions. Teller and Thomas (2008) have found that retailers choose their store locations based on a wide range of factors, including population characteristics, attributes of trading areas, the adjacency to other stores operated by the same firm or competing firms, legal restrictions, and more. Likewise, consumers choose locations where they like to shop based on a wide range of factors, including the overall appeal of the area, the distance and transportation access to the store, the availability of a choice of retailers or brands, the ease of parking, and more. Retailers usually follow these three steps in choosing a store location (Berman and Evans 2012):

1. Know the customer. This is a very critical part in site selection which will need lots of researches about the shopper profiles like people's shopping behaviors and their characteristics.
2. Determine which type of location is needed. Will it be in an unplanned business district or in a planned shopping center within the geographic area? The types of location will be defined in the next paragraphs. Then seek out locales with the highest sales potential and reasonable leasing.
3. Research existing shopping centers to see if they are compatible for customer attraction. Evaluate alternate sites and their geographic (trading) areas. Generally, this step will involve finding factors like characteristics of residents and existing retailers, stores' visibility, access, population density, household income, and traffic convenience for a specific area or location.

Steps 1–3 usually make lots of preparations, including preliminary market researches and evaluations.

9.2 Understand Customers

Understanding target customers is a vital and high priority in retailing business. The gender, age, income, and other similar demographic characteristics of customers have a significant relationship in evaluating site locations. The values of those variables summed or aggregated in certain range could indicate where potential business market locates in and which type of location consumers will select to visit (Cliquet 2007).

9.2.1 Shopping Behavior

Levy and Weitz (2013) have defined three typical shopping behaviors for most of people.

Convenience shopping: consumers are primarily concerned with minimizing their effort to get the product or service they want, like the gas stations, generally located in neighborhood strip centers and freestanding locations.

Comparison shopping: consumers already have a general idea about the type of product or service they want, but they do not have a well-developed preference for a brand. However, they are willing to expend effort to compare alternatives. This is the reason why furniture retailers and shopping malls often locate next to one another.

Specialty shopping: consumers know what they want and will not accept a substitute. They are brand and/or retailer loyal. Customer group with this type of shopping behavior is preferred by some special retailers like organic vegetables.

9.2.1.1 Demographic Data of Customers

A retailer could use GIS to learn the demographics of customers at its best locations and set up a computer model to find potential locations with the most desired attributes, such as learn which of its stores have trading areas containing households with a median annual income of more than $50,000, or pinpoint its geographic areas of strength and weakness.

Scott (2007) has found some demographic variables play an important role in site selection. Age and gender in market analysis will indirectly impact whether a customer will return to an organization to experience the same product or service. The median income and average cost could indicate consumption capacity in designated geographical area. The next two maps will show how GIS combines the retailing business with demographic big data. Those population characteristics which are of special interest by most of retailers are listed below (Table 9.1; Fig. 9.1):

Table 9.1 Customer demographic data

Total size and density	Age distribution	Education level
Occupation distribution	Total disposable income	Trends
Per capita disposable	Composition of household	Incomes

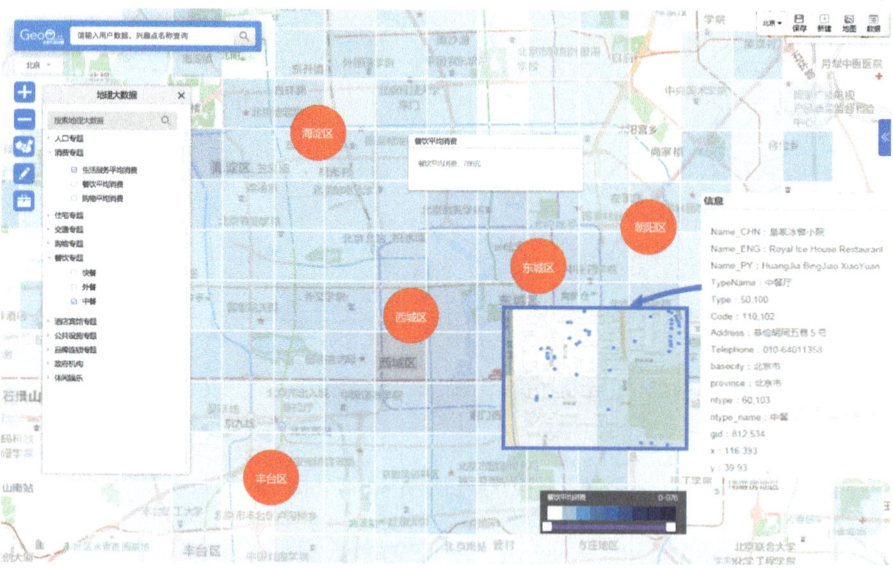

Fig. 9.1 This map shows the geographic distribution of Chinese restaurants and average catering cost of customers for each grid in Shanghai. When zoomed to highest resolution, the multi-scale map shows all the information, and the grid size is 250 m × 250 m. Each restaurant has its own attributes, like the name, address, latitude, longitude, and more

9.2.2 Determine Types of Location

Not all types of locations are suitable for each different retailing business, so the primary consideration in the choice of a specific site is to decide which type of location is best for retailers.

9.2.2.1 Traditional Types of Location

According to the traditional classification in the world, there are four different location types: Isolated Store (freestanding), Unplanned Business District, Planned Shopping Center, and Other Location Opportunity (Levy and Weitz 2013; Berman and Evans 2012). Each one has its own characteristics as to the size of the trading area, the occupancy cost, the convenience and traffic in association with the location, the composition of competitors, parking, and other factors (Van 2008).

9.2.2.2 Isolated Store

An isolated store is a freestanding retail outlet located on either a highway or a street. There are no adjacent retailers with which this type of store shares traffic (Yudelson 2009).

The advantages of this type of retail are no competition in close proximity, and rental costs are relatively low with larger space and easy parking (Wood and Sue 2007). There are also various disadvantages like: initial customers may be difficult to attract because many people will not travel very far to this type of store. Besides, advertising expenses may be high, and the cost such as outside lighting, security, grounds maintenance, and trash collection is not shared by other retailers.

Isolation is good for convenience-oriented shopping store such as gas stations and fast-food restaurants.

9.2.2.3 Unplanned Business Districts

An unplanned business district is where two or more stores situate in close proximity, but the total arrangement is not due to prior long-range planning (Table 9.2).

9.2.2.4 Planned Shopping Center

The planned shopping centers consist of a group of architecturally unified commercial establishments on a site that is owned or managed, designed and operated as a unit, based on balanced tenancy, and accompanied by parking facilities. Its location, size, and mix of stores are related to the trading area served (Jerath and Zhang 2010).

Table 9.2 Unplanned business districts

Central business district (CBD)	CBD is the downtown business area, with greatest density of public transportation and pedestrian traffic, office buildings, and stores. Its core is often very small, but formats here are various, and shoppers may be drawn from the whole urban area. However, limited parking and longer driving times can discourage suburban shoppers from patronizing stores in a CBD. The chain headquarters stores are often situated in CBDs like CCTV, People's Daily, GM, and many international financial banks
Secondary business district (SBD)	SBD can be also called Main Streets (High Streets in the UK). It's a traditional shopping area in smaller towns or not a primary business district, usually bounded by the intersection of two major streets. Services and goods sold here are very similar to those in CBDs. However, the SBD has smaller stores and trading area. They are near to residential areas with lower occupancy costs and a higher proportion of convenience-oriented items
Neighborhood business district	This type of business district is situated on the major streets within a single residential area. It contains several small stores, such as a dry cleaner, a barber shop, a beauty salon, or a restaurant, and a leader typically like a supermarket

Their positive strengths are more appealing for one-stop, family shopping and cooperative planning of goods and services based on long-range planning with sharing of common costs. The limitations associated with them are many kinds of regulations that have reduced each retailer's flexibility, like the operation hours. The competition within the center is very harsh with the domination by large anchor stores (Kramer 2010). This type of location is best suited for comparison shopping behavior and big chain retailers (Table 9.3).

9.2.2.5 Other Location Opportunities

Some stores opened in a people-aggregated area and not belonging to previous ones are also other location alternatives for many retailers. For examples, the pop-up temporary stores along the roadside and the stores in airports or railway stations are all nontraditional sites. They are popular in customers with convenience shopping behavior and fast-food restaurants (Brooks et al. 2008).

9.2.2.6 Business Districts in China

Among four types of location, both isolated store and other location opportunities have relative smaller trading areas. Instead of conducting this classification based on concepts of human geography mentioned above, the definition of business districts in GISuni follows three classifications, including CBD, SBD, and neighborhood business district. But they are aggregation of grids which have similar market segments and have been done by integrating various variables based on empirical and quantitative methods.

Table 9.3 Several planned shopping centers

Shopping mall	Enclosed, climate-controlled, lighted shopping centers with stores on one or both sides of enclosed walkways (Teller 2008). There are two classified categories—regional malls (1000,000 square feet) and superregional malls (1000,000 square feet)
Power center	Consists primarily of collections of big-box retail stores, featuring a general merchandise discount store, category specialists, off-price retailers, and warehouse clubs. Many are located near an enclosed shopping mall. It often includes several freestanding anchors and only a minimum number of small specialty store tenants
Lifestyle center	An open-air configuration of specialty stores, restaurants, and entertainment (Yan and Eckman 2009). They take more considerations of environment and architecture designs like street furniture which just cater to the lifestyle of consumers. They are particularly attractive to specialty retailers, like home products, books, or music
Theme/festival center	Typically employ a unifying theme carried by individual shops in their architectural design. The biggest appeal of these centers is to tourists, so they are most likely located in a place of historical interest and anchored by restaurants and entertainment facilities. The significant examples are Disneyland Parks and entertainment hotels in Las Vegas
Outlet center	Contains manufacturers' and retailers' outlet stores. On average, rent rates are lower than those at shopping malls. Tourism represents 50 percent of the traffic generated for many outlet centers. Thus, many are located with convenient access to popular tourist attractions

Central business district (CBD): satisfies citizens' numerous demands, like entertainment, shopping, and business. The presence of central business districts usually accompanies with the existence of department stores, shopping malls, business buildings, and financial centers. As a consequence, the total number of them would be taken into account to help classify business districts.

Secondary business district (SBD): serves a smaller range of area and has less attraction compared to the central business district. Hence, the cluster of various services seems to be weakened, and some daily demands should be satisfied by the secondary business district.

Neighborhood business district: its existence is to satisfy residents' daily demands, and hence, different types of stores and services would be considered at this stage.

9.2.2.7 Special Business Districts

Based on traditional collected information, it is difficult to specify those business districts with special features (MacEachren et al. 2011). For example, Qinhan Hutong is a famous walking street/alley in Shanghai. However, it is hard to capture this information based on the cluster of sparse POIs.

Today, with the growing popularity of location-based social media applications via the locator software like GPS built into mobile phones, the dynamics like the locations, distances, and directions to nearby POIs or stores are certainly providing a new way to inspect the flow of people, and hence, those special business districts could be explored (Zafarani et al. (2014). And the reviews also allow the retailers to design more personalized shopping experience and stimulate loyalty for customers (Mathioudakis and Koudas 2010).

Among most of the social media data, Weibo is a massive social networking site tuned toward fast communication and has played a prominent role in sociopolitical events. These data could help identify those special business districts.

9.2.3 *Analyze Trading Areas*

After a retailer investigates markets and determines what type of location is desirable, it could select out alternative site locations. Studying their proposed trading area could reveal opportunities and the retail strategy necessary to succeed. The trading area is the geographic area that encompasses most of the customers who would patronize a specific retail site, typically reflecting the boundaries within which it is profitable to sell and/or deliver products (Gambini et al. 2005). Researches by Berman and Evans (2012) suggested that each trading area has three parts:

The primary trading area encompasses 55–75% of a store's customers. It is the area closest to the store and possesses the highest density of customers to population and the highest per capita sales. There is little overlap with other trading areas.

The secondary trading area contains an additional 15–25% of a store's customers. It is located outside the primary area, and customers are more widely dispersed.

The fringe trading area includes all the remaining customers, and they are the most widely dispersed (Fig. 9.2).

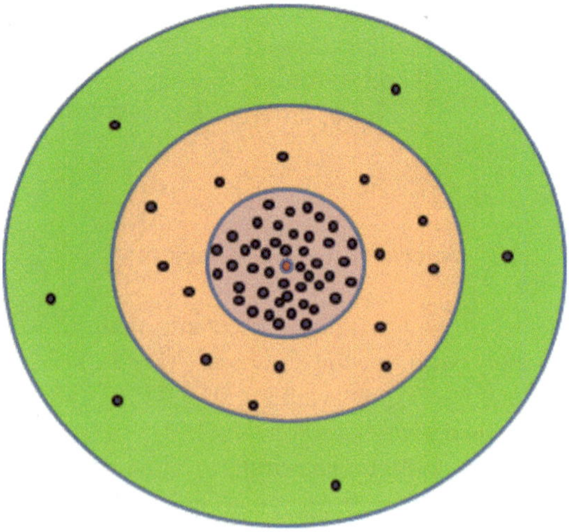

Fig. 9.2 This map shows the three types of trading areas mentioned above. The innermost circle is primary trading area. In reality, the driving time usually substitutes for the percentage of customers to account for the trading areas' division

Table 9.4 Chief evaluating factors

Traffic	Number of people flow
	Number of vehicle flow
	Traffic congestion
Transportation	Delivery distance
	Access from major roads
	Availability of mass transit
Store features	Size and shape of the buildings
	Size and shape of the lots
	Number of parking spots
	Distance from parking to store
	Condition of the lots and buildings
	Store visibility
Terms of occupancy	Leasing terms
	Operations costs
	Zoning restrictions

A thorough analysis of trading areas provides several benefits (Ozuduru 2009):

Demographic data of consumers and socioeconomic characteristics are examined. Other factors like the impact of the Internet, transportation, and supplier location legal restrictions are also reviewed.

The retailer can look at geographical coverage of proposed or existing locations and learn whether a proposed new store will service additional customers or take business from its existing stores. The best number of similar stores for a chain to operate in a given area is calculated for not too much overlap.

Geographic weaknesses are highlighted. Suppose a shopping center discovers that most of the people residing south of it do not shop there because the pedestrian overpass needs people to detour a lot. Then the manager should take actions to make up this weakness.

Berman and Evans (2012) have mentioned that when one store has created a stronger image, it may become a destination store and generate a trading area much larger than its adjacent competitors. The parasite store does not create its own real trading area which depends on people drawn to the location for other reasons, like snack bars and restaurants in a shopping center. Customers patronize these shops for convenience also with some sort of comparison.

Over the years, many researchers (e.g., Reilly 1953; Brown 1989; Craig et al. 1984; Huff 1964) have examined trading area size in a variety of settings and introduced additional factors and advanced statistical techniques to explain the consumer's choice of shopping location by including shopping center and transportation conditions, the effects of competition among stores, and shopping attractions for customers in human geography (Table 9.4).

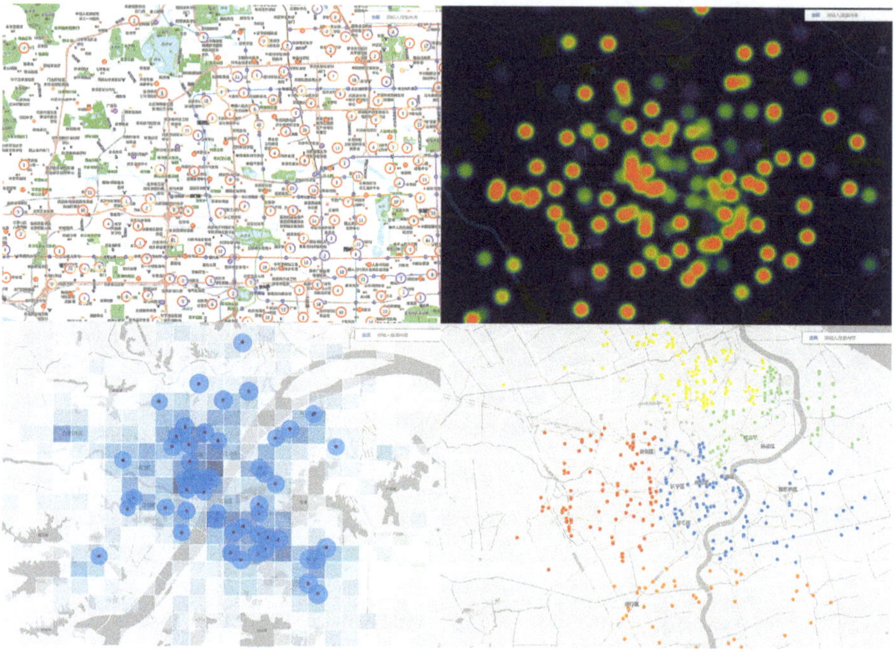

Fig. 9.3 These maps illustrate how the business software—GeoQ—realizes geographic visualization of mass big data and analyzes existing store sites and their trading areas

However, these factors are not the same in all location decisions. For example, the competition is not always considered as very bad. The clusters of retailers can add draw or agglomeration power to attract customers. This is called synergy effect (Fenker 2012) (Fig. 9.3).

Map (a) shows a cyclic graph, a synthesis of percentages of men, women, and children for one retailing site per month. The customer database is "geo-coded," and the number of each group who patronize the store is calculated and summed.

Map (b) is a heat map used for displaying similar areas with total dollar purchases aggregated. The red color means a higher customer spending. From the map, we can see that the cluster of stores doesn't always mean a higher potential sale.

Map (c) delineates the primary trading area of each retailing site which could account for nearly 75 percent of the customer base. The overlaps indirectly reflect the level of competitions among stores. This can help managers adjust their own stores.

Map (d) generates the distribution of retailing sites vs. customers to further understand the correlation between the purchase behavior and store locations. The points with the same color correspond to the true service area of each store based on registered members' information.

9.3 Commercial Methodology

Most of GIS analysis relies on grid by grid calculation. It's difficult to get accurate classifications of the whole market by only using human geography which needs lots of in-site survey work and artificial interpretation. In reality, trading areas do not usually follow circular patterns, and many of their characteristics are rather hard to calculate. They are influenced by store type, store size, competitor locations, travel time, traffic barriers, and media availability. The probability of people shopping at a location depends on customers' sensitivity to travel time which is not clear because it will vary from product types to their own shopping preferences. However, it's a common principle that each retailing business does have similar market environments. It's very powerful to use market segmentation methodology to properly describe successful business areas associated with geographical data.

9.3.1 Market Segmentation

Market segmentation is a strategy which involves dividing a broad market into subsets of consumers or businesses that have common needs, interests, or types of locations and then designing and implementing strategies to target them (McDonald and Dunbar 2012). There are several traditional types of market segmentation (Swart and Roodt 2015):

Geographic segmentation: segmentation according to geographic criteria—nations, states, regions, countries, cities, neighborhoods, or postal codes. The geo-cluster approach combining demographic data with geographic data could create a more accurate or specific profile.

Consumer segmentation: segmentation according to demography is based on variables such as age, gender, occupation, and educational level.

Behavior segmentation: it divides consumers into groups according to their knowledge of, attitude toward, usage rate, response, loyalty status, and readiness stage to a product.

Psychographic (lifestyle) segmentation: this is measured by studying activities, interests, and customer opinions (Miguéis et al. 2012).

The big data provided by LBS technology is commonly used here. However, such massive data are usually time series, from which useful information are hard to be extracted. GISuni has developed its own commercial segmentation methods on the basis of classified POIs (point of interest) with a supplement of social media data to improve precisions.

9.3.2 POI Segmentation

POI is a specific point location that is useful or interesting, like schools or supermarkets. Unlike problems of completeness, logical consistency, positional and temporal accuracy in SCD, and mobile and social media data, POI information can be widely accessible with guaranteed time consistency, so it's the core data of this market segmentation method. As of today, there are 18 classes and nearly 313 subclasses of POIs classified by GISuni (Table 9.5).

Through creating blocks (grids as 300 m × 300 m), various POIs could be integrated in each block so that subsequent calculations could be conducted at this level to do market segmentation.

From whole perspective, the market can be segmented into several parts by different business districts, not researching relevant site locations at first (Table 9.6).

From specific retailer's perspective, the business environment could be simulated by analyzing the type, number, and percentage of POIs around their own stores.

Table 9.5 18 Classes of POI

Classes of POIs	Types of POIs
Department store	Shopping service
Shopping mall	Shopping service
Brand store	Shopping service
Restaurant	Catering service
Sports Entertainment	Sports service
Business building	Business building
Finance	Financial service
Supermarket	Life service
Vegetable market	Life service
Pharmacy	Life service
Convenience store	Life service
Post office	Life service
Bank	Life service
Hair studio	Life service
Community center	Educational service
Education	Educational service
Subway station	Public service
Bus station	Public service

Table 9.6 Business districts defined by SQL

Central Business District	Department store > 0 & shopping mall > 0 & brand store > 0 & restaurant > 0 & sports entertainment > 0 & business building > 0 & finance > 0 & [(department store + shopping mall) > 3 or (business building + finance) > 10] & (subway + bus station) > 30
Secondary Business District	(Department store + shopping mall > 2) & restaurant > 0 & sports entertainment > 0 & supermarket > 0 & (subway + bus station) > 0
Neighborhood Business District	(Department store > 0 or shopping mall > 0) & (supermarket > 0 or vegetable market > 0) & (pharmacy + convenient store + post office + bank + hair studio + community center + education) > 10 & (subway + bus station) > 0

Besides, many retailers would like to be adjacent to shopping centers or supermarkets. The adjacency in space could be reflected as distance to different types of POIs. Supposed most existing site locations were successful, applying Cluster and Outlier Analysis for these patterns to remaining areas will help retailers find similar business aggregation areas.

9.4 Results

Shanghai, an economic, financial, trade, and shipping center in China, is the final research area, and existing stores of two famous fast-food retailers are test points for market segmentation. They totally have nearly 6000 stores in China in 2015 and are called as A and B retailers for short in the following paragraphs.

9.4.1 Segmentation Result for Business Districts

After calculating 18 classes of POIs in Shanghai, three different business districts are found there (Fig. 9.4):

9.4.2 Supplement of Social Media Data

According to Weibo reviews and check-in numbers within each block, excluding airports, bus stations, and railway stations, 20 most popular blocks in Shanghai have been collected. Comparing this information to the previous geographical segmentation, these blocks, which were not concerned as three pre-calculated business districts, were assigned as special business districts (Fig. 9.5).

Given a look to all existing B retailer's stores dropped into four types of business district classified, the statistic result follows next (Table 9.7).

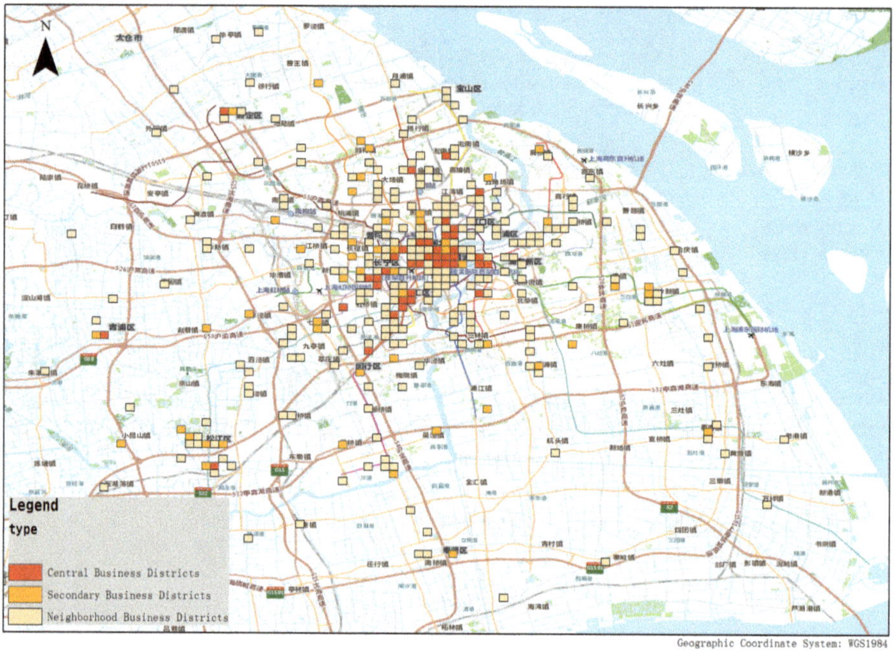

Fig. 9.4 Three business districts redefined in Shanghai

Fig. 9.5 The map above presents a couple of special business districts, which were defined based on social media data

Table 9.7 Statistic result in business districts

District type	Retailers number	Earliest opening year	Latest opening year
Central business district	47	1994	2014
Secondary business district	22	1995	2014
Neighborhood business district	56	1996	2014
Special business district	2	2008	2013

Table 9.8 Statistic result of POIs in the same blocks

Types of POI	Total number in the block	Percentage (%)	Total number in Shanghai	Percentage (%)
Shopping service	12,706	31	76,830	27
Catering service	7686	18	35,056	12
Life service	5617	14	42,596	15
Business building	3765	9	31,816	11
Enterprise	4472	11	40,802	15
Financial service	1543	4	6580	2
Public service	1032	2	7498	3
Educational service	1016	2	7823	3
Sports service	1287	3	7648	3
Residential service	788	2	5162	2
Healthcare service	682	2	5535	2
Sum	41,560		281,450	

One hundred twenty-seven out of 179 B retailer's stores are located within classified business districts, which means more than 70% of stores can be explained by defining the business district.

It could be found that they tend to open stores in the central business district first and then expands its services to secondary business districts and neighborhood districts. Special business districts can also attract their attention like locations near schools.

9.4.3 Segmentation Result for Specific Retailers

Table 9.8 reveals the total number of POIs dropped into same blocks as their stores, and the percentage has indicated patterns of most successful site locations.

A sector diagram shows the components of POIs around existing stores more clearly (Fig. 9.6):

Comprehensive similar aggregation areas are generated by overlaying similar aggregation areas of each type of POIs with percentage above as weights (Fig. 9.7).

Total numbers of their stores in Shanghai are 247 and 179, and among which, 83% of A retailer's stores and 90% of B retailer's stores have dropped into comprehensive similar aggregation areas.

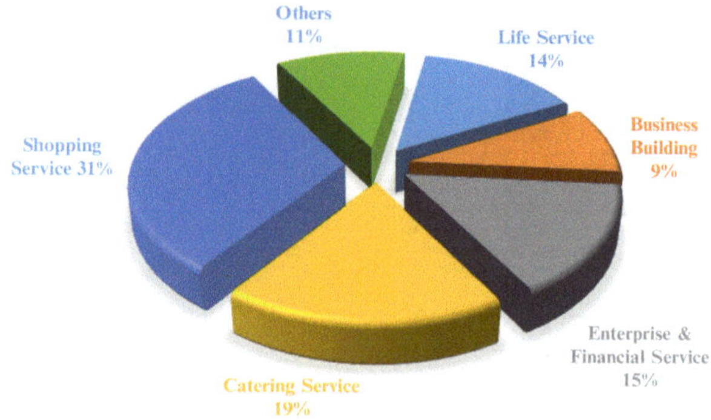

Fig. 9.6 This sector diagram illustrates proportions for different types of POI around existing stores. Shopping service, catering service, life service, business building, and enterprise and financial service are five most important components

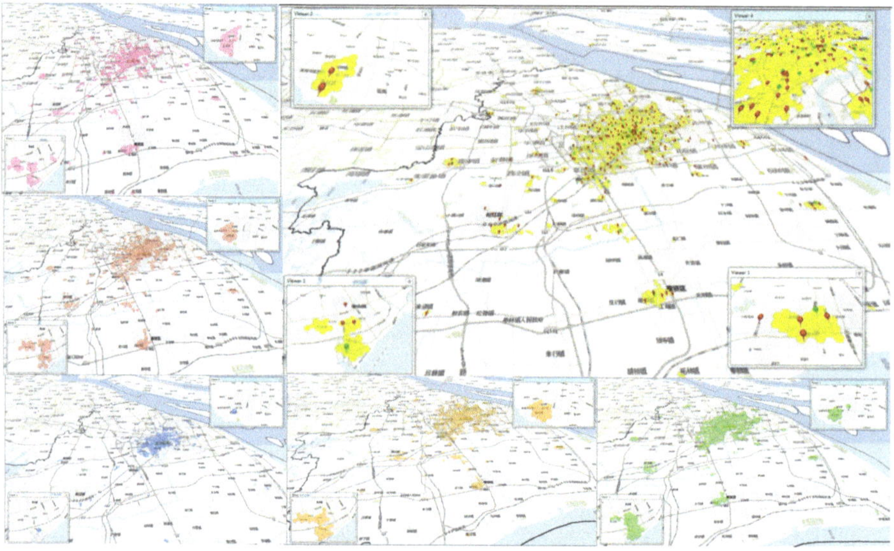

Fig. 9.7 These maps illustrate similar aggregation areas of POIs in order of (**a**) shopping service, (**b**) catering service, (**c**) life service, (**d**) business building, (**e**) enterprise and financial service. (**f**) The *yellow* parts on the map are comprehensive similar aggregation areas of POIs, *red pins* are A retailer's stores, and *green pins* are B retailer's stores

9.4.4 Calculated by Average Distance

The processes above may ignore some isolated areas which are strongly interested by retailers. Apart from total numbers of POIs, the average distance of all POIs to nearest stores is 183 meters, and the median of grid's length is 150 meters. So the adjacent relationship can be defined as:

Strong dependent: average distance <150 m
Medium dependent: 150 m < average distance <183 m
Weak dependent: average distance >183 m (Table 9.9)

Table 9.9 Statistic result of strong dependent POIs

Types of POI	Total number within 300 m	Average distance to existing (m)
Shopping service	1162	114
Catering service	2065	139
Life service	200	126

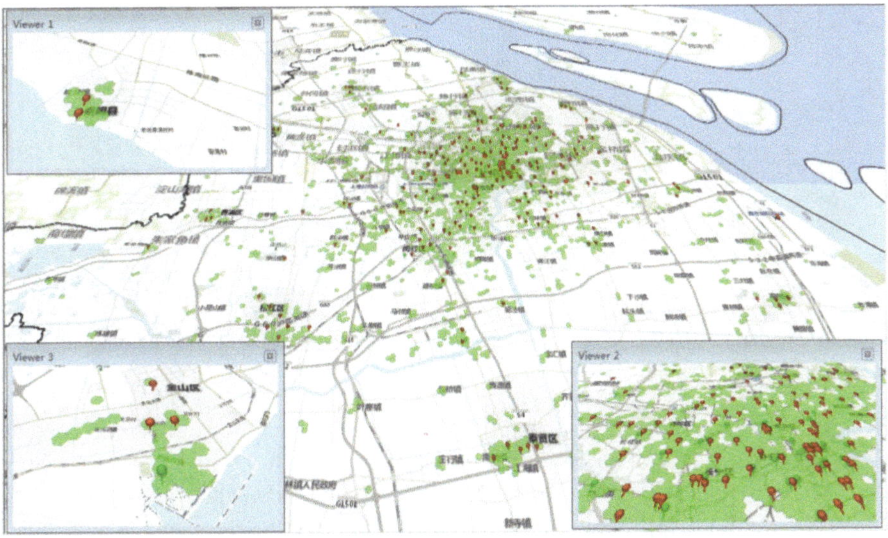

Fig. 9.8 The *green parts* on the map are comprehensive similar aggregation areas of POIs, *red pins* are A retailer's stores, and green pins are B retailer's stores

Similar aggregation areas for each type of POI and comprehensive aggregation areas are generated by Cluster and Outlier methodology based on distances (Fig. 9.8).

Two overlaid comprehensive aggregation areas of POIs are calculated by numbers and distances (Fig. 9.9):

This time, 97% of A and B retailers' stores have dropped into overlaid comprehensive similar aggregation areas. Besides, compared to new opened stores in 2015, 28 out of 33, nearly 85% of them are located there.

9.5 Conclusions and Future Researches

Despite the fact that parts of this chapter are concerned with discussing the accuracy and practicability of traditional methodologies, its contribution is to provide marketing experts an innovative and nonexclusive segmentation approach stemming

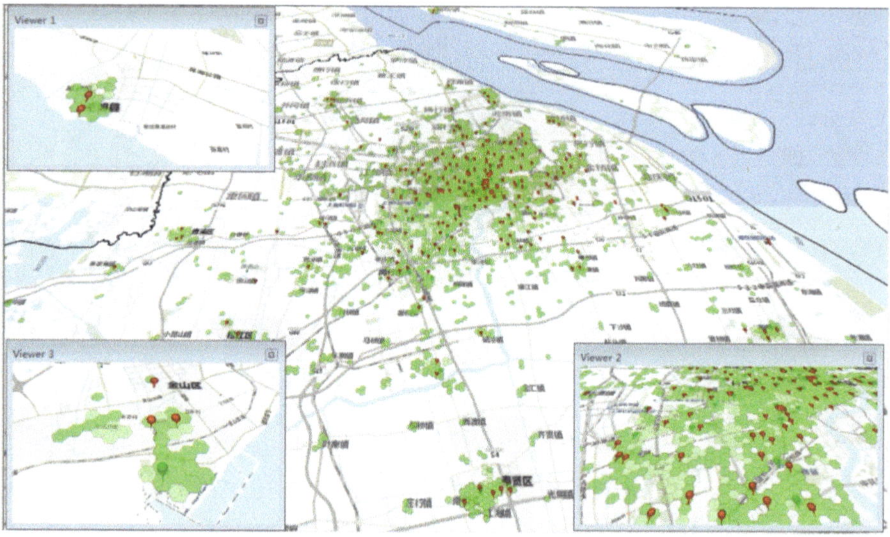

Fig. 9.9 Overlaid comprehensive similar aggregation areas of POIs

from POIs and supplements of social media data that meet the commercial need to understand and interpret ambiguous market information as realistically as possible (Purohit & Sheth 2013). Different from traditional methods, which cannot be easily expanded, methods explored in this chapter tend to be more accessible and feasible. The real case study has demonstrated this solution is effective to reflect correlations between geographic data and successful locations, satisfying most of considerations in retailing business. Business district classifications can help examining retail sites' selections, and most of retail stores tend to choose those business districts, which may be explained by the economic status and traffic flow within those districts. Also, POIs can significantly contribute to explain the selection of retail locations (Li & Liu 2012) since retailers always have their own competitive and complementary units, which could be covered by the aggregation of POIs. Even though the best solution of retail site selection is ambiguous and unable to be determined, POIs do provide intuitive insights in order to promote the decision of site selection. However, compared to traditional in-site surveys, the aggregation area of POIs delineated by this method is impossible to perfectly cover all existing stores even after introducing adjacent relationships and Weibo data to improve the precision. The reasons accounting for mismatching points may be:

1. The choice of grid's size and cluster methods will generate differential in segmentation results.
2. Some areas, like tourist spots and airports, will not be divided into similar business aggregation areas because POIs there are very few, even sparse. But the people flow is big enough to attract them.

Due to difficulties in distinguishing all locations for successful sites, there exist certain directions for future researches. First of all, the social media data not only just indicates spatial positions but also contains massive temporal and personal preference information in comments. This can be dug further to find missed areas, thus increasing the accuracy of market segmentation. Secondly, some special factors like competitiveness and customers' sensitiveness to travelling distance can also be included to narrow the range of final result. Only by adding more and more powerful indicators step by step into analysis can this method be really responsive and accountable for decision makings. Besides, the regularity varies for different business formats, so principles of Cluster and Outlier should be adjusted associated with relevant requirements.

References

Berman, B., & Evans, J. R. (2012). *Retail management: A strategic approach* (12th ed.). Boston: Pearson.

Brooks, C., Patrick, J. K., et al. (2008). Trip chaining behavior in multi-destination shopping trips: A field experiment and laboratory replication. *Journal of Retailing, 84*(1), 29–38.

Brown, S. (1989). Form and Function in an Unplanned Retail Park: Evidence from the Abbey Trading Centre. *Area, 21*(2), 137–144.

Calabrese, F., & Ratti, C. (2006). Real time Rome. *Networks and Communication Studies, 20*, 247–258.

Cliquet, G. (2007). *Geomarketing methods and strategies in spatial marketing*. Washington, DC: ISTE Special Priced Titles.

Cressie, N., & Wikle, K. C. (2010). *Statistics for spatio-temporal data*. Hoboken, N.J.: Wiley.

Duggal, N. (2008). *Use of GIS in retail location analysis*. Saarbrucken: VDM Verlag.

Fenker, R. (2012). Using property scoring to find the right location. *Retail Property Insights, 19*(1), 32–34.

Gambini, R., Huff, L. D., et al. (2005). Geometric properties of market areas. *Papers in Regional Science, 20*(1), 85–92.

Hernandez, T. (2007). Enhancing retail location decision support: The development and application of geovisualization. *Journal of Retailing and Consumer Services, 14*(4), 249–258.

Huff, D. L. (1964). Defining and Estimating a Trading Area. *Journal of Marketing, 28*(3), 34.

Jerath, K., & Zhang, Z. J. (2010). Store within a store. *Journal of Marketing Research, 47*(4), 748–763.

Kramer, A. (2010). *Dollars & Cents of Shopping Centers/The SCORE 2010*. Washington, DC: Urban Land Institute.

Levy, M., & Weitz, B. A. (2013). *Retailing management*. New York: McGraw Hill.

Li, Y. R., & Liu, L. (2012). Assessing the impact of retail location on store performance: A comparison of Wal-Mart and Kmart stores in Cincinnati. *Applied Geography, 32*, 591–600.

MacEachren, A. M., et al. (2011). *Senseplace2: Geotwitter analytics support for situational awareness. IEEE Conference On Visual Analytics Science and Technology (VAST)* (pp. 181–190). Providence, RI: IEEE.

Mathioudakis, M., & Koudas, N. (2010). Twitter monitor: Trend detection over the twitter stream. In *Proceedings of the 2010 ACM SIGMOD International Conference on Management of data* (pp. 1155–1158). Indianapolis, IN, USA.

McDonald, M., & Dunbar, I. (2012). *Market segmentation: How to do it and how to profit from it, revised 4th edition*. Oxford: Butterworth-Heinemann.

Miguéis, V. L., Camanho, A. S., et al. (2012). Customer data mining for lifestyle segmentation. *Expert Systems with Applications, 39*, 9359–9366.

Mover, M. S. (1972). Management science in retailing. *Journal of Marketing, 36*, 3–9.

Ozuduru, B. (2009). *Trade area modeling in retail location analysis*. Saarbrucken: VDM Verlag.

Purohit, H., & Sheth, A. (2013). Twitris v3: From citizen sensing to analysis, coordination and action. In *Proceedings of the 2013 International Conference on Weblogs and Social Media*. Cambridge, Massachusetts, USA.

Rachel, J. C. (2007). Significance and variety of Geographic Information System (GIS) applications in retail, hospitality, tourism, and consumer services. *Journal of Retailing and Consumer Services, 14*(4), 247–248.

Ratti, C., Frenchman, D., et al. (2006). Mobile Landscapes: Using Location Data from Cell Phones for Urban Analysis. *Environment and Planning B: Planning and Design, 33*(5), 727–748.

Reilly, W. J. (1953). *The law of retail gravitation* (2d ed.). New York: Pilsbury Publishers.

Scott, P. (2007). *Geography and retailing*. Edison: Aldine Transaction.

Swart, P. M., & Roodt, G. (2015). Market segmentation variables as moderators in the prediction of business tourist retention. *Service Business, 9*, 491–513.

Teller, C. (2008). Shopping streets versus shopping malls—Determinants of agglomeration format attractiveness from the consumers' point of view. *International Review of Retail, Distribution and Consumer Research, 18*(4), 381–403.

Teller, C., & Thomas, R. (2008). The evolving concept of retail attractiveness: What makes retail agglomerations attractive when customers shop at them? *Journal of Retailing and Consumer Services, 15*(3), 127–143.

Van, U. C. (2008). *Malls & department stores*. Berlin: Braun.

Wood, S., & Sue, B. (2007). Convenience store location planning and forecasting—A practical research. *International Journal of Retail and Distribution Management, 35*(4), 233–255.

Yudelson, J. (2009). *Sustainable retail development: New success strategies*. New York: ICSC.

Yan, R. N., & Eckman, M. (2009). Are lifestyle centers unique? Consumers' perceptions across locations. *International Journal of Retail & Distribution Management, 37*(1), 24–42.

Zafarani, R., Abbasi, M. A., et al. (2014). *Social media mining: An introduction*. Cambridge University Press.

Chapter 10
Redefinition of the Social Space Based on Social Atlas Analysis: A Case Study of Dongguan, China

Yungang Liu and Haiyu Su

Abstract Social area analysis is a conventional approach for studying social divisions and social structures and involves deducing and examining existing theories in America. However, because it is confined to the designated demography index and not spatial features, social area analysis is incapable of discovering the diversity of social spaces. This paper presents and analyzes a series of social maps, using data from the Sixth Population Census and POI data from electronic maps and then discusses the sociospatial features of Chinese cities. From the maps, it is found that with demographic properties, social facilities, and organization, an improved index system of social spaces emerges. Using these insights, a study of the social area is conducted. The result is that 11 types of social areas are found in Dongguan, respectively, (1) old town centers, (2) high-grade residential and commercial areas, (3) general residential areas, (4) sub-commercial centers, (5) administration centers, (6) industrial community areas, (7) low-density suburban areas, (8) historical culture areas, (9) new residential areas, (10) agricultural areas, and (11) old age communities. Furthermore, these areas are classified into three kinds of social spaces, rural communities, rural migrant worker communities, and urban communities, and a triad structure society model of Dongguan is uncovered in the space. In the end, this research demonstrates that the method of social atlas can relieve the limitations of social area analysis. It contributes to understanding a very complex and diverse social space and reconstructing the research framework to adapt to particular social situations in China.

Keywords Social atlas • Social space • Social element • Social area • Dongguan

---

Y. Liu (✉) • H. Su
School of Geography and Planning, Guangdong Key Laboratory for Urbanization and Geo-simulation, SUN Yat-sen University, Guangzhou, Guangdong 510275, China
e-mail: liuyung@mail.sysu.edu.cn

Z. Shen, M. Li (eds.), *Big Data Support of Urban Planning and Management*,
Advances in Geographic Information Science, DOI 10.1007/978-3-319-51929-6_10

10.1 Introduction

As Chinese society steps into a transition period, social development is becoming a more significant issue, which makes understanding urban social spaces increasingly important. A number of empirical studies in China on social spatial structures or social differentiation are based on the social area analysis developed by Shevky and bell (1955), such as the research in Shanghai by Wei yu (1986) and Li zhigang (2006); in Guangzhou by Xu xueqiang (1989), Zheng jing (1995), and Zhou chunshan (2006); in Beijing by Gu chaolin (2003) and Feng jian (2003); in Nanjing by Xu di (2009); etc. Although these studies effectively present social spaces in different cities through cluster analysis, they deviate from studies of the social elements in spaces because they are overly focused on an analysis of integrated features and principal components. Furthermore, another concern that needs further examination is whether the conventional social area theory from America is suitable for interpreting social spaces in China. As such, social area analysis has substantial limitations that make it difficult to explore the specificity and diversity of social spaces in Chinese cities.

In the past, induction was the basic approach to discover new models or establish theories in geography studies. The course of theorizing the concentric zone model by Burgess and the social area analysis by Shevky is no exception. In those processes, marking the relevant social elements and phenomena on maps is the first action, and then with a series of maps, some specific spatial features or patterns are discovered and concluded. In the same way, re-induction should be the basis of an approach to determine the features of social spaces in China instead of deducting them from Western theory directly. Therefore, the research of social atlas is one of the basic works of social study in China. In this research, with a case in Dongguan, we attempt to discuss the sociospatial features of a Chinese city by presenting a series of social maps and analyzing the social areas based on mapping.

10.2 Review of Social Atlas in Social Studies

10.2.1 Comprehensive Description of Social Elements

In the first study, the social atlas is applied to primitively present the spatial features of all sorts of social elements and the relationship between social elements and the physical environment with specific visualization methods. In his research on poverty in London in 1903, Booth (1903) marked the residence of different social classes using colors varying from light to dark while labeling the site of shops and factories with dots on the same map. From such methods, a connection between facility and social stratum was found, and the study became the original example of the application of social atlas in social space studies. Afterwards, Shepherd and

Phelps continued these methods and conducted research in London and central Lancashire, respectively. In Australia, Davis and Spearritt (1974) took the lead in compiling the atlas in Sydney in 1974, which was then updated by Horvath et al. (1989) in 1989. Later, social atlas, as a basic work, received increasing attention from different institutions, and a series of social atlas updates (see Perth (Nagle, 2002), Melbourne (Lazzaro, 2003), Darwin and Palmerston (Elliott, 2003), Adelaide (McGrath, 2008), Sydney (Smith, 2008), and others) were completed in succession under the guidance of the Australian government. In recent years, with the rapid development of open data and data processing technology, more kinds of social data have become available for presentation in space. For instance, the topic of population changes, crime, traffic accidents, commuting, cultural life, the concept of value, and so on could increasingly be included in the social atlas.

10.2.2 Thematic Interpretation of Social Issues

Some research focuses on the application of the social atlas for the interpretation of specific issues. For example, Poulsen and Spearritt (1981), based on the data from 1971 to 1976, analyzed the relation between social capital and the election results in space from social politics, history, and policy through the methods of social and political atlases. Seager (1997 and 2003); Seager and Olson (1986), a feminist researcher, proved that the opportunities of women in different countries are inequality by atlases. Shinagawa and Jang (1998) and Brewer and Suchan (2001), in the same way, proved that the spatial segregation between different races, including White, Black, American Indian, Asian, Latinos, and Alaskan natives, among others, has become a universal and serious problem in America. Benson (2002) presented the change of poverty in space and illuminated the relevant causes through a series of atlases of income, health, birth, education, culture, and life as well. Moreover, Tennant et al. (2003) and Glover et al. (1999, 2006) published three editions of the social health atlas of South Australians in succession, which showed the health welfare change of young adults and children and the matching degree of health services in space. Oswalt and Rieniets (2006), after his studies of shrinking cities in the world, published an atlas of shrinking cities to show that a large number of industrial cities were troubled by the demographics of shrinking populations. Recently, Tsvetkov (2013) mapped the stereotypes in the nation showing prejudice in today's fast-changing world in his atlas publication about prejudice.

10.2.3 Theoretical Verification of Social Divisions

Social atlases are less likely to be used to research social divisions due to the widespread use of social area analysis starting in the 1950s. But their use has not completely faded away. For example, Morrill (1990) conducted a validation studies in

1990, based on a social atlas, with a series of maps of social elements, such as age, fertility, livelihood, and so on. Kurasawa and Asakawa (2004) examined the social areas and structure of Tokyo by social atlas and found that it is different from that of the American cities.

10.3 Research Methods

10.3.1 Definition of Social Space

According to the social space production theory of Henri Lefevre (1991), social spaces can be divided into three respects: material space, institutional space, and living space, which correspond to social facilities, organization, and demographic property, respectively—the three compound elements in the entity space (Fig. 10.1).

10.3.2 Data Source

The data for this study comes from two sources. One is from the Sixth Population Census of Dongguan 2010, which includes 587 statistical cells (community or village). The other is POI data from an electronic map of 2011.

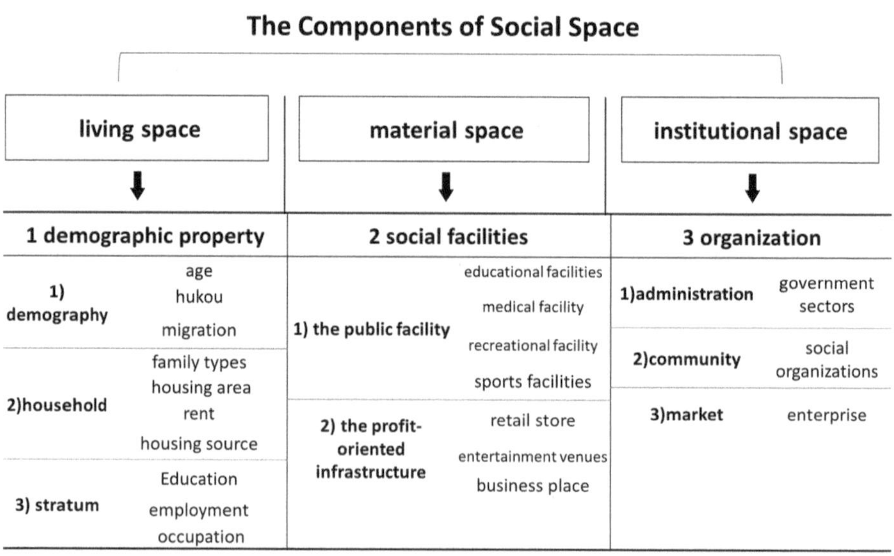

Fig. 10.1 The definition of social space

10.4 Research Region

Dongguan city is located in Guangdong Province in South China, close to Guangzhou and Shenzhen. The total area of Dongguan is 2465 km^2. In 2010, the population is eight million, making it the third largest city in Guangdong. Additionally, the portion of immigrant population is 80%, which is well above the average of the Pearl River Delta region (Figs. 10.2, 10.3, 10.4, 10.5, 10.6, 10.7, and 10.8).

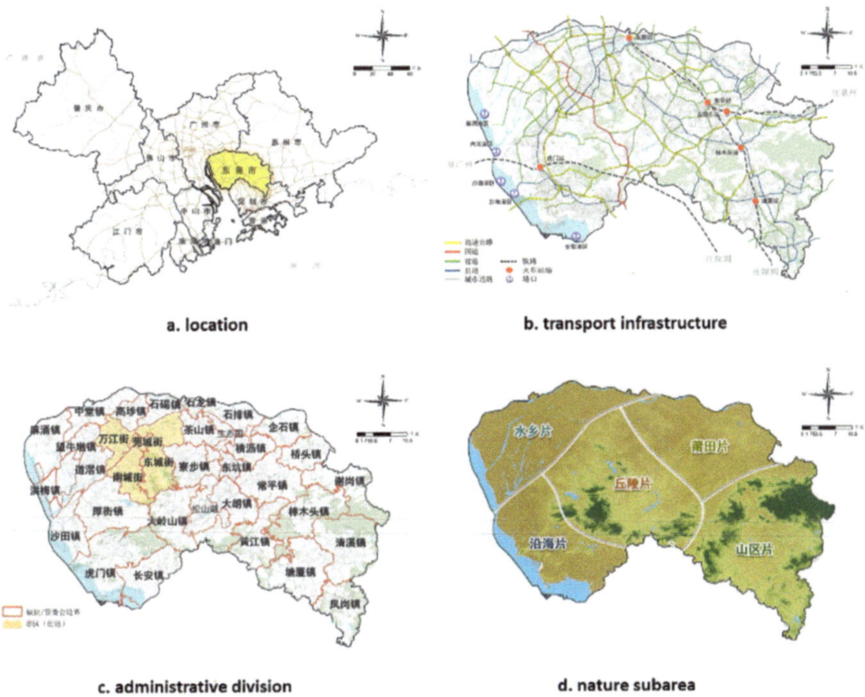

Fig. 10.2 The location, administrative division, and natural conditions of Dongguan

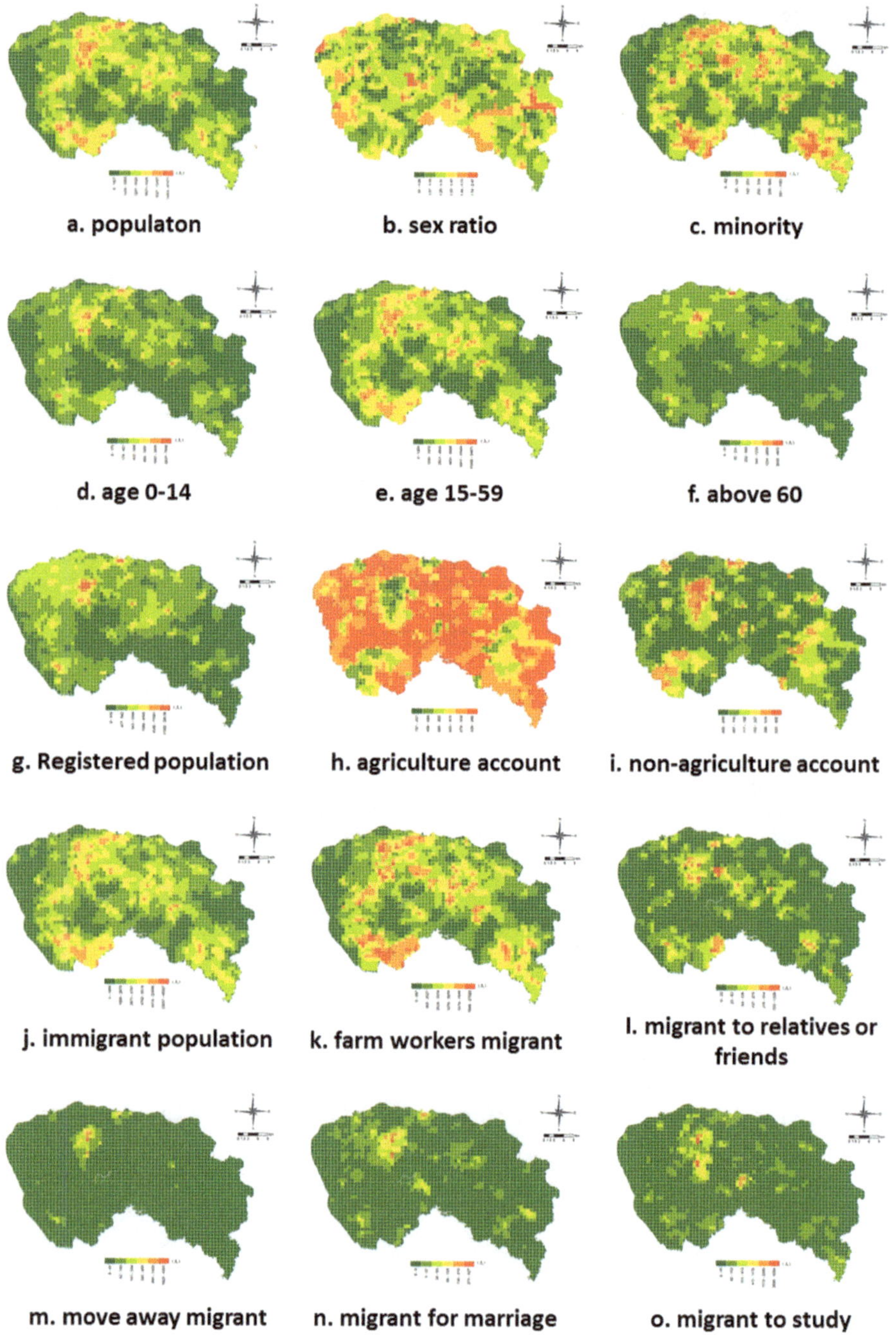

Fig. 10.3 The demographic maps of Dongguan

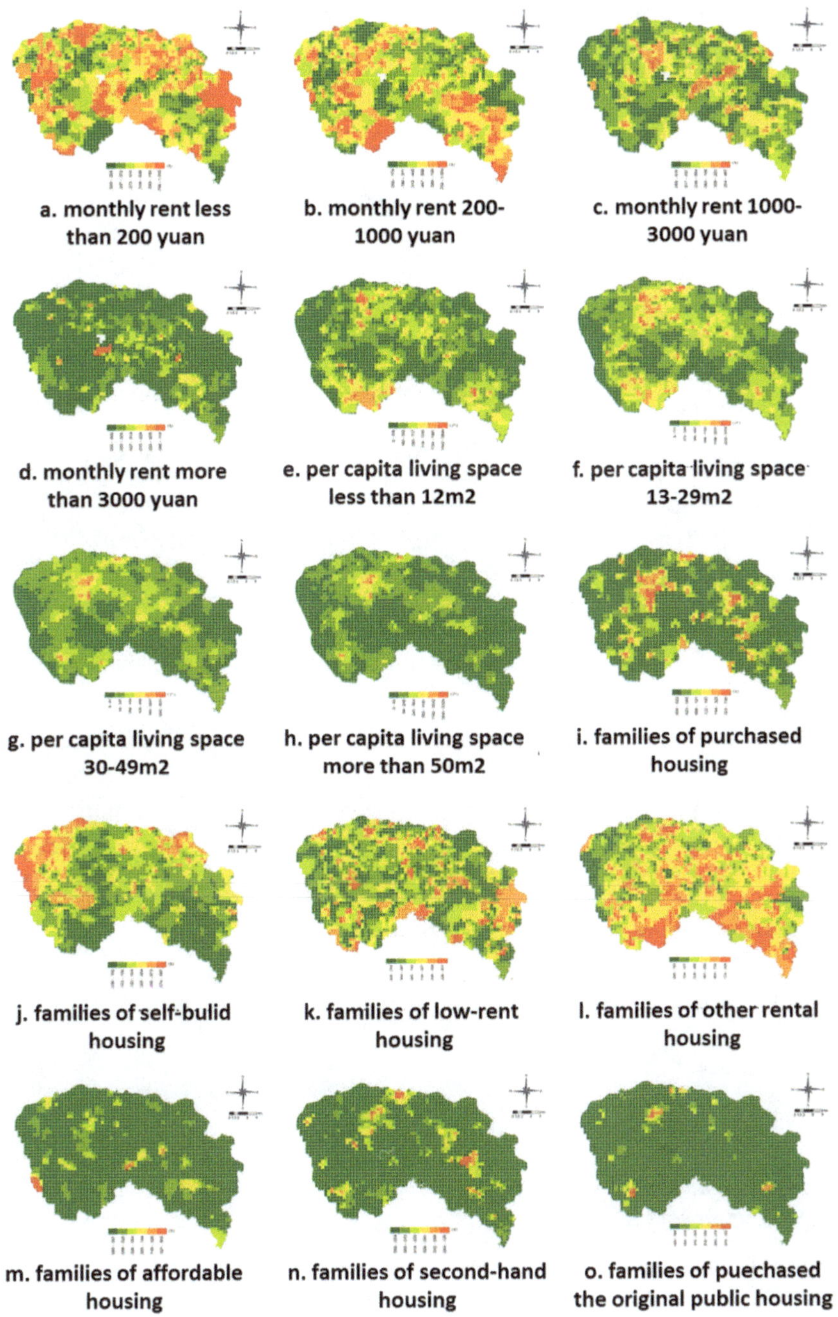

Fig. 10.4 The household maps of Dongguan

10.5 Features of Dongguan Social Atlas

10.5.1 Spatial Properties of Demography

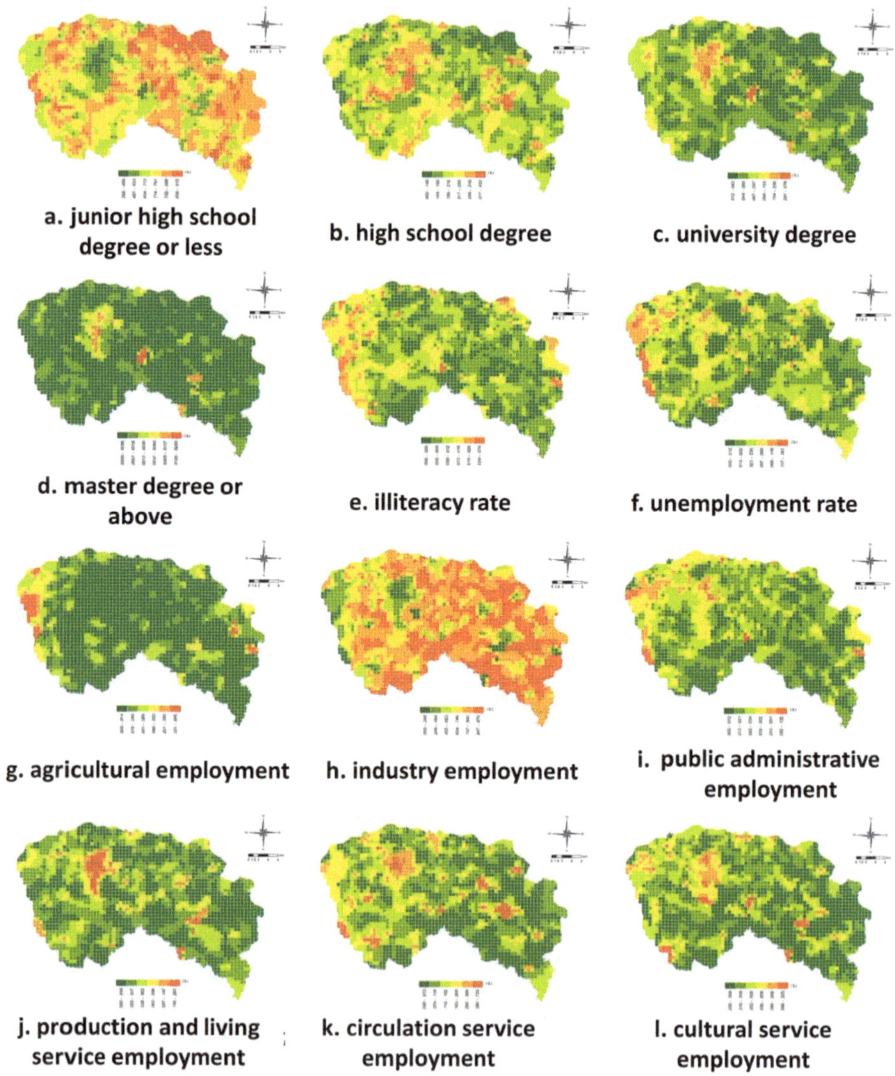

Fig. 10.5 The stratum maps of Dongguan

10.5.2 Spatial Properties of Facility

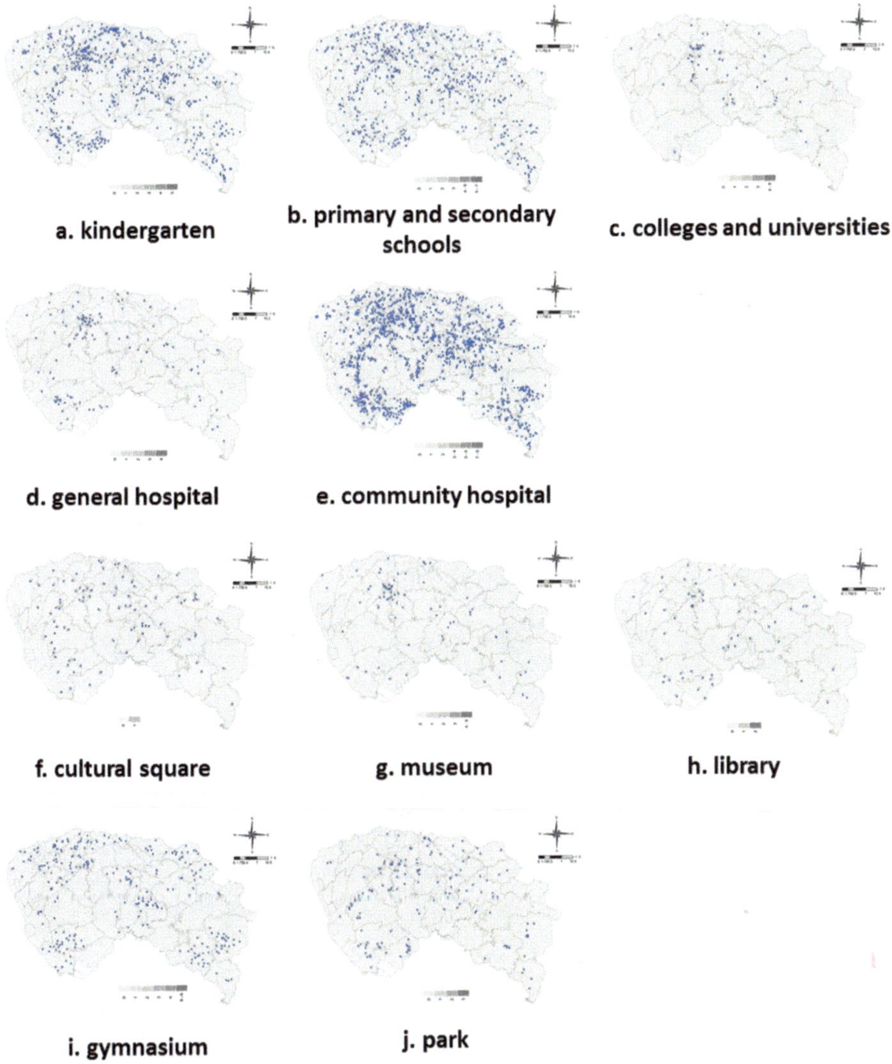

Fig. 10.6 The public facility maps of Dongguan

10.5.3 Spatial Properties of Organization

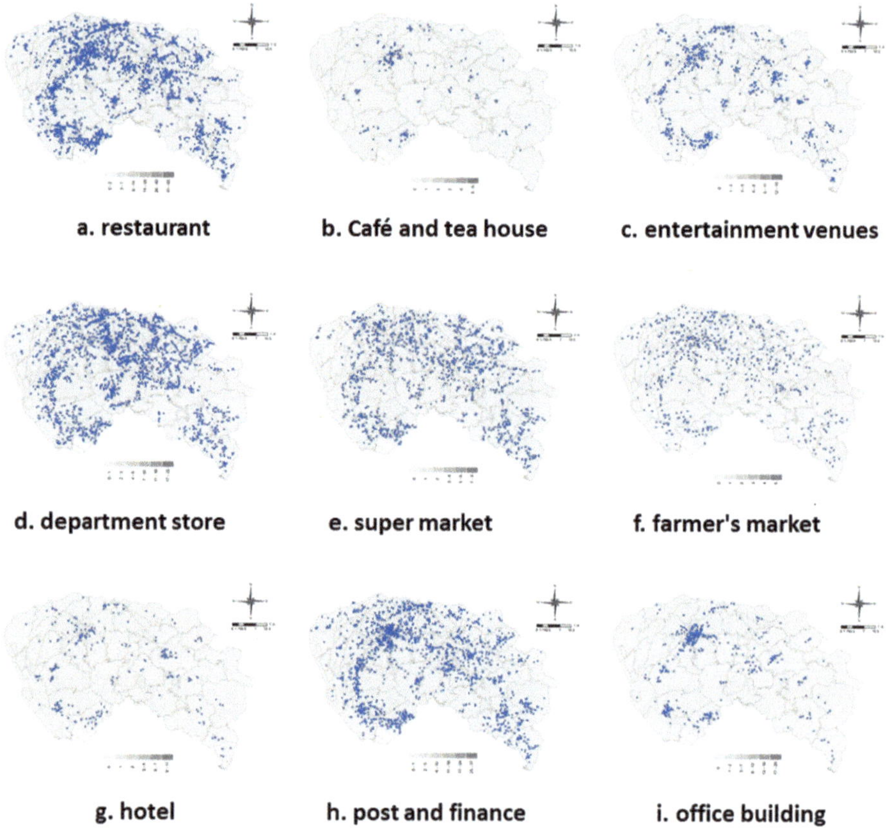

Fig. 10.7 The profit-oriented infrastructure maps of Dongguan

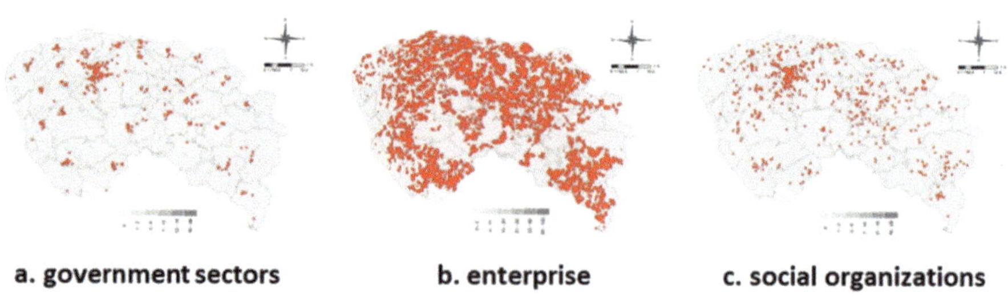

Fig. 10.8 The organization maps of Dongguan

10.6 Social Areas of Dongguan

10.6.1 Index System

Based on social facilities, organization, and demographic property, we conducted a study of social divisions by cluster analysis.

Since the demographic property elements contain too many indexes, it was necessary to do dimensionality reduction first. According to the results of factor analysis, the KMO is 0.89 with a confidence level of 99%. The 68 indexes could be reduced to five factors: (1) elites, (2) migrant workers, (3) elderly people, (4) commercial service people, and (5) farm workers (Table 10.1).

Furthermore, the social facility elements could be reduced to two types: public facilities and profit-oriented infrastructure. Organization elements include government sectors, enterprise, and social organizations. From the above, it is possible to set up a new index system for social area studies (Table 10.2).

10.6.2 Social Areas

Using the above index system, the social areas of Dongguan are present in the clustering analysis of K-means. As the result, the social space of Dongguan is divided into eleven areas: (1) the old town center, (2) high-grade residential and commercial areas, (3) general residential areas, (4) sub-commercial centers, (5) administration centers, (6) industrial community areas, (7) low-density suburban areas, (8) historical culture areas, (9) new residential areas, (10) agricultural areas, and (11) old age communities (Fig. 10.9, Table 10.3).

Table 10.1 The result of factor analysis of social demography

	The factor	Eigenvalue(variance explained)
1	Elites	23.326(34.303%)
2	Migrant workers	14.845(21.832%)
3	Elderly people	11.678(17.173%)
4	Commercial service people	4.029(5.925%)
5	Farm workers	2.081(3.060%)

Table 10.2 The indicator system for social divisions

	The big class	Subclass
Demographic property	Elites	
	Migrant workers	
	Elderly people	
	Commercial service people	
	Farm workers	
Social facilities	Public facility	Educational facilities
		Medical facilities
		Recreational and sports facilities
	Profit-oriented infrastructure	Retail stores
		Entertainment venues
		Business places
Organization	Government sectors	
	Enterprise	
	Social organizations	

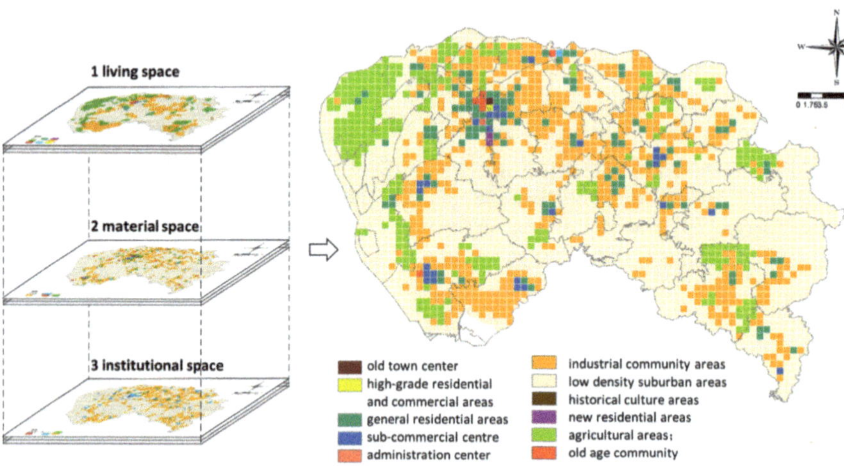

Fig. 10.9 The social areas of Dongguan

10.6.3 Social Spatial Structure

For further studies, we placed the 11 categories of social areas into three types of space, which are the rural society, the farmer workers' society, and the urban society (Fig. 10.10, Table 10.4).

Table 10.3 The clustering results of social areas in Dongguan

	1	2	3	4	5	6	7	8	9	10	11
Elites	9.22	32.81	0.57	0.45	2.89	−0.09	−0.05	0.05	18.26	−0.09	0.40
Migrant workers	0.42	1.06	0.68	0.89	0.29	0.92	−0.35	−0.89	−0.13	−0.16	−0.62
Elderly people	10.25	1.72	0.77	1.06	−0.06	0.03	−0.16	18.46	−2.80	0.12	12.86
Commercial service people	−11.51	19.46	0.27	4.17	3.27	−0.02	−0.05	−4.31	−9.05	−0.10	−1.52
Farm workers	−2.55	2.93	−0.22	−0.60	−0.29	−0.10	−0.31	−2.17	0.98	2.09	−1.52
Educational facilities	8.98	1.42	1.82	2.47	1.42	0.56	−0.35	6.15	1.89	−0.17	3.62
Medical facilities	3.49	4.29	1.56	3.73	1.09	0.84	−0.42	3.49	1.49	−0.32	2.29
Recreational and sports facilities	10.47	11.82	1.69	1.66	1.01	0.25	−0.25	11.82	−0.34	−0.08	3.71
Retail stores	0.96	0.33	1.25	2.35	0.64	1.17	−0.47	1.59	0.80	−0.38	1.44
Entertainment venues	1.10	2.10	1.94	5.94	2.93	0.42	−0.33	5.43	1.68	−0.31	1.29
Business places	6.57	6.23	2.13	5.36	4.57	0.43	−0.36	6.23	4.57	−0.31	3.46
Government sectors	8.39	2.85	2.22	3.35	23.17	−0.06	−0.19	1.00	0.39	−0.17	3.98
Enterprise	24.44	11.67	1.78	2.26	4.00	0.05	−0.19	2.29	0.16	−0.14	4.00
Social organizations	−0.52	0.17	0.10	−0.26	−0.71	1.28	−0.39	−0.61	−0.47	−0.30	−0.25

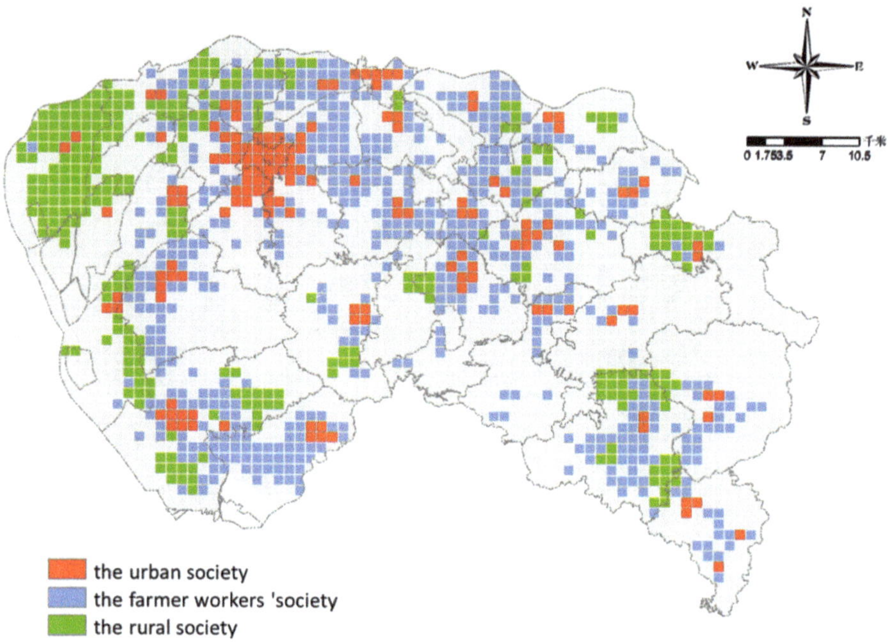

Fig. 10.10 The social space structure of Dongguan

Table 10.4 The features of social space structure of Dongguan

	The rural society	The farmer workers' society	The urban society
Distribution	Edge of the city	The villages and towns	The center of city and town
Subjects	Farmers	Farmer workers	Urban residents
Housing	General (self-build housing)	Bad (rental housing)	Good (purchased housing)
Family size	Big	Small	Big
Income	The lowest	Low	High
Employment	Industry + agriculture	Industry	Service
Educational levels	Low	Low	High
The profit-oriented infrastructure	Bad	General	Good
The public facility	General	Bad	Good
Organization type	Enterprise + government sectors	Enterprise	Enterprise + government sectors + social organizations

10.7 Conclusion and Discussion

In conclusion, this is the first research attempt to re-understand the social spaces in Chinese cities by the methods of social atlas. As a result, eleven types of social areas are found in Dongguan. Furthermore, these could be placed into three kinds of social spaces: rural communities, rural migrant worker communities, and urban communities.

Compared with the conventional methods, the social atlas is an improved research method for social study, as it efficiently helps to understand a very complex social space and reconstruct the research framework of social space in China. Social atlas, as a method, should be given more attention, and continuing research is necessary.

Foundation Item: Under the auspices of National Natural Science Foundation of China (No. 41571130, 41271165), the Fundamental Research Funds for the Central Universities, Collaborative Innovation Center for Geopolitical setting of Southwest China and Borderland Development (YNNU).

References

Booth, C. (1903). *Life and labour of the people in London*. New York: Macmillan Publishing Company.

Benson, T. D. (2002). *Malawi: An atlas of social statistics*. Washington, DC: International Food Policy Research Institute.

Brewer, C. A., & Suchan, T. A. (2001). *Mapping census 2000: The geography of US diversity*. Redlands: Esri Press.

Davis, J. R., & Spearritt, P. (1974). *Sydney at the census, 1971: A social atlas*. Canberra: Australian National University.

Elliott, R. (2003). *2001 census of population and housing: Darwin and Palmerston: A social atlas*. Canberra: Australian Bureau of Statistics.

Feng, J., & Zhou, Y. X. (2003). The social spatial structure of Beijing metropolitan area and its evolution: 1982–2000. *Geographical Research, 22*(4), 465–483. (in Chinese).

Gu, C. L. (2003). Study on urban social areas in Beijing. *Acta Geographica Sinica, 58*(6), 917–926. (in Chinese).

Glover, J., Harris, K., & Tennant, S. (1999). *A social health atlas of Australia* (2nd ed.). Adelaide: The University of Adelaide.

Glover, J., Hetzel, D., Glover, L., et al. (2006). *A social health atlas of South Australia* (3rd ed.). Adelaide: The University of Adelaide.

Horvath, R. J., Harrison, G. E., & Dowling, R. M. (1989). *Sydney: A social atlas*. Sydney: Sydney University Press.

Kurasawa, S., & Asakawa, T. (2004). *New social atlas of metropolitan* Tokyo, 1975–90[M]. Tokyo: University of Tokyo Press. (in Japanese).

Lazzaro, V. (2003). *Melbourne- a social atlas: 2001 census of population and housing*. Canberra: Australian Bureau of Statistics.

Li, Z. G., & Wu, F. L. (2006). Sociospatial differentiation in transitional Shanghai. *Acta Geographica Sinica, 61*(2), 199–211. (in Chinese).

McGrath, M. (2008). *Adelaide-a social atlas: 2006 census of population and housing*. Canberra: Australian Bureau of Statistics.

Morrill, R. L. (1990). Regional demographic structure of the United States. *The Professional Geographer, 42*(1), 38–53.
Nagle, C. (2002). *Perth: A social atlas: 2001 census of population and housing*. Canberra: Australian Bureau of Statistics.
Oswalt, P., & Rieniets, T. (2006). *Atlas of shrinking cities*. Ostfildern: Hatje CantzVerlag.
Poulsen, M., & Spearritt, P. (1981). *Sydney: A social and political atlas*. Sydney: George Allen & Unwin Pty.
Seager, J., & Olson, A. (1986). *Women in the world: An international atlas*. New York: Simon and Schuster.
Seager, J. (1997). *The state of women in the world atlas* (2nd ed.). London: Penguin.
Seager, J. (2003). *The atlas of women*. London: Womens Press.
Shevky, E., & Bell, W. (1955). *Social area analysis: Theory, illustrative application, and computational procedures*. Redwood City: Stanford University Press.
Smith, G. (2008). *Sydney-a social atlas: 2006 census of population and housing*. Canberra: Australian Bureau of Statistics.
Shinagawa, L. H., & Jang, M. (1998). *Atlas of American diversity*. Walnut Creek: Altamira Press.
Tsvetkov, Y. (2013). *Atlas der vorurteile*. München Bayern: Knesebeck.
Tennant, S., Hetzel, D., & Glover, J. (2003). *A social health atlas of young south Australia* (2nd ed.). Adelaide: Public Health Information Development Unit.
Xu, X. Q., & Hu, H. Y. (1989). A factorial ecological of social spatial structure in Guangzhou[J]. *Acta Geographica Sinica, 44*(4), 385–399. (in Chinese).
Xu, D., Wang, Z., Zhu, X. G., et al. (2009). The urban social areas in Nanjing: Analysis of the data based on the Fifth National Population Census. *Geographical Research, 28*(2), 484–498. (in Chinese).
Yu, W. (1986). A study and planning of urban social spatial structure. *City Planning Review, 06*, 25–28. (in Chinese).
Zheng, J., Xu, X. Q., & Chen, H. G. (1995). An ecological reanalysis on the factors related to social spatial structure in Guangzhou City. *Geographical Research, 14*(2), 15–26. (in Chinese).
Zhou, C. S., Liu, Y., & Zhu, H. (2006). Analysis on social areas of Guangzhou City during the economic system transformation. *Acta Geographica Sinica, 61*(10), 1046–1056. (in Chinese).

Chapter 11
The Spatial and Temporal Evolution of Innovative Function of Science and Technology of Beijing Based on the Analysis of Enterprise Data

Juan Li, Miaoyi Li, Anrong Dang, and Zhongwei Song

Abstract With the orientation of the capital's four functions defined as political center, cultural center, international communication center, and innovative center of science and technology, the development of innovative ability of science and technology in Beijing has been upgraded to a national strategy. This chapter analyzes the spatial evolution of Beijing's high-tech industry based on enterprise data and further explains how government policies influence the spatial evolution. Results show that spatial distribution of high-tech industries in Beijing spread from city center to the periphery gradually under the guidance of government policies, which demonstrates that top-down policies play a key role in the innovation-driven stage of city development. According to the concept mode of spatial evolution of Beijing's high-tech industry, future strategy should focus on fostering multiple leading cores to guide the cooperation among all districts, further integrating them into a well-collaborative network, in order to improve the comprehensive ability in the innovation of science and technology.

Keywords Innovative function of science and technology • High-tech industry • Enterprise data • Beijing

J. Li • A. Dang
School of Architecture, Tsinghua University, Beijing, China

M. Li
Joint International FZUKU Lab SPSD, Fuzhou University, Fuzhou City, China

Z. Song (✉)
Quantum Data (Beijing) Technology Co. Ltd, Beijing, China
e-mail: zhongwei@liangzisj.com

Z. Shen, M. Li (eds.), *Big Data Support of Urban Planning and Management*, Advances in Geographic Information Science, DOI 10.1007/978-3-319-51929-6_11

11.1 Introduction

As the process of globalization continues, innovative ability is becoming one of the main determinants of a country's international competiveness. Many countries have made the development of innovative ability in science and technology a national strategy. In October 2015, America released the new "Strategy for American Innovation" in substitution for a previous version in 2009, recognizing the need for the federal government to "invest in the building blocks of innovation, to fuel the engine of private sector innovation, and to empower a nation of innovators."[1] In 2010, the European Commission announced "Europe 2020: A Strategy for Smart, Sustainable and Inclusive Growth," and one of the three mutually reinforcing priorities put forward by the EU was to achieve smart growth by "developing an economy based on knowledge and innovation" (European Commission 2010). In June 2014, Japan unveiled the "Japan Revitalization Strategy" which was a revision of its former plan, "The New Growth Strategy: Blueprint for Revitalizing Japan."[2] In the new plan, "promoting innovation in science and technology and becoming the world's leading intellectual property-based nation" is the one of the main strategies to achieve industry revitalization for Japan. In addition, Russia and India et al. also make great efforts to develop innovation-driven economies (Yuan 2014). With this inevitable developing trend, China also attaches great importance to innovation in science and technology. China's 13th Five-Year Plan for National Economic and Social Development released in 2016 has made it clear that implementing an innovation-driven strategy is of great priority in the next stage. All the major initiatives taken by different countries remind us to pay attention to the improvement of innovative ability.

There is no doubt that innovative ability is critical to economic vitality (Zhang 2012; Chen 2005). After a few decades of rapid growth, the Chinese economy has entered into a transitional period with "new normal" as the main characteristic, as proposed by president Xi.[3] Demographic benefit from the large population which has made a great contribution to economic development during the past 30 years has started to fade out; at the same time, negative impact from rising aging rates is being felt. To motivate economic growth under these adverse conditions, upgrading the industrial structure is among the most urgent steps required. Innovation in science and technology has been proven to play a key role in promoting the upgrade (Wang 2007; Li 2015). Li (2015) studied the influence of technological innovation on employment size and structure of Beijing and concluded that technological innovation would increase employment indirectly and promote the upgrading of employment structure, especially increasing the ratio of employees in tertiary industry and ratio of employees with higher education.

---

[1] https://www.whitehouse.gov/the-press-office/2015/10/21/fact-sheet-white-house-releases-new-strategy-american-innovation

[2] http://www.kantei.go.jp/jp/singi/keizaisaisei/pdf/honbunEN.pdf

[3] http://news.xinhuanet.com/world/2014-11/09/c_1113175964.htm

In addition to enhancing the upgrading of industrial structure, the improvement of innovative ability also promotes the comprehensive competiveness of an individual city (Zhang 2011; Li and Lu 2002). Sassen (2001) claims that global cities "now function not only as command points in the organization of the world economy, but also as key locations and marketplaces for leading industries" and "as sites for the production of innovations in those industries." Therefore, a city's global influence in the world system is determined to a great extent by its innovative ability in science and technology.

According to Porter (1998), economic development can be divided into four phases: resources-driven phase, investment-driven phase, innovation-driven phase, and wealth-driven phase. During the innovation-driven phase, the main driving force comes from domestic research institutions and enterprises. A lot of studies also emphasize the key effect of enterprises on technological innovation (Li 2015; Zhang 2012; Wang 2007; Zhao et al. 2015). Yuan (2014) concluded that there are three major effects that enterprises have on technological innovation. First, the majority of technological innovation comes from enterprises. To maintain strong competitiveness in markets, enterprises will take the initiative to promote innovation in every process including the development of new products, the improvement of production procedure, and the adjustment of organization structure. Second, enterprises are the pivotal bridge between technology supply and market demand. Enterprises put technological achievements of universities and research institutes into industrial production, while also giving feedback on new market demands discovered in their communication with users. And lastly, the emerging and growth of new enterprises are an important support resource for more innovation.

A large number of existing studies focus on evaluation systems of innovative ability (Sun 2008; Zhang 2012; Zhao et al. 2015). Zhao et al. (2015) developed an evaluation system including three first-level indicators—competitiveness, leadership, and influence—of scientific and technological innovation, which can be subdivided into 9 second-level indicators and a further 64 third-level indicators. Similarly, Sun (2008) developed a three-level evaluation system with 4 first-level indicators, 10 second-level indicators, and 23 third-level indicators. Both of the above evaluation systems adopt similar indicators of which the fundamental contents are the same, such as ratio of R&D employees, per capita GDP, Internet penetration rate, and so on. Zhang (2012) also used most of the same indicators, but, instead of getting conclusions directly from data, she calculated indexes through different combinations of data. Since, as earlier mentioned, enterprises are the main resources for innovation, it is important to assess the innovative ability so as to provide reasonable suggestions for enterprises to get improvement. However, very little has been done to explore the spatial evolution of enterprises, which is believed to be of great importance. In this chapter, we attempt to fill this gap based on enterprise data.

In 2014, President Xi proposed to enhance Beijing's main functions as capital, which focus on its being four things—political center, international communication center, cultural center, and innovative center of science and technology. After that, a lot of policies not only from the local government but also from the central government have been put forward to encourage the development of scientific and technological innovation industry in Beijing to improve innovative ability.

Therefore, we analyze the spatial evolution of Beijing's high-tech enterprises with an aim to explore how these government policies impact the distribution of high-tech industry.

Following the introduction in Sect. 11.1, Sect. 11.2 explains the data and study area we use. Section 11.3 makes a general analysis of Beijing's high-tech industry to find out its main characteristics. Section 11.4 analyzes the spatial evolution of Beijing's high-tech industry. Then, Sect. 11.5 combines what we have concluded in Sects. 11.3 and 11.4 to analyze the impact that policies have on the spatial distribution of the high-tech industry. And, finally, we present a summary in Sect. 11.6.

11.2 Data and Study Area

According to the classification criteria of the National Bureau of Statistics of the People's Republic of China,[4] the scientific and technological innovation industry can be represented as a high-tech industry which includes the high-tech manufacturing industry and the high-tech service industry, respectively. The high-tech manufacturing industry, characterized by relatively higher investment in R&D, contains six subcategories of industries, whereas the high-tech service industry contains eight main subcategories of industries which provide service for the society by means of advanced technology. Table 11.1 lists all subcategories of the high-tech

Table 11.1 List of high-tech industry categories and subcategories

High-tech industry	Subcategories
Manufacturing industry	Pharmaceutical manufacturing
	Aviation, spacecraft, and related equipment manufacturing
	Electronic and telecommunication equipment manufacturing
	Computer and office equipment manufacturing
	Medical equipment and instrument manufacturing
	Photographic equipment manufacturing
Service industry	Information service
	E-commerce service
	Inspection and testing service
	High-tech service of professional and technical service
	R&D and design service
	Transformation of scientific and technological achievements service
	Intellectual property and related legal service
	Environmental monitoring and governance service

[4] http://www.stats.gov.cn/tjsj/tjbz/

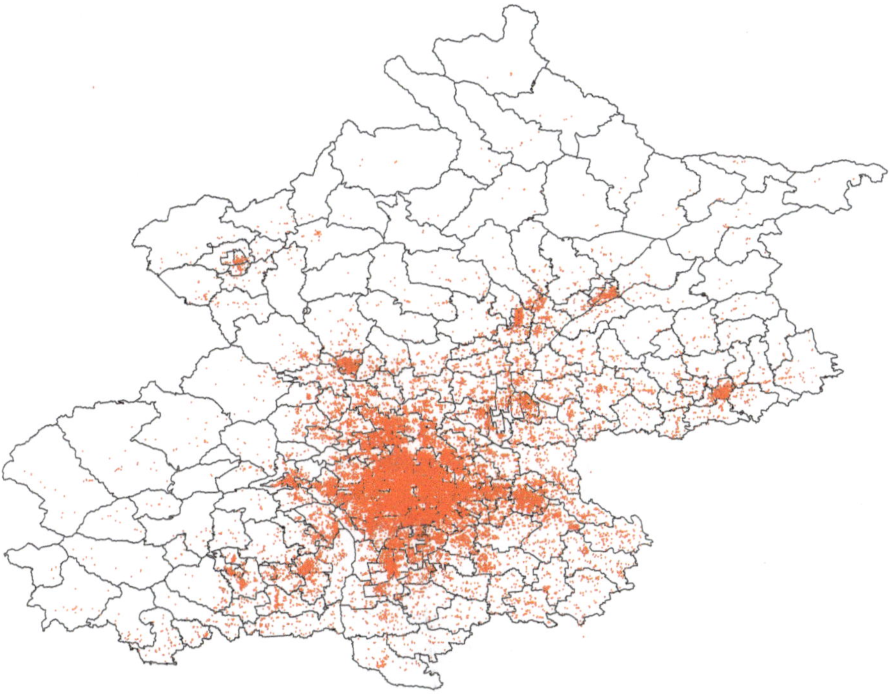

Fig. 11.1 Enterprises of the high-tech industry in 2015

industry. Based on the registered information from State Administration for Industry & Commerce of the People's Republic of China, all enterprises belonging to the 14 categories of high-tech industry are gathered and geocoded. Figure 11.1 shows the spatial distribution of all high-tech enterprises in Beijing in 2015. Attributes of high-tech enterprise data include industry category, date of establishment, date of revocation, registered capital, address of enterprise, investment received, and count of patents and software copyrights.

In this study, we use the administrated area (*Shixiaqu*) of Beijing including 16 districts as study area and 318 subdistricts (*Jiedao*) as basic analysis units. To figure out how the high-tech industry evolves, we analyze the enterprise data in Beijing from 2000 to 2015 and draw conclusions based on the statistical results at 5-year intervals.

11.3 The Profile of the High-Tech Industry

11.3.1 The High-Tech Industry

As the capital, the innovative ability of science and technology of Beijing is of great importance to the whole country. Significant efforts have been made to improve Beijing's innovative ability, through which great benefit to both economic and urban development has been gained. After the explicit claim that being the innovative center of science and technology is one of the four functions of Beijing, more new government policies focusing on the advancement of its high-tech industry have been put forward, and the performance of the high-tech industry is promising.

Figure 11.2 shows the total number of high-tech enterprises and the number of emerging enterprises in each period. The total count of high-tech enterprises keeps increasing at an accelerated growth rate. As of 2015, the high-tech enterprises account for 35% of all enterprises in Beijing. In 2014, the State Council of PRC issued the Registered Capital of the Registration System Reform Plan[5] and reduced the requirements for registered capital of enterprises. Except for some specific industries, the former regulation that the minimum registered capital of limited liability company was 30,000 RMB, of one-person limited liability company was 100,000 RMB, and of company limited was 5 million RMB were canceled. Limitations about the initial investment ratio of shareholder and time limit for full payment of investment when setting up a new company have also been abolished. Lower requirements provide a lower entry threshold for entrepreneurs, which motivates the establishment of more new enterprises. The new reform plan especially encourages entrepreneurs in the high-tech industry. Among all emerging enterprises in 2015, high-tech enterprises account for up to 40%. Evidence from Fig. 11.2 also shows that the number of emerging high-tech enterprises in 2015 is more than twice what it was in 2010.

At the same time, the enthusiasm of small and medium high-tech enterprises has been greatly aroused as a result of the reform plan. Table 11.2 lists the registered capital of surviving high-tech enterprises[6] in 2015 in Beijing which indicates that

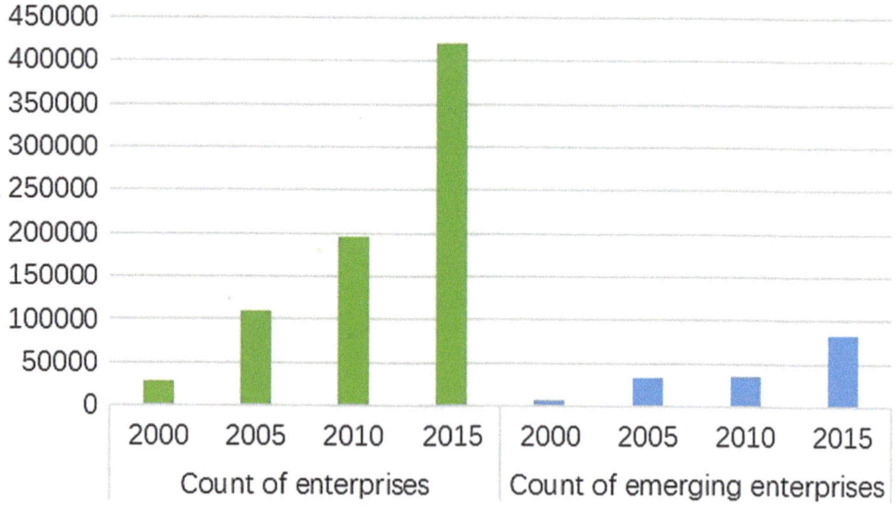

Fig. 11.2 Total number of high-tech enterprises and emerging enterprises

[5] http://legal.china.com.cn/2014-02/18/content_31513129.htm

[6] Surviving enterprises refer to enterprises that exist at the end of the year.

Table 11.2 Registered capital of surviving high-tech enterprises in 2015 in Beijing

Registered capital (10000 RMB)	Count of high-tech enterprises	Proportion
0–50	104,944	0.250
50–200	170,672	0.406
200–500	45,033	0.107
500–1000	42,568	0.101
1000–2000	33,996	0.081
2000–5000	11,625	0.028
Above 5000	11,126	0.026

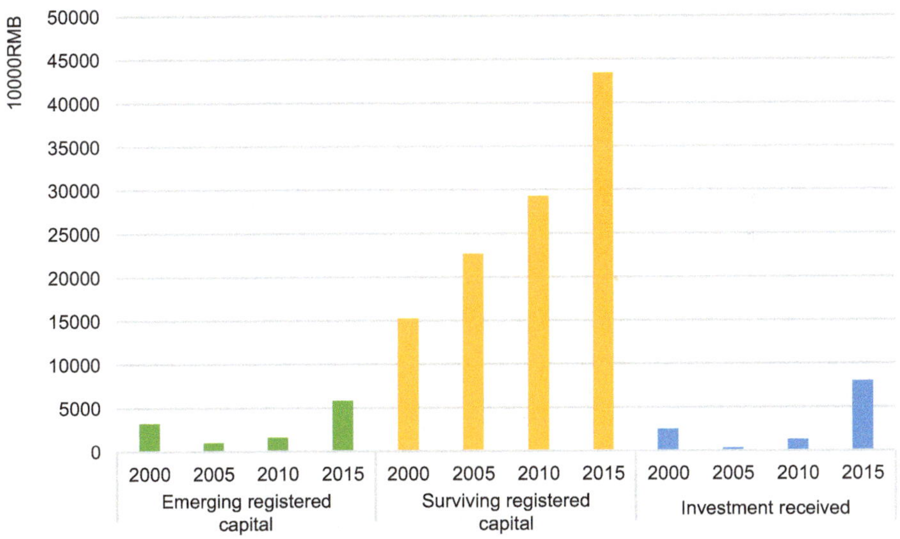

Fig. 11.3 Emerging registered capital, surviving registered capital, and investment received of high-tech enterprises

the registered capital of over 65% of the high-tech enterprises is less than 2 million RMB. Small and medium high-tech enterprises have become the main components of high-tech industry.

Even though the individual registered capital of a lot of small and medium high-tech enterprises was small, the total amount of emerging registered capital was considerable in 2015 owing to the large number of emerging high-tech enterprises, which reversed the downturn trend in 2005 and 2010 as Fig. 11.3 shows. The emerging registered capital in high-tech industry in 2015 is 3.6 times the amount in 2010 and accounts for 21% of all industries in Beijing. Figure 11.3 also shows the surviving registered capital[7] of high-tech enterprises and the investment received form the whole country. In 2015, surviving registered capital of high-tech enterprises accounts for 19.3% of that of all enterprises in Beijing. Due to the large

[7] Surviving registered capital refer to the total capital that surviving enterprises have at the end of the year.

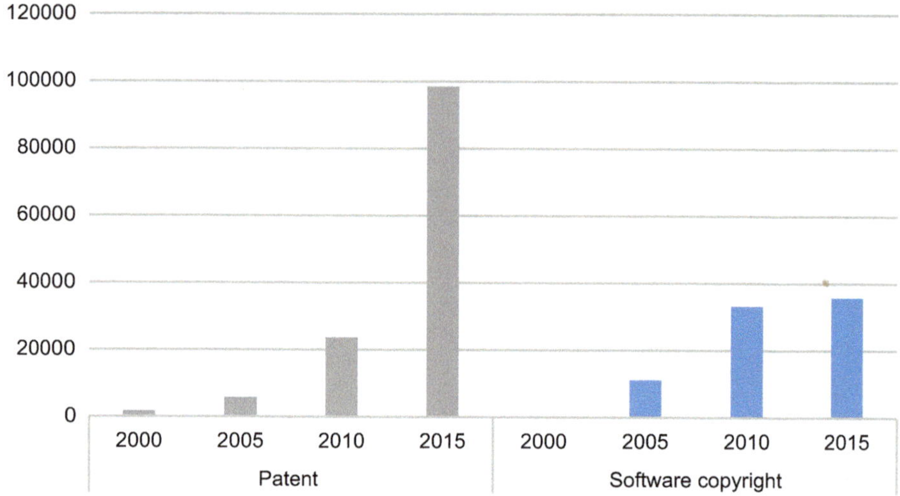

Fig. 11.4 Amount of patents authorized and software copyrights applied of high-tech enterprises

number of small and medium enterprises, the potential capacity of high-tech industry for much more investment is improved significantly, and the investment received in 2015 is six times what was received in 2010, and accounts for 25% of all investment received in Beijing. The above data also indicates that the high-tech industry has become one of the main contributors to the economic development of Beijing.

Figure 11.4 shows the amount of patents authorized and software copyrights applied, which are the direct representation of innovative ability in science and technology. In 2015, amount of patents authorized of high-tech enterprises accounts for 60% of all patents authorized in Beijing, and software copyrights applied of high-tech enterprises accounts for 87.6% of all applied in Beijing. The innovation from high-tech enterprises have become the main resource of Beijing's innovative ability. In addition, the number of patents authorized keeps increasing rapidly, indicating that the innovative ability of Beijing is improving as expected.

The most authoritative assessment of the innovative ability of Beijing comes from the Capital Innovation Index of Science and Technology. In February 2016, the "Capital Innovation Index of Science and Technology 2015" was released jointly by the Capital Institute of Science and Technology Development Strategy (CISTDS) and Beijing Municipal Commission of Science and Technology. The index system consists of three-layer indicators. The first layer contains four indicators which are innovative resource, innovative environment, innovative service, and innovative performance. The second layer subdivides the four indicators in the first layer into 15 indicators, and the third layer further subdivides them into 64 indicators. The evaluation results show that the Capital Innovation Index of Science and Technology implies a significant growing trend and the index score rises from 79.77 in 2010 to 88.72 in 2014,[8] from which we can conclude that the innovative ability is being

[8] http://www.most.gov.cn/dfkj/bj/tpxw/201603/t20160304_124437.htm

enhanced continuously. Evidence from the authoritative assessment confirms what we have found from the enterprise data.

11.3.2 High-Tech Manufacturing Industry and High-Tech Service Industry

Following the general analysis of all high-tech enterprises, we take a closer look at the two categories of high-tech industries—the high-tech manufacturing industry and high-tech service industry. Figure 11.5 shows the number of high-tech manufacturing enterprises and high-tech service enterprises. As is shown, the high-tech manufacturing enterprises account for a very small percentage of the high-tech industry and increase only a little in number for 10 years. In contrast, the high-tech service enterprises increase rapidly, and the growth rate is also rising.

Figure 11.6 indicates that the number of emerging enterprises in the high-tech manufacturing industry is gradually decreasing while the developing trend of the high-tech service industry shows the complete opposite.

To establish itself as a more international city and have more influence in the world, Beijing has committed to upgrading its industrial structure by transferring the secondary industry including manufacturing out of the city and promoting the development of the tertiary industry in the city. Therefore, for the high-tech industry, it is reasonable to invest significant effort into developing the high-tech service enterprises, as it turns out the high-tech service industry has almost entirely represented the high-tech industry up till now. Correspondingly, the high-tech service industry has made major contributions to the improvement of innovative ability of science and technology, as Fig. 11.7 shows.

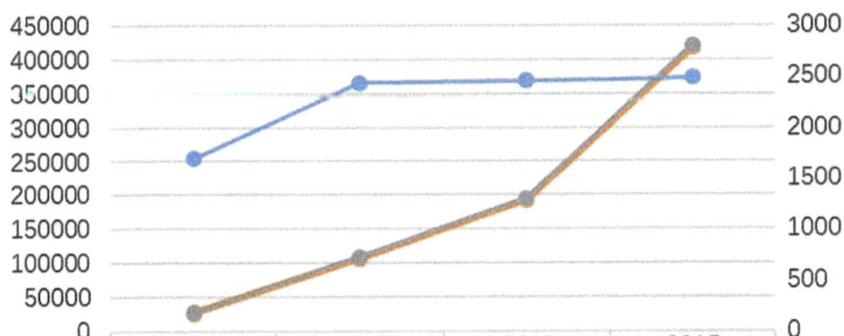

	2000	2005	2010	2015
High-tech service industry	26191	106836	192949	417479
All high-tech industry	27889	109277	195407	419964
High-tech manufacturering industry	1698	2441	2458	2485

Fig. 11.5 Number of high-tech manufacturing enterprises and high-tech service enterprises

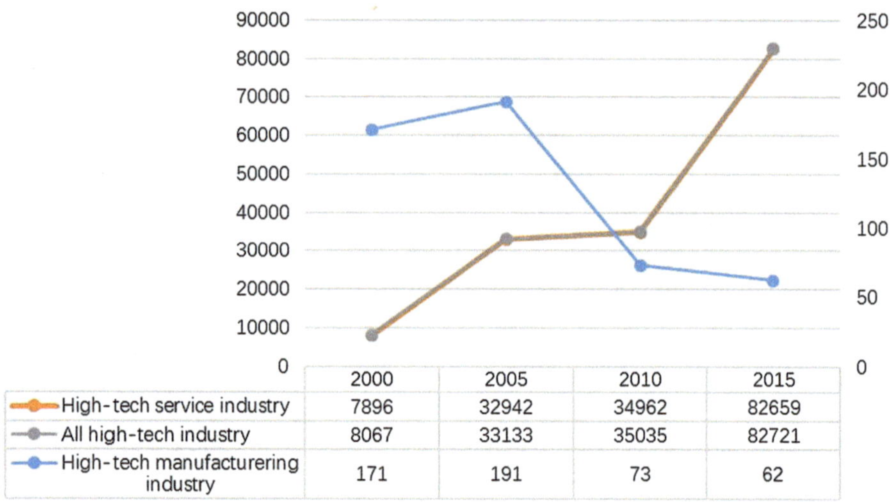

Fig. 11.6 Number of emerging enterprises in the high-tech manufacturing industry and the high-tech service industry

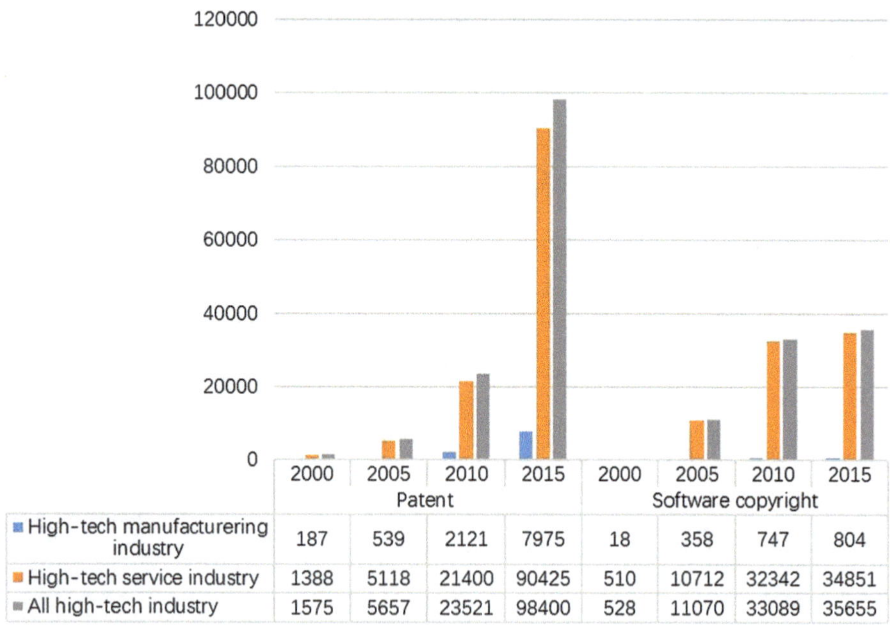

Fig. 11.7 Amount of patents authorized and software copyrights applied of high-tech manufacturing enterprises and high-tech service enterprises

11.3.3 Subcategories of the High-Tech Industry

As previously stated, the high-tech industry is subdivided into 14 subcategories. Figure 11.8 shows the number of enterprises within each subcategory. The number of enterprises of all eight subcategories of the high-tech service industry grew quickly from 2000 to 2015, and the majority came from transformation of scientific and technological achievements service and information service, followed by high-tech service of professional and technical service, and R&D and design service. Those four subcategories of the high-tech service industry account for 98% of all high-tech enterprises in Beijing and definitely consist of the leading scientific and technological innovation industries of Beijing. The number of enterprises of intellectual property and related legal service is much smaller than the top four subcategories, but the growth rate is remarkable. On the other hand, the number of enterprises in the six subcategories of the high-tech manufacturing industry didn't increase significantly. The number of enterprises from electronic and telecommunication equipment manufacturing industry even started to decrease from 2005. The only subcategory of the high-tech manufacturing industry that keeps growing is medical equipment and instrument manufacturing, which together with electronic and telecommunication equipment manufacturing accounts for the majority of the high-tech manufacturing industry.

Figure 11.9 shows the emerging enterprises of each subcategory, which provides more details about how they change over time. The four leading subcatego-

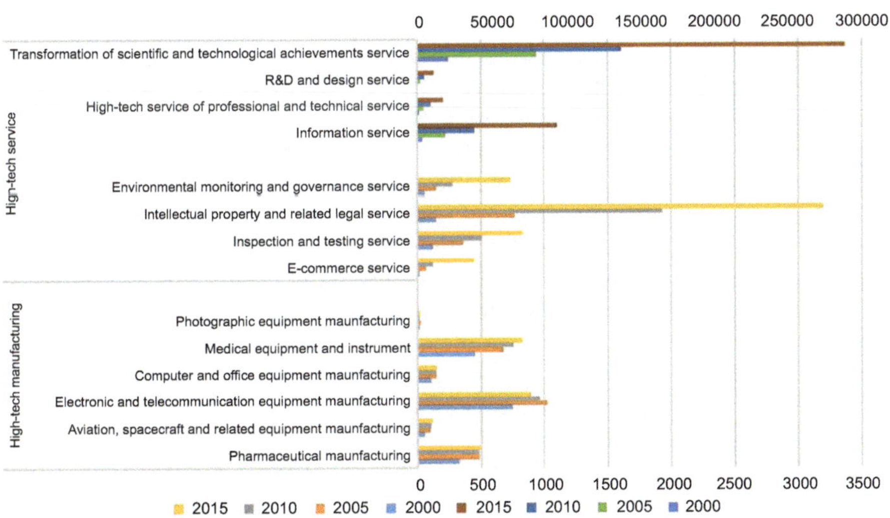

Fig. 11.8 Amount of enterprises of 14 subcategories of the high-tech industry

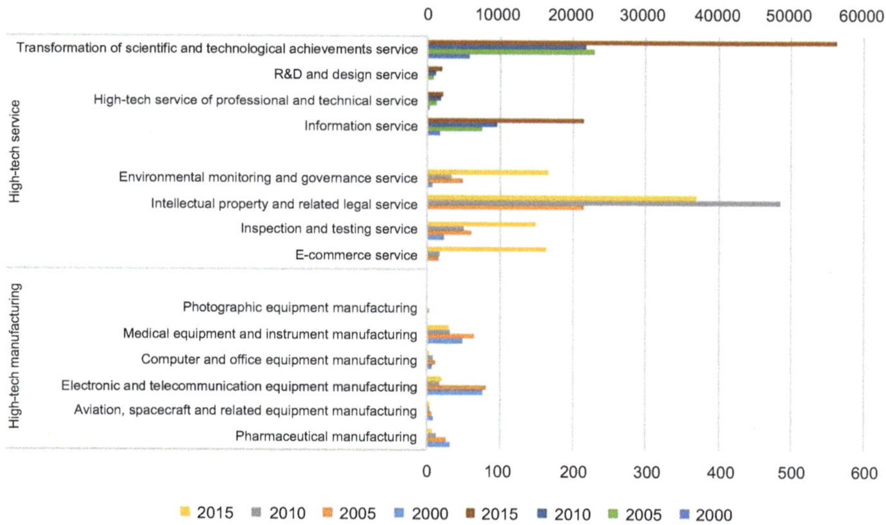

Fig. 11.9 Number of emerging enterprises of 14 subcategories of the high-tech industry

ries—transformation of scientific and technological achievements service, information service, high-tech service of professional and technical service, and R&D and design service—have without a doubt gained a large number of new enterprises. The number of enterprises of intellectual property and related legal service also indicates significant growth during the last 15 years. The focus of the intellectual property and related legal service is to guarantee the legality of procedures for acquiring, utilizing, and maintaining all kinds of intellectual property, including patents, brands, copyrights, business secrets, and so on, which are some of the key achievements of innovative activities.[9] A well-developed intellectual property service is a necessity for high-tech industries to carry out innovative activities securely and successfully. Therefore, as the prosperity of the high-tech industry continues, the demand for intellectual property service grows greater and greater.

The other three subcategories—environmental monitoring and governance service, inspection and testing service, and e-commerce service—which experienced sudden growth in 2015 are also worthy of attention. In recent years, the haze problem in China has gained worldwide attention, and environmental pollution has become a huge obstacle on the way forward for Beijing. The desire of the central government to develop the economy efficiently, and of local residents to live a healthy life, urges an effective action to control the worsening air pollution. The number of enterprises in environmental monitoring and governance service thus

[9] http://www.sipo.gov.cn/ztzl/ywzt/hyzscqgz/fwjg/zcxx/201302/t20130217_785582.html

increased rapidly in a short time. However, the long-standing tendency to give absolute priority to economic development to the extent of sacrificing the natural environment has posed much more challenges than we can imagine. On one hand, for the enterprises, awareness of environmental protection is inadequate, and their knowledge of how to protect control pollution is severely insufficient. On the other hand, for the government, a well-founded policy system of monitoring the environment and workable measures to guide enterprises are difficult to establish. The performance of environmental governance gains little success as a result. Hence, it is believed that there is much room for the environmental monitoring and governance service to improve.

Inspection and testing service industry is the quality foundation serving economic and social development and plays an important role in enhancing quality, promoting industrial development, protecting consumers' rights, protecting the environment, and improving public security. As the issues of quality and safety draw more and more attention, demand for inspection and testing service grows fast, which in turn fosters more and more enterprises providing these services.

The increase in the number of enterprises delivering the e-commerce service is also promoted by social demand. The rise of e-commercial platforms such as Taobao, Tmall, and JD contributes more and more to economic development. Taking Tmall as an example, on 11 November 2016, its transaction volume reached up to 120.7 billion RMB in just 1 day.[10] People are already used to the new consumption pattern which has penetrated into every aspect of daily life. With this kind of business model accounting for higher and higher proportions in economic activities, the demand for e-commerce service increases correspondingly.

At the same time, we can detect a downward trend in the emerging enterprises of all subcategories of the high-tech manufacturing industry compared to the significant upward trend in the high-tech service industry.

Figure 11.10 shows the amount of patents authorized of each subcategory. The four leading subcategories still contribute the most, especially the transformation of scientific and technological achievements service. Among all high-tech manufacturing industries, the amount of patents of electronic and telecommunication equipment manufacturing increases fastest, even though the number of its enterprises is decreasing. In the Information Age, electronic and telecommunication equipment as infrastructure including optical fiber and optical cable, communication devices, audio-visual equipment, etc. is one of the prerequisites to guarantee that the innovative activities can be carried out successfully. Meanwhile, with great advances in technology, electronic and telecommunication equipment are gradually becoming a necessity of daily life. The huge demand derived from various aspects ranging from industrial development to everyday life has motivated the vitality of innovation significantly.

[10] http://news.xinhuanet.com/info/2016-11/14/c_135827080.htm

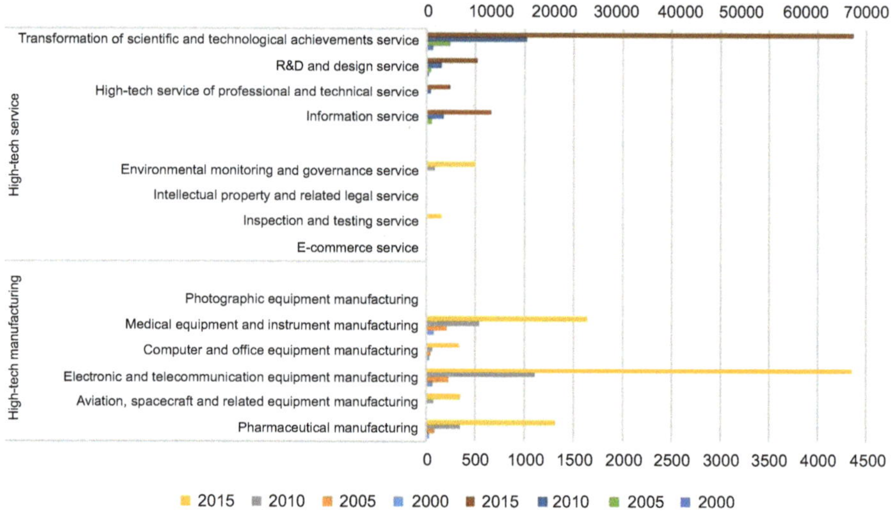

Fig. 11.10 Amount of patents authorized of 14 subcategories of the high-tech industry

11.4 Spatial Distribution of the High-Tech Industry

On the basis of spatial attributes of all enterprise data, we can generate the spatial distribution of high-tech enterprises and try to figure out how it changes over time. Figure 11.11 shows the spatial distribution of high-tech enterprises in Beijing in 2000, 2005, 2010, and 2015, respectively. In 2000, most high-tech enterprises concentrated in some subdistricts (*Jiedao*) of Haidian, Chaoyang, and Fengtai districts, but the most concentrated subdistrict was Zhongguancun which belongs to Haidian District. In 2005, more high-tech enterprises started to gather in the other subdistricts of Haidian, Chaoyang, and Fengtai districts and further spread into the periphery of the three districts. New areas with significant growth are Daxing and Tongzhou districts. The most concentrated subdistrict in 2005 was still Zhongguancun, however. In 2010, spatial distribution of high-tech enterprises continued to spread to peripheral districts. More enterprises were located in Changping, Shunyi, south of Huairao, and east of Fangshan districts. Apart from Zhongguancun, Shangdi and Beixiaguan subdistricts also became the most concentrated areas, and all three of them belong to Haidian District. By 2015, in addition to a few subdistricts of Yanqing, Fangshan, Huairou, and Miyun districts, all other subdistricts indicated great increase in the number of high-tech enterprises. The most concentrated areas in 2015 were Shangdi, Zhongguancun, and Xincun subdistricts, and Xincun belongs to Fengtai District. Figure 11.12 shows the number of high-tech enterprises in 16 districts. As is shown therein, the high-tech enterprises in Haidian, Chaoyang, and Fengtai districts account for the majority of those in Beijing, and the proportion is up to 60% of the total.

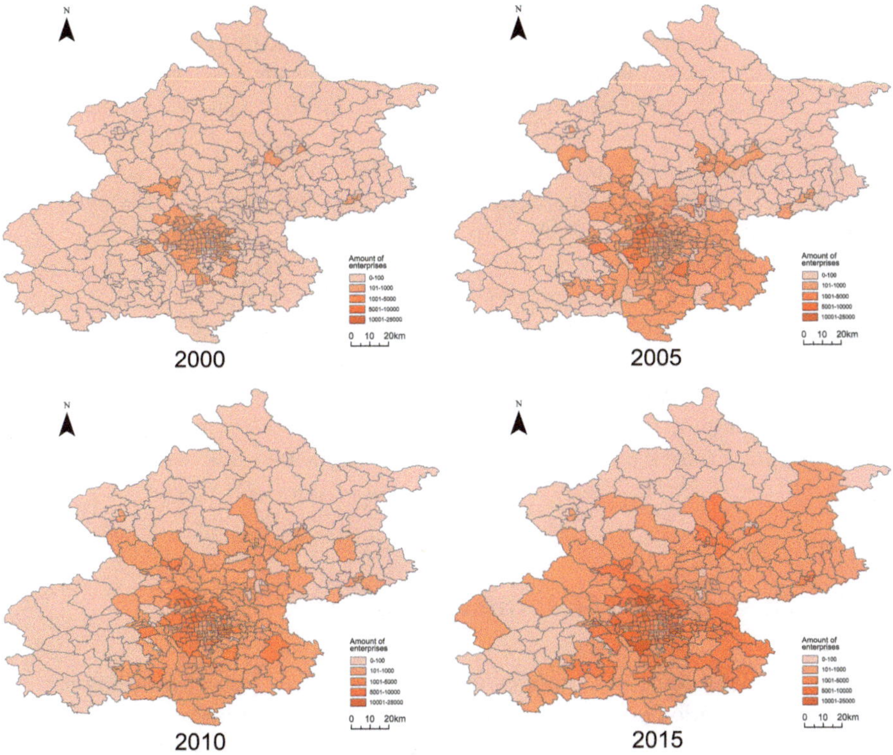

Fig. 11.11 Spatial distribution of the high-tech industry in Beijing

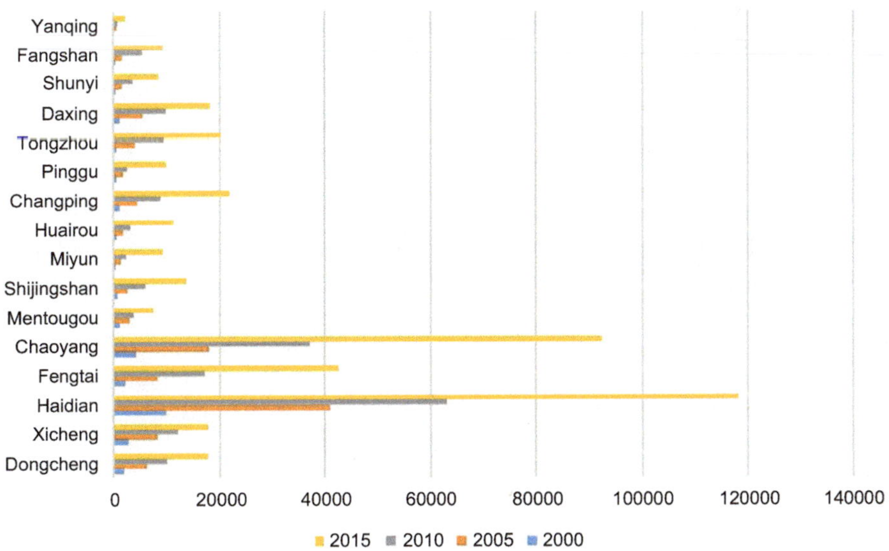

Fig. 11.12 Number of high-tech enterprises in 16 districts

Table 11.3 shows the most concentrated area for each subcategory of the high-tech industry. For high-tech manufacturing enterprises, the most concentrated areas are Tongzhou, Daxing, Fangshan, Haidian, and Chaoyang districts. For high-tech service enterprises, the most concentrated areas are Haidian, Chaoyang, and Fengtai districts.

Figure 11.13 shows the amount of patents authorized of high-tech enterprises in 16 districts. All districts had increased numbers of patents, but Haidian District was the most outstanding district both in the total amount and growth rate.

Figure 11.14 shows the spatial distribution of patents authorized, which indicates that patents authorized didn't spread in equilibrium like enterprises themselves did. Even though most subdistricts grew significantly in number of high-tech enterprises, the amount of patents authorized remained concentrated in a few subdistricts, of which most were located in Haidian District. Since Haidian District has the most universities and research institutions, it started to develop the high-tech industry in Beijing from the very beginning. Its long history and most advantaged resources have made Haidian District the innovation core of Beijing.

11.5 The Impact of Government Policies

When the driving force of economic development is transferring from investment driven into innovation driven, the direct intervention of government is crucial to the success of the transition (Yuan 2014). A series of policies made by government are needed to improve the innovative ability and to support entrepreneurship of high-tech enterprises in order to overcome the challenges in the transitional period, and this is believed to be one of the most dominant factors that determine the spatial distribution of high-tech enterprises. Therefore, we try to understand how the process was carried out in Beijing.

Generally, the history of the development of high-tech enterprises in Beijing happened in four stages,[11] and during each stage, certain policies have played an important role in driving the progress of the high-tech industry.

1. First Stage (1983–1988)

In the early 1980s, as the concept that "science and technology are the primary productive forces" became widely accepted, a great number of people who previously worked at universities or research institutes tried to participate in the competition of market economy by starting high-tech enterprises and to explore approaches of transforming the scientific and technological achievements into products. Benefitting from their proximity to a number of universities and research institutions, high-tech enterprises first concentrated at Zhongguancun subdistrict. By the end of the first stage, the so-called Electronics Street, of which the Zhongguancun

[11] http://www.zgc.gov.cn/sfqgk/55179.htm

Table 11.3 Spatial distribution of the most concentrated areas for the14 subcategories of the high-tech industry

High-tech industry	Subcategories	The most concentrated districts	Number of enterprises
High-tech manufacturing industry	Pharmaceutical manufacturing	Tongzhou	18
		Daxing	16
		Haidian	16
	Aviation, spacecraft, and related equipment manufacturing	Haidian	15
		Chaoyang	9
		Tongzhou	9
	Electronic and telecommunication equipment manufacturing	Haidian	61
		Chaoyang	38
	Computer and office equipment manufacturing	Haidian	14
		Daxing	13
	Medical equipment and instrument manufacturing	Haidian	27
		Chaoyang	22
		Daxing	20
	Photographic equipment manufacturing	Daxing	4
		Fangshan	2
		Tongzhou	2
High-tech service industry	Information service	Chaoyang	28,314
		Haidian	15,218
		Fengtai	9012
	E-commerce service	Chaoyang	114
		Haidian	97
	Inspection and testing service	Chaoyang	179
		Haidian	155
	High-tech service of professional and technical service	Chaoyang	2436
		Haidian	1736
	R&D and design service	Chaoyang	1374
		Haidian	902
		Fengtai	621
	Transformation of scientific and technological achievements service	Haidian	75,548
		Chaoyang	32,519
		Fengtai	23,888
	Intellectual property and related legal service	Haidian	882
		Chaoyang	587
		Xicheng	449
	Environmental monitoring and governance service	Haidian	134
		Fengtai	68
		Chaoyang	46

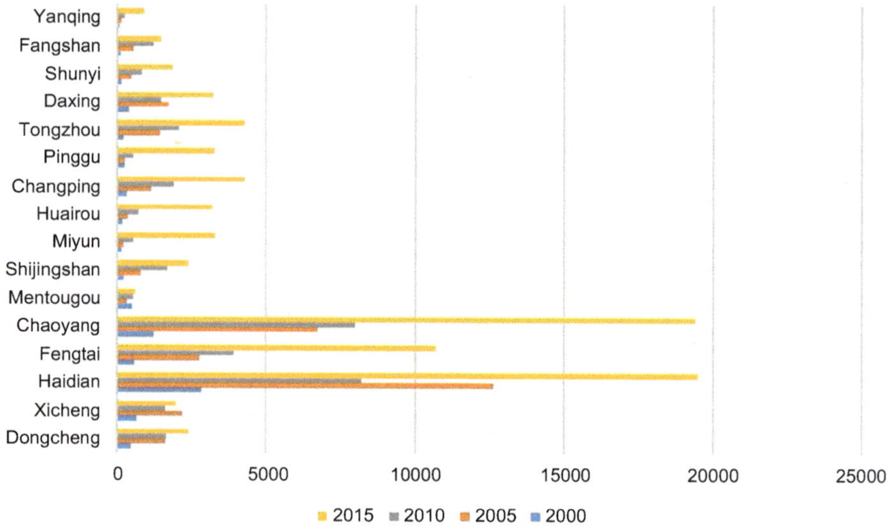

Fig. 11.13 Amount of patents authorized of high-tech enterprises in 16 districts

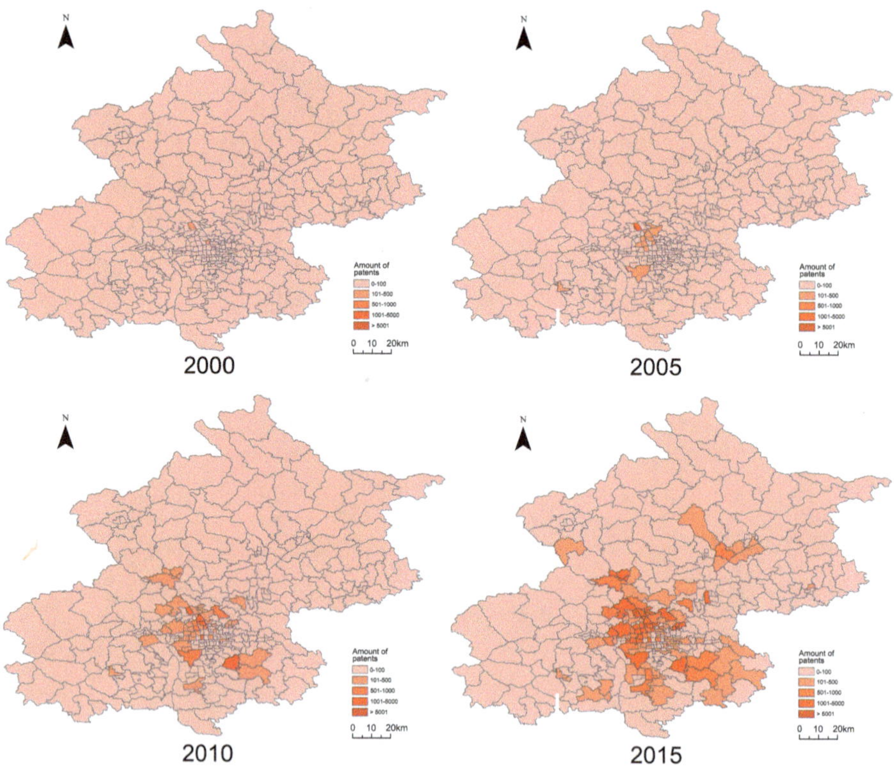

Fig. 11.14 Spatial distribution of patents authorized of the high-tech industry in Beijing

Street was the representative, had a reputation for being the most innovative area in Beijing.

2. Second Stage (1988–1999)

In 1988, the "Provisional Regulations on Beijing New Technology Industrial Development Trial Zone" was released by the State Council of PRC, and it claimed to delineate an area of about 100 km^2 in Haidian District with Zhongguancun sub-district as the core area, aiming to establish the first national-level high-tech industry park in China. Several kinds of preferential policies would be provided for high-tech enterprises registered in the trial zone. In 1994, the State Science and Technological Commission approved the decision which integrated Fengtai and Changping districts into the trial zone. Further, in 1999, the extent of the trial zone was adjusted again, and the "Electronics City" located in Chaoyang District and Yizhuang Park located in Daxing District was included. At last, the spatial pattern of "one district with five sub-parks" covering the whole trial zone came into being.

3. Third Stage (1999–2009)

In July 1999, advice on enhancement of the development of Zhongguancun Science and Technology Park put forward by Beijing municipality and the Ministry of Science and Technology of PRC was approved by State Council, as well as the developing plan of Zhongguancun Science and Technology Park. Two months later, the Beijing municipality officially changed the name of Beijing New Technology Industrial Development Trial Zone Committee to Zhongguancun Science and Technology Park Committee. By 2006, the spatial pattern of "one district with ten sub-parks" was formed including Haidian Park, Fengtai Park, Changping Park, Desheng Park, Yonghe Park, "Electronics City," Yizhuang Park, Tongzhou Park, Shijingshan Park, and Daxing Biomedicine Industrial Base. Figure 11.15 shows the spatial distribution of the "one district with ten sub-parks."

4. Fourth Stage (2009 to present)

In March 2009, the State Council approved the plan of constructing the Zhongguancun National Demonstration Zone, aiming to establish a scientific and technological innovation center which could exert international influence. In 2011, the "Development Plan Outline for Zhongguancun National Demonstration Zone (2011–2020)"[12] was launched by the State Council. It claimed to optimize the spatial layout of high-tech industries by promoting "one town" and "two belts." "One town" is Zhongguancun Science Town, which focused on building a well-organized platform for universities, institutions, and enterprises to promote collaborative innovation. One of the "two belts" is Northern Development Belt for R&D Services and High and New-tech Industries, focusing on building an industrial cluster to promote R&D and information service and facilitate the commercialization of high-tech achievements. The other is Southern Development Belt for High-tech Manufacturing and Emerging Strategic Industries and is focused on developing the key technology

[12] http://www.zgc.gov.cn/zgcsnghgy/gy/69400.htm

Fig. 11.15 Spatial distribution of "one district with ten sub-parks" (*Source*: http://www.zgc.gov.cn/sfqtp/sfqgh/70657.htm)

manufacturing by integrating Yizhuang and Daxing districts. Figure 11.16 shows the spatial structure of "one town" and "two belts." The Southern Development Belt is almost accomplished as expected in view of the analysis we have revealed before that the most concentrated areas for high-tech manufacturing enterprises are located in Tongzhou, Daxing, Fangshan, Haidian, and Chaoyang districts.

In October 2012, the State Council agreed to adjust the area and layout of Zhongguancun National Demonstration Zone by extending the area from "one district with ten sub-parks" to "one district with sixteen sub-parks". The adjusted zone included Dongcheng Park, Xicheng Park, Chaoyang Park, Haidian Park, Fengtai Park, Shijingshan Park, Mentougou Park, Fangshan Park, Tongzhou Park, Shunyi Park, Daxing-Yizhuang Park, Changping Park, Pinggu Park, Huairou Park, Miyun Park, and Yanqing Park. Each district of Beijing owns one sub-park, and

Fig. 11.16 Spatial structure of "one town" and "two belts" (*Source*: http://www.zgc.gov.cn/sfqtp/sfqgh/70656.htm)

Fig. 11.17 shows the spatial distribution of all sub-parks. Figure 11.18 shows the process of expansion of Zhongguancun National Demonstration Zone which is consistent with the process of spatial expansion of high-tech enterprises. Government policies indeed played a positive role in determining the spatial distribution of the high-tech industry.

Analysis of previous results reveal that high-tech enterprises have spread into the whole region of Beijing, but the sources of innovation are still concentrated in a few districts like Haidian, Chaoyang, and Fengtai, which are all located in the central district of Beijing. Even though they have made great contributions to the socioeconomic development, they are so dominant that most developing resources pool here,

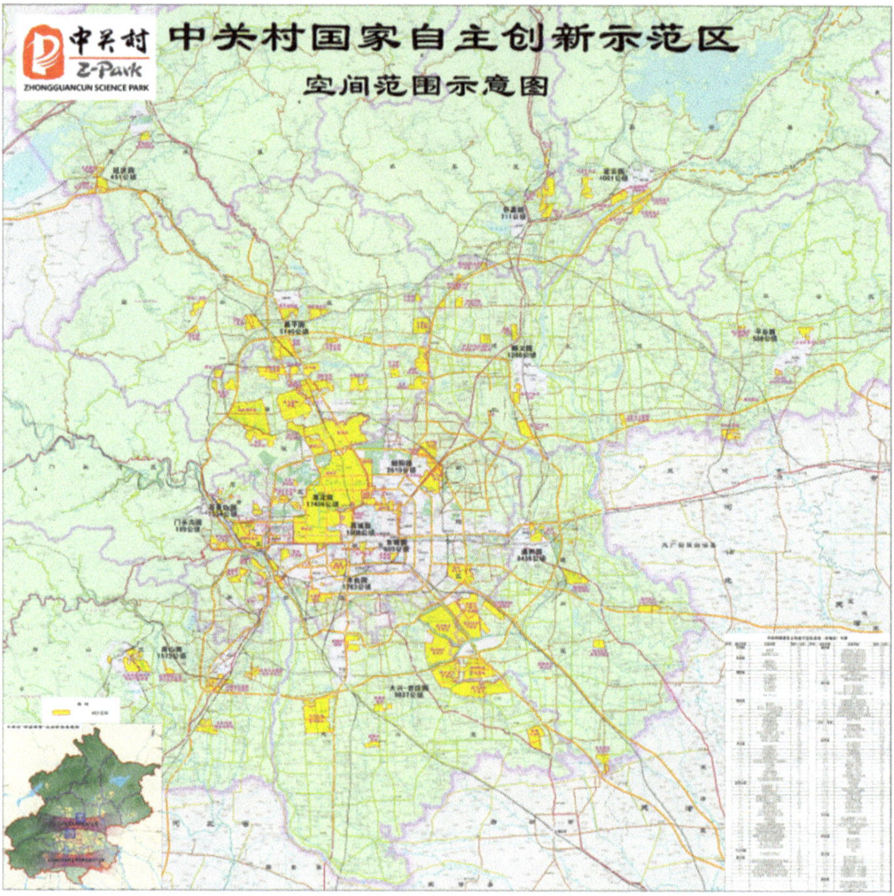

Fig. 11.17 Spatial distribution of "one district with sixteen sub-parks" (*Source*: http://www.zgc.gov.cn/document/20130228165218843366.jpg)

putting other districts at a disadvantage. Furthermore, several strong points functioning very close to each other cannot exert their impact on the country effectively; hence, a platform based on the whole Beijing area must be organized to promote the collaboration of innovation activities.

Therefore, during the fourth stage, the newest policies aimed at two tasks. One is to continue to integrate all districts into a whole and promote the comprehensive innovation ability of Beijing. Beijing Technology Innovation Action Plan (2014–2017)[13] has proposed to facilitate the collaborative relationship between 16 sub-parks and the districts they are located in and to build a city-level coordination mechanism for managing major industrial projects. The other is to foster multiple leading cores in city regions. The development of "three science towns"—

[13] http://zhengwu.beijing.gov.cn/ghxx/qtgh/t1352269.htm

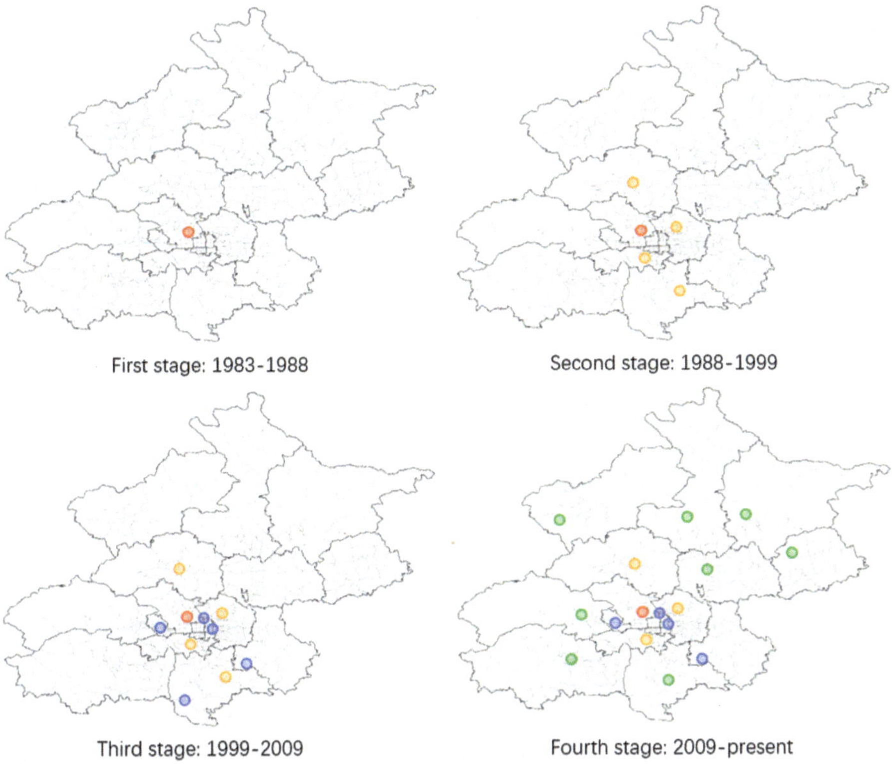

Fig. 11.18 Process of expansion of Zhongguancun National Demonstration Zone

Zhongguancun Science Town, Huairou Science Town, and S&T Future Town—has been emphasized in the Construction Plan on Enhancing the National Scientific and Technological Innovation Center in the 13th Five-Year Plan in Beijing.[14] Each science town is to maximize its own advantages and avoid vicious competition with each other.

Following analysis of the overall process of the development of Beijing's high-tech industry and related policies, a concept model such as in Fig. 11.19 can be generated to explain it. In general, the spatial evolution of Beijing's high-tech industry experienced three phases in turn—the emerging phase, the spreading phase, and the collaborating phase. First, a well-developed innovative environment like Haidian District induces high-tech enterprises to cluster and form the original innovative core. Then, demand for the upgrading of industrial structure accelerates the process of spreading, in which, on one hand, the high-tech service industry generates the majority of innovative activities and contributes the most to improving innovative ability, and on the other hand, the original innovative core dominates the development of the high-tech industry. Finally, to improve the vitality of innovation, both

[14] http://zhengwu.beijing.gov.cn/gh/dt/t1454520.htm

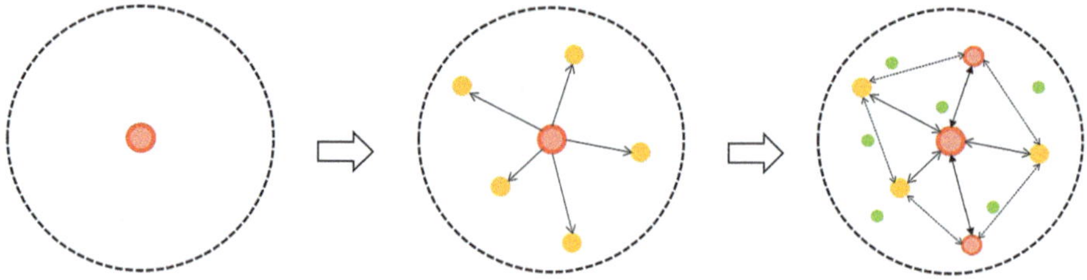

Fig. 11.19 The concept mode of spatial evolution of Beijing's high-tech industry

multiple leading cores and a well-collaborative network among districts are key factors. Beijing is currently experiencing the third phase.

11.6 Conclusion

Based on enterprise data, we analyzed Beijing's high-tech industry and its spatial evolution. Further, we observed how government policies influenced the spatial evolution. The results show that numbers of both high-tech enterprises and patents authorized increased significantly and at an accelerating rate. The innovative ability of science and technology was thus greatly improved, which in turn contributes significantly to socioeconomic development. More importantly, the high-tech industry became the main resource of scientific and technological innovation for Beijing.

Among all high-tech industries, the high-tech manufacturing industry accounts for a very small percentage of innovative activities, whereas the high-tech service industry makes up the majority of these innovative activities; this confirms that Beijing almost accomplishes the upgrading of its industrial structure. A detailed analysis of the subcategories of high-tech industries reveal that the driving forces of the development of high-tech industries come not only from the demand for upgrading of industrial structure but also from the social need and demand for improving industrial chain.

The spatial distribution of high-tech industries in Beijing spread from city center to the periphery gradually, and during this process, the guidance of government policies played a key role, for example, by providing preferential policies, delineating specific parks for high-tech industries, and so on. Thus, it is clear that top-down policies are necessary in the innovation-driven stage. While the full coverage of the city by enterprises will not necessarily lead to the improvement of innovative ability for all districts, the current status where only one district dominates the high-tech industry is to the disadvantage of comprehensive innovative ability. According to the concept mode of spatial evolution of Beijing's high-tech industry, the future strategy should focus on fostering multiple leading cores to guide cooperation among all districts, further integrating them into a well-collaborative network, in order to improve comprehensive ability in the innovation of science and technology.

References

Chen, J. (2005). *Empirical analysis on economic development promoted by science and technology* (Master thesis). Beijing: Capital University of Economics and Business.

European Commission. (2010). *Europe 2020: A strategy for smart, sustainable and inclusive growth: Communication from the commission*. Publications Office of the European Union. http://ec.europa.eu/europe2020/index_en.htm

Li, M. (2015). *Research on the influence of technological innovation about employment number and structure in Beijing* (Doctoral dissertation). Beijing: Beijing Jiaotong University.

Li, G., & Lu, M. (2002). The preliminary research on the mode and policy tendency of Beijing being constructed into a World City. *Scientia Geographica Sinica, 22*(3), 263–269.

Porter, M. E. (1998). *Clusters and the new economics of competition* (vol. 76, no. 6, pp. 77–90). Boston: Harvard Business Review.

Sassen, S. (2001). *The global city: New York, London, Tokyo*. Princeton: Princeton University Press.

Sun, M. (2008). *The grey comprehensive evaluation of Fujian province's scientific & technological innovation capacity* (Master thesis). Xiamen: Xiamen University.

Wang, W. (2007). *Impact and strategies of scientific and technological innovation on economic development of capital* (Master thesis). Beijing: Capital Normal University.

Yuan, X. (2014). *Research on the planning of innovation-driven science and technology city* (Doctoral dissertation). Beijing: Tsinghua University.

Zhang, Y. (2011). On Beijing effort for international tech-innovation hub and international urban strategy. *China Open Journal, 5*, 50–53.

Zhang, H. (2012). *Comparison of S&T innovation function development of Beijing and Shanghai* (Master thesis). Shanghai: East China Normal University.

Zhao, Z., Liu, Y., et al. (2015). Strategies and estimation system of Beijing constructing scientific and technological center. *China Development Observation, 6*, 77–81.

Part III
Mobile Device Data for Integrating Land Use and Transportation Planning

Chapter 12
Spatial Development Analysis of the Southern Area of Beijing Based on Multisource Data

Wenqi Lin, Liang Ma, Qiao Chu, and Yong Gao

Abstract The Action Plan of Promoting the Development of Southern Area of Beijing has been implemented by the local government since 2009. How this action plan influences the southern area of Beijing and what lessons we can draw from this experience are required to be assessed before schemes can be discussed for the next stage. This chapter is proposed to provide multidimensional analysis on the development path and existing problems of the southern area of Beijing. Permanent population, property price, infrastructure construction and utilization, and relative popularity of space units to different transport modes are considered based on multisource data, such as taxi GPS data, transit smart card data, secondhand property pricing data, and demographic data of Beijing. The Action Plan of Promoting the Development of Southern Area of Beijing is proved to stimulate the growth of the southern area in this chapter. However, the southern area is still below average in many aspects, such as the spatial structure and the economy vitality, which does not meet the targets stated in related documents.

Keywords Spatial analysis • Data-oriented analysis • Decision-making support system • The Action Plan of Promoting the Development of Southern Area of Beijing

W. Lin (✉)
Key Laboratory of Urban-Rural Eco Planning & Green Building, Ministry of Education, Tsinghua University, Beijing 100084, China
e-mail: linwq@tsinghua.edu.cn

L. Ma • Q. Chu
Beijing Tsinghua Tongheng Urban Planning & Designing Institute, Beijing 100085, China

Y. Gao
Beijing Branch of Shenzhen Urban Transport Planning Center, Beijing 100083, China

Z. Shen, M. Li (eds.), *Big Data Support of Urban Planning and Management*, Advances in Geographic Information Science, DOI 10.1007/978-3-319-51929-6_12

12.1 Introduction

The southern area of Beijing, including the former Chongwen District, the former Xuanwu District, Fengtai District, Daxing District, and Fangshan District, accounts for 20.5% of the total area of Beijing (Zhang 2010). For various reasons, the southern area is still inferior to the northern area of Beijing in many aspects, such as the condition of transport and municipal infrastructure, the economy aggregate, and the fiscal revenue. By the end of 2008, the total gross domestic product (GDP) of the five districts in the southern area is a fifth of that of the five districts in the northern area. The GDP per capita, the total fixed assets investment, and the retail sales consumer goods (RSCG) are a third of the northern area's level, respectively (Yang 2012). The unbalanced development of Beijing has become an important issue for this city. However, compared with the limited land resources for potential development in the northern area, the southern area of Beijing is a plain, which can provide a large amount of flat area for future development and for the transition of the city. Moreover, with the government taking the policy of the Integration of Beijing, Tianjin, and Hebei as a national development strategy, the southern area, located at the common boundary of these three provinces (or municipalities), may become a gateway of Beijing to participate in the corporation and integration.

To mitigate the congestion in the northern area and compensate the inequality in development, the Tenth Congress of Beijing Government proposed to build the southern area into a vibrant region of New Beijing. During the Two Sessions in 2008, Beijing Development and Reform Commission announced the Report on China's Economic, Social Development Plan for Beijing (2008), which drew attention to the infrastructure construction of the southern area. In September 2009, the local government issued the Action Plan of Promoting the Development of Southern Area of Beijing (or the South Beijing Development Plan). The Implementation Outline of the South Beijing Development Plan (2010–2020) was then released, which determined the spatial development pattern of the southern area: "one axis, two wings, three districts, and four cores" (Beijing Development and Reform Commission 2009). By 2013, the first stage of the South Beijing Development Plan was brought to its end. This led to the announcement of the Second Stage of the Action Plan of Promoting the Development of Southern Area of Beijing (2013–2015) (Beijing Development and Reform Commission 2013).

Two stages of the South Beijing Development Plan have been taken into actions so far. How this action plan influences the southern area of Beijing, whether the discrepancy between the southern area and the northern area is mitigated, whether the southern area meets the target of being a vibrant region of New Beijing, and what lessons we can draw from this experience are required to be assessed before schemes can be discussed for the next stage.

This chapter is proposed to provide multidimensional analysis on the development path and existing problems of the southern area of Beijing. Permanent population, property price, infrastructure construction and utilization, and relative popularity of space units to different transport modes are considered based on multisource data, such as taxi GPS data, transit smart card data, secondhand property pricing data, and demographic data of Beijing. Conclusions and suggestions are presented, which may support the plans and policy adjustments in the next stage.

12.2 Methodology and Data

Analyzing the spatial development pattern and identifying the existing problems of cities with multisource data have become an important method for policy assessment. Data which is commonly used in current urban researches includes point of interest (POI) data, map data, GPS data, transit smart card data, mobile signaling data, location-based service (LBS) data, video monitoring data, environmental and meteorological data, and social activity data (Wang et al. 2014). Wang et al. analyze shopping centers in Shanghai with mobile signaling data. Shopping centers like East Nanjing Road, Wujiaochang, and Anshan Road are classified into different levels according to the spatial distribution of their customers (Wang et al. 2015). Wang et al. apply space syntax method to analyze the evolution of the spatial structure of Beijing and reveal the mechanism of interactions among factors such as the land use and the social economy (Wang et al. 2008). Fine-grained spatial and temporal data has been commonly used in current urban researches based on urban geographic theories. Identifying the functions of space units with data mining methods also attracts increasingly more attentions in these years (Leng and Zhao 2014; Bram and Mckay 2005). However, most researches are still based on single source of data. There are few comprehensive researches and empirical studies considering multisource data.

In this chapter, longitudinal comparison and lateral comparison are taken to assess the South Beijing Development Plan. The former reveals the changes in southern area before and after the South Beijing Development Plan. The latter compares the current conditions of the southern area after the implementation of the South Beijing Development Plan with that of the northern area. Considering the availability of data and their sensitivity to the targets of the analysis in this chapter, the longitudinal comparison mainly established on three aspects: the spatial distribution of permanent population based on mobile signaling data, the relative popularity of bus stations, and the relative heat of bus stations based on transit smart card data. The lateral comparison then considers four factors: the urban land price based on online secondhand property pricing data, the time-phased relative popularity of bus stations based on transit smart card data, the relative popularity of space units based on taxi GPS data and transit smart card data, and the road network structure based on space syntax method with demographic data.

The main indicators and related calculation methods are as follows:

1. Spatial distribution of permanent population. The spatial distribution of population, especially the permanent population, directly indicated the spatial development of cities. Before 2010, the census data was mostly used to show the

spatial distribution of population. With the development of information and communication technology, the mobile signaling data is now commonly used instead of census data (Gao et al. 2014). The census data is collected according to the subdistrict or the neighborhood committee. The mobile signaling data is gathered according to the base station. Therefore, to guarantee the comparability of two sets of data, the mobile signaling data is added up to subdistrict level, and both datasets are uniformed in order to indicate the change in the spatial distribution of permanent population.

2. Relative popularity of bus station or space unit. The level of individual trip generation reveals the level of activity of regional transport, which indirectly reflects vitality of cities. Series of reform measures have been put forward to improve public transport in Beijing since June 2005. The transit smart card is utilized by 85% of trips with public transport for payment in Beijing. This could provide the analysis of changes in passenger flow with a rich source of data. The amount of passengers starting their trips on the bus station is chosen to reflect the level of activity of this station. Considering the change in total passenger volume of public transport before and after the action plan, the data in different years are uniformed to reveal the relative popularity of bus stations. According to the level of activity of bus station, its relative popularity is then calculated:

$$K_{ij} = \frac{Q_{ij}}{\sum_i Q_{ij}}$$

where Q_{ij} is the number of passenger starting trips at bus station i,
K_{ij} is the relative popularity of bus station i at time period j,
i is i^{th} bus station,
and j is time period j.

Residents may choose different modes of transport according to their personal preferences in many aspects, for example, the time cost, monetary cost, comfort level, and safety. The level of activity of taxi trip generation is therefore chosen as an additional indicator. This could also reflect the relative popularity of space units among different types of residents.

3. Secondhand property price. The secondhand property price could reflect the difference in land price among regions and also the efficiency of land use indirectly. Based on the online property pricing data, the differences between the southern area and other regions in absolute value and the change of property price are calculated, which then shows the discrepancy of the land price.
4. Degree of integration of road network. Space syntax is a theory about the architectures and urban space, which could be applied to quantitatively analyze the urban space form (Hillier and Hanson 1984; Li 2009). The degree of integration is one of the most commonly used indicators in space syntax theory. Evaluating the potential to be a trip destination of a space unit, this indicator could show the accessibility and centrality of the space unit. Therefore, it could also reflect the relationship between a single space unit and the whole urban space. The degree

of integration excludes the interference from the amount of space units with mathematical treatment. This enables this indicator to be a standardized parameter to compare different levels of urban space.

12.3 Empirical Research on Southern Area of Beijing

This section aims to empirically assess the current conditions and the changes of spatial development of the southern area of Beijing with quantitative methods mentioned above. Longitudinal comparison for conditions of southern area in different time periods and lateral comparison for current conditions of different areas in Beijing are taken in this section to provide the implementation effect assessment of the South Beijing Development Plan.

12.3.1 Longitudinal Comparison for Different Periods

12.3.1.1 Spatial Distribution of Permanent Population

The sixth census was carried out in 2010, which is also the publication year of the South Beijing Development Plan. The sixth census data is therefore used to reflect the spatial distribution of the permanent population before the plan (Fig. 12.1 (a)).

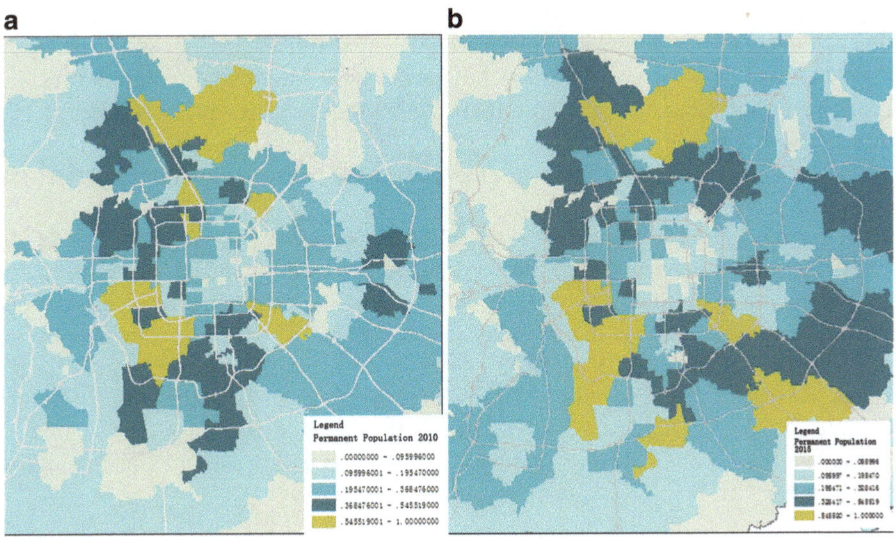

Fig. 12.1 Spatial distribution of permanent population in Beijing. (**a**) 2010. (**b**) 2015

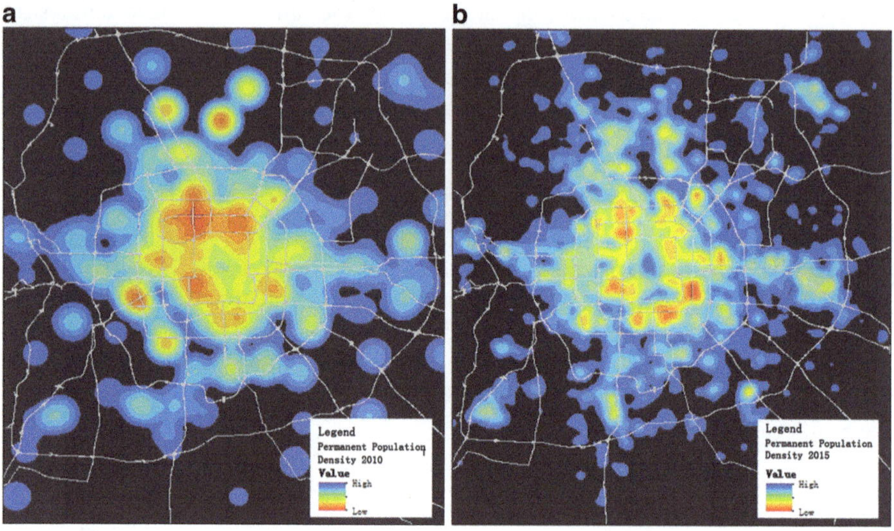

Fig. 12.2 Heat map of residential area of permanent population in Beijing. (**a**) 2010. (**b**) 2015

The condition of that after the plan is reflected based on the mobile signaling data in 2015, shown in (Fig. 12.1 (b)).

As shown in Fig. 12.1, the permanent population around Southeast Fifth Ring Road has increased since the action plan was implemented. This may result from the rapid development of Yizhuang Economic and Technical Development Zone and the policy of the Integration of Beijing, Tianjin, and Hebei. The permanent population of the area near Southwest Fourth Ring Road also rises because of the increasing land use intensity, especially the newly built residential communities in Liangxiang and Changyang. Given that, "two wings" in the layout strategy mentioned above start to be formed.

Figure 12.2 shows the popular residential area for permanent population in 2010 and 2015.

In Fig. 12.2, the location of the residential area of permanent population in 2010 is represented by the centroid of the subdistrict. The location in 2015 is represented by the location of the base station. Although the difference in representation method may result in the discrepancy of heat maps in two observation years in microlevel, it is comparable in macro-level to consider the relative popularity of zones in a city. As shown in Fig. 12.2, the popular area spread to the area out of the Fifth Ring Road in 2015, compared with 2010. High-density zones start to appear in the southern area of Beijing, such as Yizhuang in southeast and Changyang in southwest.

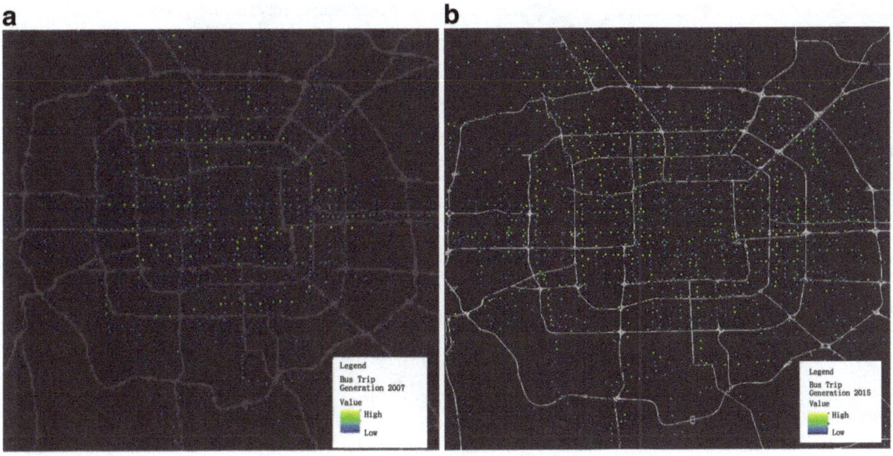

Fig. 12.3 Spatial distribution and popularity of bus stations in Beijing. *Metro stations and suburban bus stations excluded. (**a**) 2007. (**b**) 2015

12.3.1.2 Relative Popularity of Bus Station

One weekday transit smart card data in 2007 and 2015 are respectively selected to calculate the popularity of bus stations, as shown in Fig. 12.3.

Compared with Fig. 12.3(b), the popularity of bus stations in the southern area of Beijing has been improved, benefiting from the increased vitality caused by the South Beijing Development Plan. Before the implementation of the action plan, popular bus stations in the South Beijing mainly concentrate in the area near the central area of the city, such as Liuliqiao, Yangqiao-Fangzhuangqiao, and Qianmen. After the implementation, the popularity of bus stations near Nanyuan Airport, Wangxinghu Park, and Yizhuang has increased. Particularly, the amount of bus stations between Metro Line 9 and Metro Line 10 within the Fourth Ring Road and their relative popularity have significantly increased. Bus stations in the south of Shijingshan District have also improved their performance as a whole.

12.3.1.3 Relative Heat of Bus Station

The relative heat reflects both the density and the relative popularity of bus stations. It reveals the relationship between the demand and the supply of public transport. The amount of passengers getting on the bus is uniformed to indicate the discrepancy of the level of activity of public transport in the city (Fig 12.4).

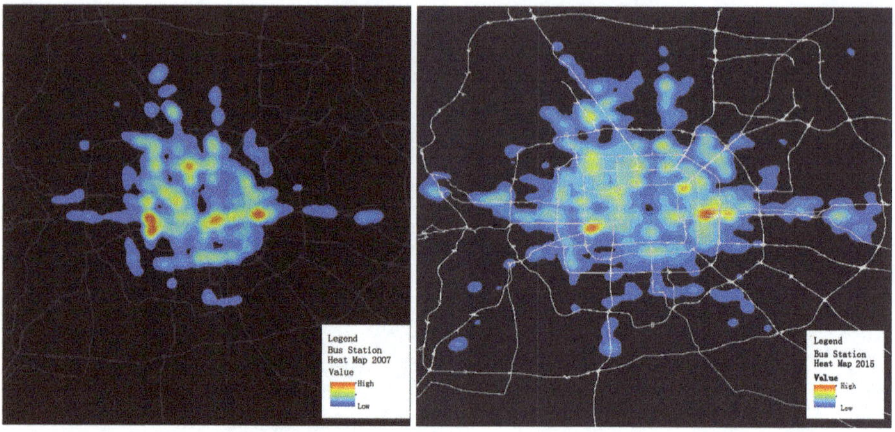

Fig. 12.4 Bus stations heat map in Beijing. *Metro stations and suburban bus stations excluded. (**a**) 2007. (**b**) 2015

The high value area in the south has moved north, while the medium value area has spread to the southwest, since the implementation of the action plan. Key developmental areas such as Nanyuan Airport and Yizhuang and areas along the metro lines have improved to be low-medium value areas. Area between the South Fourth Ring Road and South Fifth Ring Road is a weak point for the South Beijing considering the construction and utilization of public transport infrastructure. As for the north, the amount of high value areas has decreased. Except for the World Trade Center-Sihuiqiao, areas on each side of the east-west axis have been respectively cooled down to some extent.

12.3.2 Lateral Comparison for Different Areas

12.3.2.1 Secondhand Property Price

Based on online data, the price of each residential community in March, July, and November 2015 is extracted to calculate the average price of each space unit in these 3 months (Fig. 12.5).

The average price of secondhand property in light color area is higher than that in dark color area in Fig. 12.5. The average price is indicated to decrease from the central to the suburban. This coincides with the characteristics in most cities in China. Within the Fifth Ring Road, the average price of the space unit in the northern area is generally higher than that in the southern area when Chang'an Avenue is taken as the north-south axis. Except for some space units within the South Second Ring Road, the average price in most of the space units in the southern area is below the average. In particular, some space units around South Third Ring Road even have lower average price than that of the space units around North Fifth Ring Road.

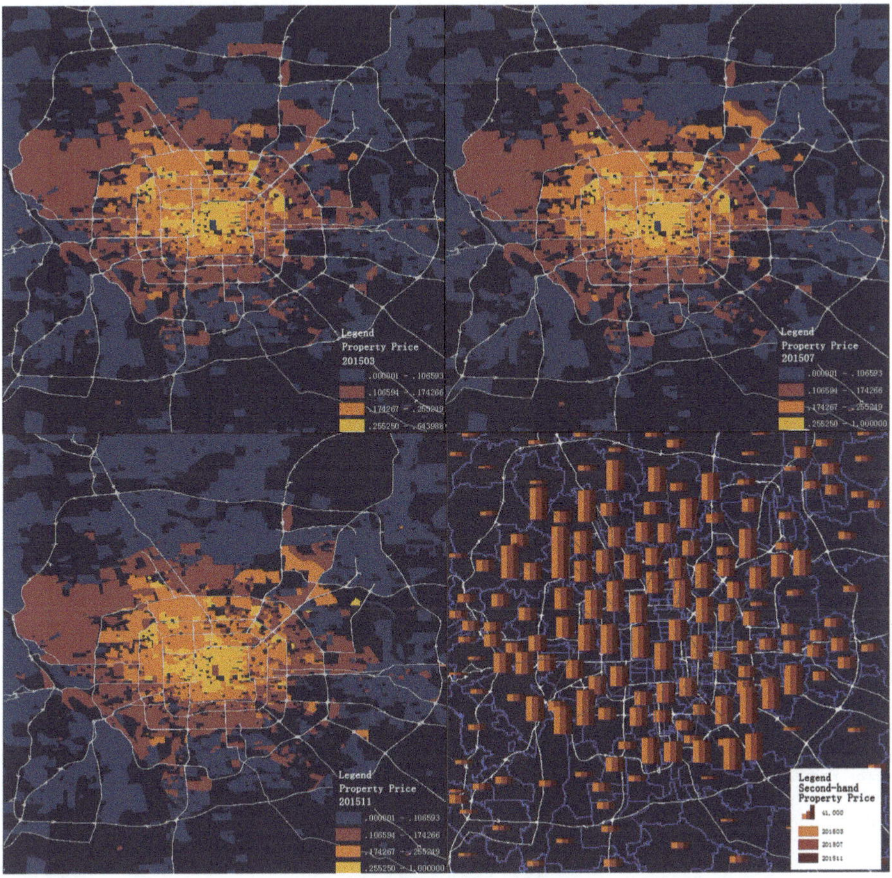

Fig. 12.5 Average price of secondhand property in Beijing (2015)

Compared with 2014, the annual change in secondhand property price in December 2015 is shown in Fig. 12.6.

It is indicated in Fig. 12.6 that the average prices of space units mainly decreased in December 2015, most of which reduced by within 10%. In general, the change is more obvious in the southern area than that in the northern area. The regions nearer to the key development areas have more significant change.

12.3.2.2 Time-Phased Relative Popularity of Bus Station

In addition to the analysis in Section 12.3.1.2, the relative popularity of bus station is calculated in finer level to provide detailed analysis. Time-phased indicator is derived every 30 min for one weekday in 2015. The relative popularity of bus station is classified into five levels. The value of the indicator in each time interval is indicated in Fig. 12.7.

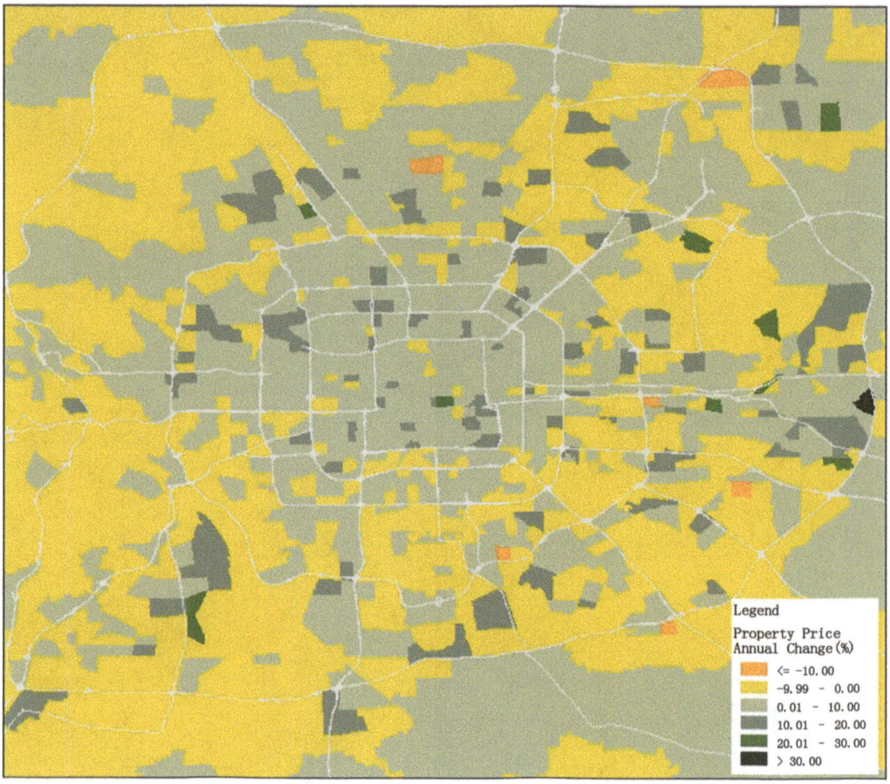

Fig. 12.6 Annual change in secondhand property price in Beijing (December 2015)

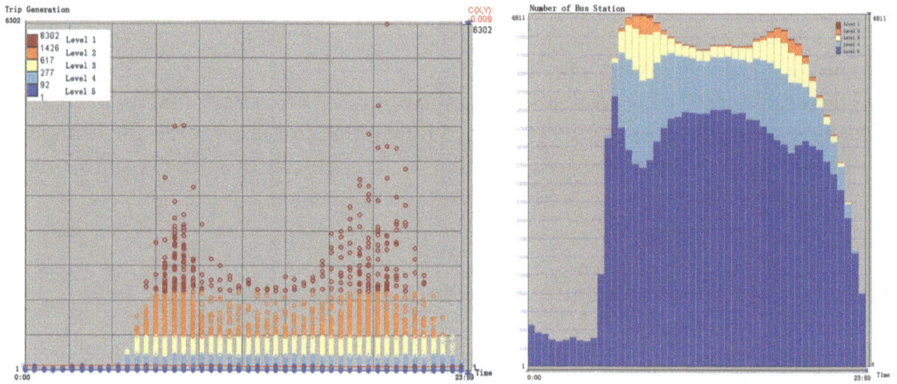

Fig. 12.7 Time-phased relative popularity of bus station and frequency distribution. (**a**) Time-phased relative popularity with levels. (**b**) Time-phased number of stations in each level

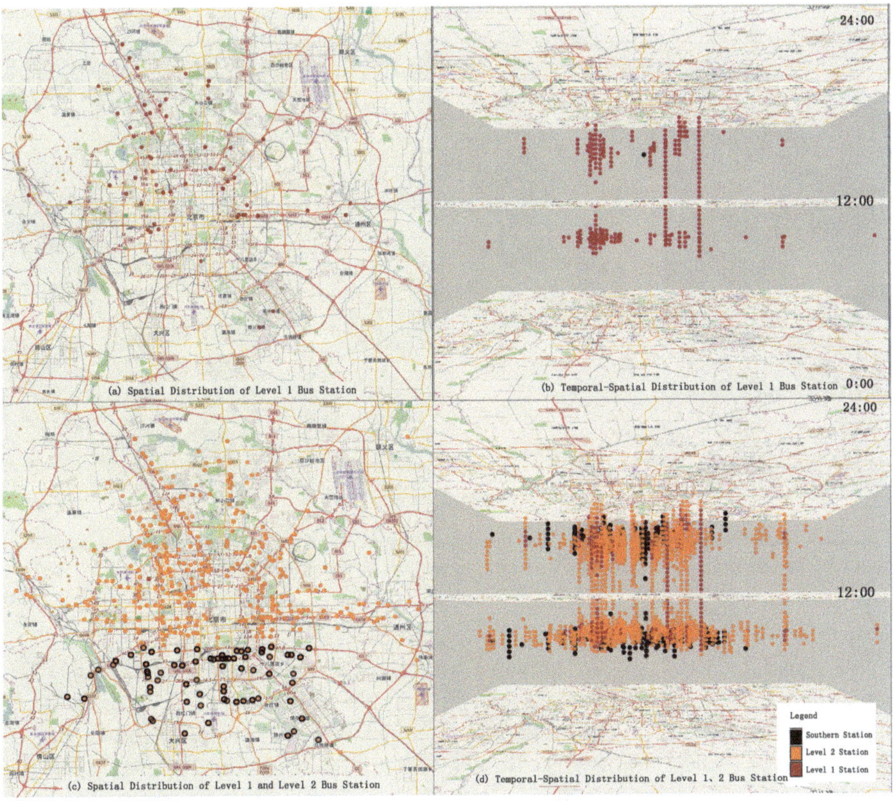

Fig. 12.8 Spatiotemporal distribution of high level bus stations in Beijing

With the increasing amount of passengers in the morning peak, the number of stations in level 1 and level 2 significantly grows. The relative popularity of bus stations in the evening peak is more discrete than that in the morning peak. More stations appear to be in high level in the evening peak. Compared with the morning peak, the evening peak continues longer.

The stations in level 1 and level 2 are selected for further analysis. The spatiotemporal distribution of these stations is illustrated in Fig. 12.8.

Figure 12.8(a) and Fig. 12.8(c) shows the spatial distribution of bus stations in level 1 and level 2. The black dots are the bus stations in the area which is north to the South Fifth Ring Road, south to the South Second Ring Road, east to the Jingkai Expressway, and west to the Jingjin Expressway. Figure 12.8(b) shows the spatiotemporal distribution of stations in level 1. Figure 12.8(d) shows the spatiotemporal distribution of stations in level 1 and level 2.

Fig. 12.9 Relative popularity of space unit for different transport modes. *Metro stations and suburban bus stations excluded. (**a**) Taxi trip generation. (**b**) Bus trip generation

As illustrated in Fig. 12.8, Nanyuan Road Guoyuan Station is in level 1 at 17:00, which is the only bus station in level 1 in the area defined by the four roads mentioned above. Except this, all other stations in level 1 are north to the Chang'an Avenue or near the Beijing West Railway Station. In addition, the stations in the southern area generally only have relatively high popularity in the morning. The majority of the stations in the southern area remain in medium or low level during 10:30–16:00 and after 20:00.

12.3.2.3 Relative Popularity of Space Unit for Different Transport Modes

Taxi GPS data and transit smart card data are extracted for the same period to calculate the relative popularity of space units for different modes of transport (Fig. 12.9).

In Fig. 12.9(a), it is obvious that more trips by taxi are generated from the northern area. A large number of taxi trips are originated at the area between the North Fourth Ring Road and the South Third Ring Road, especially the area along the part of the Third Ring Road from Sanyuanqiao to Fenzhongsi. In Fig. 12.9(b), popular area spreads to some extent. However, the northern area is still more active than the south area, which is the same as the pattern shown by the taxi trip.

Comparing the popularity of the same space unit in terms of different transport modes, the regions in the southern area such as Yuegezhuang and Liuliqiao are more popular to the bus trip than the taxi trip. A large amount of people starting trips in these regions prefer bus to taxi.

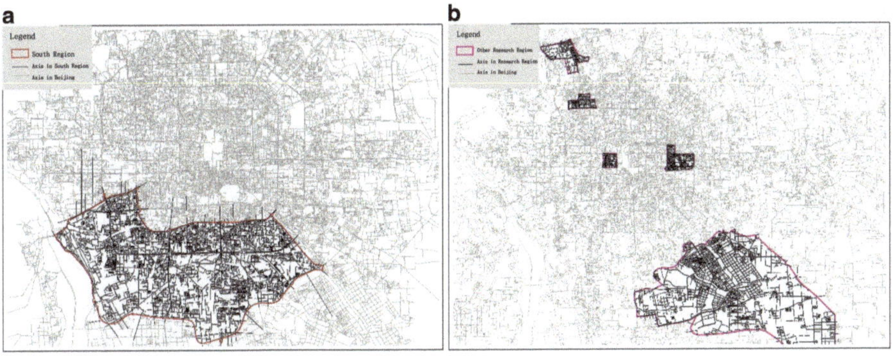

Fig. 12.10 Axial map of research regions. (**a**) Axial map of Main South Region. (**b**) Axial map of other research regions

12.3.2.4 Degree of Integration of Road Network

Road network of Beijing in 2015 is analyzed under the space syntax theory. The World Trade Center, Financial Street, Zhongguancun, and Shangdi in the northern area are selected as research regions to compare with that in the southern area: Yizhuang and the region (named Main South Region) defined by the South Third Ring Road, the Jingkai Expressway, the South Fifth Ring Road, and the Jingjin Expressway. The axial maps of the research regions are show in Fig. 12.10.

Figure 12.11 gives the scatter plot of the relative locations of the axis lines in each research region compared with the whole city. The horizontal axis is the global degree of integration (radius-n). The vertical axis is the local degree of integration (radius-3). The dots are the axis lines containing in each research region.

In Fig. 12.11, the lines are the linear regression models for the particular research region and for the whole urban space. The correlation coefficient indicates the relationship between the local degree of integration and the global degree of integration. The slope of the linear model indicates the advantage of the particular research region over the other regions in the whole urban space. High slope value shows that the axis lines, which are the streets, in the particular research region have higher local degree of integration. The intersection of the two linear models indicates the connectivity of the particular region to the whole urban space. Intersection point to the left downside shows a worse connectivity.

As shown in Table 12.1, the correlation coefficients and slopes of the Financial Street, the World Trade Center, and the Zhongguancun are relatively high, which indicate a satisfactory local accessibility and a good relationship between particular regions and the whole urban space. However, all the coefficients of the Yizhuang and the Main South Region are in low level. The situation of Main South Region is worse than that of Yizhuang. This reflects the weakness of infrastructure construction and adjustment in the southern area. Imbalanced development in the southern area is also obvious.

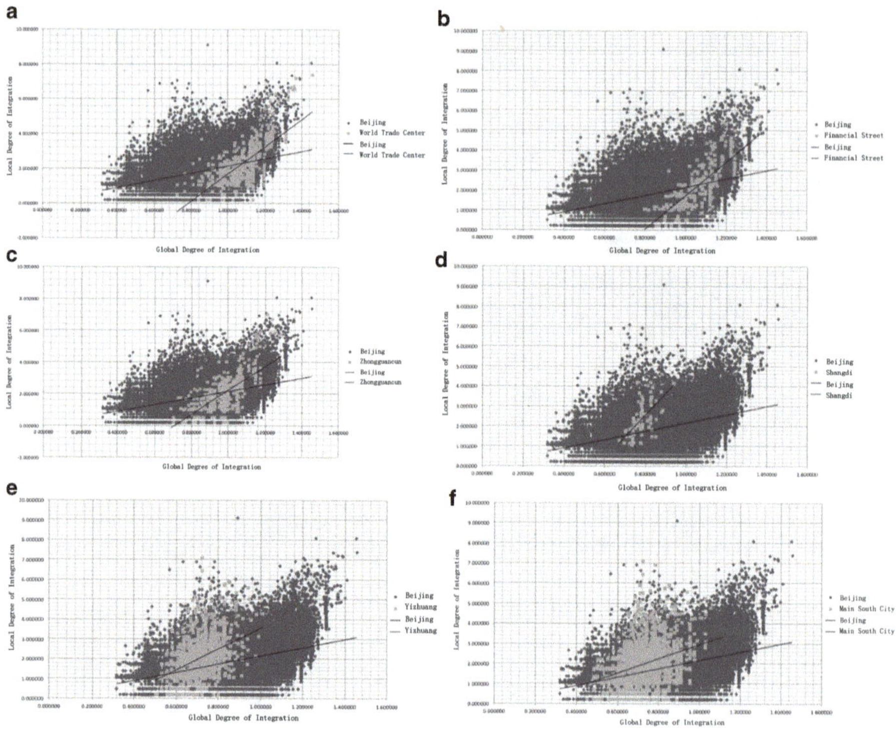

Fig. 12.11 Scatter plot of relative locations of axis lines in research regions. (**a**) World Trade Center. (**b**) Financial Street. (**c**) Zhongguancun. (**d**) Shangdi. (**e**) Yizhuang. (**f**) Main South City

Figure 12.12 gives the core of global integration, which is the lines with highest degrees of integration in the whole urban space.

As shown in Fig. 12.12, the amount of global integration core in the southern area is less than that in the northern area. The cores in the south include part of South Third Ring Road, Fengtai Road, Guangcai Road, and Jiayuan Road. The global integration cores in the Main South Region are mainly distributed in the northern part and not well extended to the entire south region.

12.3.3 Research Findings

Compared with the period before the implementation of the South Beijing Development Plan, the permanent population spreads to the southern area these years. In particular, the southeast and the southwest experience the most significant

Table 12.1 gives the value of above coefficients for the research regions. Space syntax coefficients for research regions

		Financial street	World Trade Center	Zhongguan-cun	Shang-di	Yi-zhuang	Main South Region
Correlation coefficient R^2		0.5832	0.5403	0.4993	0.3794	0.2122	0.1379
Slope k		8.4571	7.8098	7.5758	10.773	5.2967	3.5735
Intersection	Global	1.0617	0.9869	1.0795	0.7047	0.5652	0.4692
	Local	2.3538	2.1966	2.0811	1.4784	1.1634	1.000

Fig. 12.12 Core of degree of integration within Sixth Ring Road (*Blue*) and in Main South Region (*Green*)

increase in permanent population and become new popular residential areas. The relative popularity of bus stations in the southern area is also generally improved. Bus stations between South Third Ring Road and South Fourth Ring Road begin to have high popularity. The relative popularity of the stations around South Fifth Ring Road has also been improved to medium level. These indicate that the southern area has become more vibrant. Both of three indicators in the longitudinal comparison show the increase in activeness and residents. Highlighted regions in terms of indicators are also similar. The conclusions are generally consistent with the implementation effect assessment of the first stage of this action plan (Zhao 2013).

Considering the lateral comparison on the current conditions of different areas, all selected indicators show there is still a discrepancy between the southern area and the northern area. The secondhand property price and the popularity indicators reflect that the regions between the South Third Ring Road and South Fourth Ring Road, Yizhuang in the southeast, and Liangxiang and Chaoyang in the southwest have been improved significantly to a relatively high level. However, the other regions in the southern area still lag behind. The development of regions in the southern area is imbalanced.

The road network analysis based on space syntax theory indicates that the arterial roads in the southern area are not well connected to the central area of Beijing. In addition, the arterial roads do not perfectly join the roads in all levels in the southern area to improve the overall performance. The road network of Yizhuang has a higher degree of integration than the Main South Region. This partly explains the weakness of the southern area in property price and relative popularity of space units.

12.4 Conclusions and Future Work

The Action Plan of Promoting the Development of Southern Area of Beijing is proved to stimulate the growth of the southern area by the multidimensional analysis based on the multisource data. The development of "one core" and "two wings" has already made a progress, which coincides with the strategy in the plan. However, the southern area still lags behind the northern area in many aspects, such as the infrastructure construction and the space unit vitality. The congestion in the northern area is also not obviously mitigated after the development of the southern area. The development of the southern area is not balanced, and the effect of the strategy of "point to area" is not significant.

Therefore, investment in construction and support in industry are still necessary for the next stage of the development of the southern area. More attention needs to be drawn on the construction and adjustment of road network. The efficiency of the arterial roads and the connections among function zones could be the focus. The development plan also needs to improve the radiation and driving effect of key development regions. Balanced development could be a new target for the next stage.

A lack of data, especially the lack of historical data, is still one of the main issues in the spatial development analysis and in the process of identifying urban problems. This directly restricts the selection of research periods in longitudinal analysis. The collection and storage of data, particularly the raw data, is strongly recommended. This could benefit the comparison analysis and the establishment of new analytical method.

Reference

Beijing Development and Reform Commission. (2009). Report of the action plan of promoting the development of southern area of Beijing: A future vibrant area in new Beijing. *Frontline, 12*, 29–31.

Beijing Development and Reform Commission, The Second Stage Action Plan of Promoting the Development of Southern Area of Beijing (2013–2015). Available: http://zhengwu.beijing.gov.cn/ghxx/qtgh/t1303285.html.

Bram, J., & Mckay, A. (2005). Evolution of commuting patterns in the New York City metro area[J]. *Current Issues Econ Finance, 11*, 1–7.

Gao, Z., Tang, X., & Zhao, B. (2014). The application of mobile signalling data in urban planning: A case study of Chongqing urban and suburban integrated planning. Chongqing Mountainous Region Planning.

Hillier, B., & Hanson, J. (1984). *Social logic of space*. Cambridge: Cambridge University Press.

Leng, B., & Zhao, W. (2014). Region ridership characteristic clustering using passenger glow data. *Journal of Computer Research and Development, 51*(12), 2653–2662.

Li, Y. (2009). Research in assessment indicator system of urban network based on space syntax theory. Central South University.

Wang, J., Mao, Q., & Dang, A. (2008). An evolution model of Beijing— A discussion on the evolution of Urban Spatial and functional patterns based on space syntax. *Journal of Urban Planning, 21*(3), 82–88.

Wang, J., Li, C., Xiong, Z., & Shan, Z. (2014). Survey of data-centric smart city. *Journal of Computer Research & Development, 51*(2), 239–259.

Wang, D., Wang, C., Xie, D., et al. (2015). Comparison of retail trade areas of retail centers with different hierarchical levels: A case study of East Naning Road, Wujiaochang, Anshan Road in Shanghai. *Journal of Urban Planning, 3*, 50–60.

Yang, X. (2012, April). Action of South Beijing: Rebuild a new South Beijing. *Beijing Daily*.

Zhang, H. (2010). South Beijing: Move to the rapid development. *China Real Estate Market, Z1*, 110–111.

Zhao, F. (2013). Review the South Beijing at the final year. Beijing Investment, 67–70.

Chapter 13
Spatio-temporal Dynamics of Population in Shanghai: A Case Study Based on Cell Phone Signaling Data

De Wang, Weijing Zhong, Zhenxuan Yin, Dongcan Xie, and Xiao Luo

Abstract The analysis of spatial and temporal dynamic distribution of population is an important basis for recognizing people's behavior patterns and urban spatial structure, allocating urban public infrastructures, and making emergency plans of public safety. Due to the lack of data on spatial and temporal dynamic distribution of population, research related to this area is limited in China. As cell phone becomes the most popular communication terminal, the spatial and temporal distribution of cell phone users should be able to reflect that of population accurately. Using datasets of cell phone signaling records and related data of Shanghai like land use, this study attempts to build an analytical framework on internal relations among population, time, and behavior, to recognize characteristics of spatial and temporal dynamic distribution of population in Shanghai. The results show as follows: (1) the density of population appears to be monocentric distribution, and this characteristic is more significant during the day compared to that at night. The spatial distribution experiences the central aggregation process during daytime and the decentralization to suburban area during nighttime. (2) People's behaviors (like commuting, leisure, and consumption) could cause the spatial and temporal distribution of population change. The spatial mismatch between residences and workplaces as well as the high dependence on the central area result in unevenly distribution of population and form the central aggregating pattern. (3) The dependence level of consumption and leisure on central area is significantly higher than that of employment, especially in the suburban areas adjacent to the central area.

Keywords Cell phone signaling data • Daytime and nighttime population • Workplaces and residential places • Land use • Shanghai

---

D. Wang (✉) • D. Xie
Collage of Architecture and Urban Planning, Tongji University, Shanghai, China
e-mail: dewang@tongji.edu.cn

W. Zhong
Collage of Architecture and Urban Planning, Tongji University, Shanghai, China

Hanzhou City Planning and Design Academy, Hangzhou, China

Z. Yin • X. Luo
Shanghai Tongji Urban Planning and Design Institute, Shanghai, China

Z. Shen, M. Li (eds.), *Big Data Support of Urban Planning and Management*,
Advances in Geographic Information Science, DOI 10.1007/978-3-319-51929-6_13

13.1 Introduction

Spatial and temporal dynamic distribution of population is an important field in the population geography. A lot of research focus on estimating distribution of population (Sleeter and Wood 2006; Wang et al. 2014), analyzing spatial dynamic distribution of population (Qin et al. 2013), and researching coupling relation between distribution of population, industry, and land use (Yang and Ning 2015; Rao et al. 2015). However, because the data, number of population during daytime and nighttime, and number of floating population, are difficult to access, the spatial cells of existing research are regions (Bian et al. 2013), provinces (Li et al. 2015), and cities (Sleeter and Wood 2006; Wang et al. 2014), and the unit of time is years (Pan et al. 2013). There is little research analyzing distribution of population in smaller spatial cells, such as districts, during smaller unit of time, such as daytime and nighttime. In addition, existing research analyzes the distribution of population based on the census data, daily movement of persons (Foley 1954, 1952; Zhang et al. 2016), origin-destination matrix (Akkerman 1995), travel survey data (Roddis and Richardson 1998), and high-resolution remote sensing data (Sleeter and Wood 2006; Elvidge et al. 2001). There is little research focusing on individual data and analyzing the reasons that make the distribution of population change (Long et al. 2011, 2014). As a result, this field still needs to be further studied.

Nowadays, China is in the process of rapid urbanization. It is shown that 190 million people moved to cities from rural areas in the sixth national census (The Floating Population Service Management Department of national population and Family Planning Commission 2011). In addition, analysis of spatial and temporal dynamic distribution of population is an import tool to help us to understand how cities develop, to perceive mechanisms of urban functional areas, and to advise us how to allocate public facilities. On the other hand, managing public safety needs us to know spatial and temporal dynamic distribution of population more accurately. In important and random events, understanding distribution of population is the basis to rescue. As a result, analysis of spatial and temporal dynamic distribution of population is not only an important field in the population geography but an important basis to recognize human behaviors and urban spatial structures and then advise us how to allocate public infrastructures and how to make public security plans. It has both important research value and practical significance.

It is possible to access big, dynamic, accurate spatial personal data with the development of information technology. Big data gets more and more attention and has been widely used in various fields of research and management (Liu et al. 2015; Zhen and Wang 2015; Wu et al. 2015; Zhen et al. 2015; Yang et al. 2015). Because of wide coverage and low cost to collect data, the locating data provided by cell phone, the most popular communication terminal equipment, has supplied unprecedented possibility to do research for geography, urban planning, and other subjects focusing on locating data. Much research have focused on this topic in recent years (Guo et al. 2014; Liu et al. 2011). Some researchers have used cell phone data to recognize urban spatial structure (Niu et al. 2014; Calabrese et al. 2011; Reades et al. 2009), to evaluate built environment (Wang et al. 2015b) and jobs-housing relationship

(Niu and Ding 2015), to recognize commuting (Yuan et al. 2012; Ahas et al. 2010; Ding et al. 2015) and consumption (Wang et al. 2015a), and to understand individual mobility patterns (Calabrese et al. 2013; Diao et al. 2015). In the field of spatial and temporal dynamic distribution of population, some national researchers have used cell phone data to make dynamic maps of distribution of population and to recognize high population density areas (Deville et al. 2014; Vieira et al. 2010; Ahas and Mark 2005). These studies show that cell phone data is useful and important in the field of spatial and temporal dynamic distribution of population.

Because cell phone signaling data is able to provide large amount of spatial records of persons, taking Shanghai as an example, this study analyzes spatial and temporal dynamic distribution of population within city and at personal level with cell phone signaling data and aims to provide reference for the future study in this field.

13.2 Data and Framework

13.2.1 Data

This study takes 2G cell phone signaling data of China Mobile Shanghai in some 2 weeks of the first half year in 2014 as data source. Every time cell phone connects with the China Mobile phone base station, the base station records it. The data is anonymous, without any private information. Each record includes user ID, time stamp, serial number of the base station, type of issue (e.g., calling, sending or receiving message, and updating of location), and so on. Sixteen to 18 million different mobile user IDs, which account for 70% of the total permanent residents, are recorded daily, and 600–800 million cell phone records are recorded daily. There are 36,000 base stations in Shanghai. The distance between base stations in the central city is between 100 and 300 m, and the distance between base stations in the suburban areas is between 1000 and 3000 m. Mobile signaling data can continually record the locations of cell phone users. As the proportion of cell phone user is very high,[1] cell phone signaling data can properly show the spatial and temporal behavior patterns.

13.2.2 Framework

Distribution of population changes periodically. The difference between distribution of population during daytime and nighttime is a main feature. Influenced by urban self-organizing system and urban planning, urban spatial structure is represented as different urban functional areas. Commuting, consumption, leisure, and other behaviors between different urban functional areas are the main reasons to make distribution of population change. As a result, based on the time and

[1] According to *2014 Shanghai Statistical Yearbook*, the percentage of cell phone users is 132.5%.

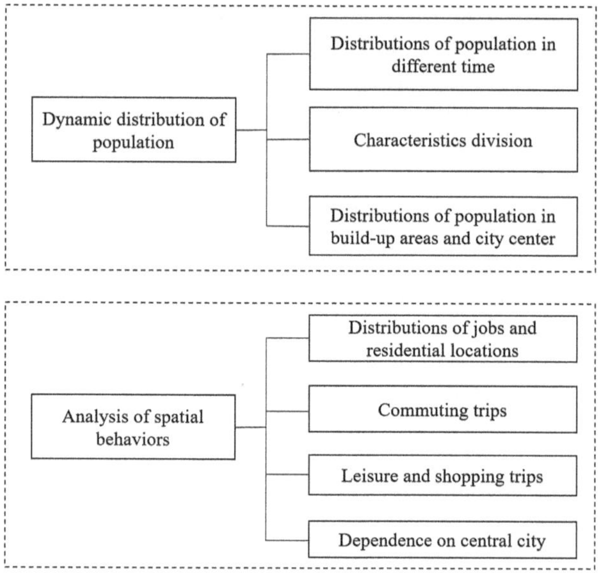

Fig. 13.1 Research framework

behaviors, this study builds a framework to analyze spatial and temporal dynamic distribution of population. Figure 13.1 shows the framework.

First, this study calculates the number of cell phone users in different periods. Based on the number of cell phone users and the spatial characteristics of behaviors, this study chooses the suitable time to represent the daytime and nighttime. At the same time, the selective analyses of distribution of population in city center and important areas are done.

Second, based on research on the spatial characteristics of commuting, leisure, and consumption, this study makes rules to recognize behaviors. Then this study analyzes the characteristics of different behaviors, such as the distribution of workplaces and residences, the distance of commuting, and having leisure and consumption trips, and the degree of dependency on central city. According to the framework, this study finds that commuting, leisure, and consumption affect the distribution of population.

13.3 Dynamic Distribution of Population

13.3.1 Distribution of Population in Different Time

Figure 13.2 shows the distributions of population in every 2 h. It can be found that distribution of population is monocentric. In the same time, the gap between densities of population in central city in different periods is huge. From the view of mobility of population, the figure shows that people aggregate during daytime and

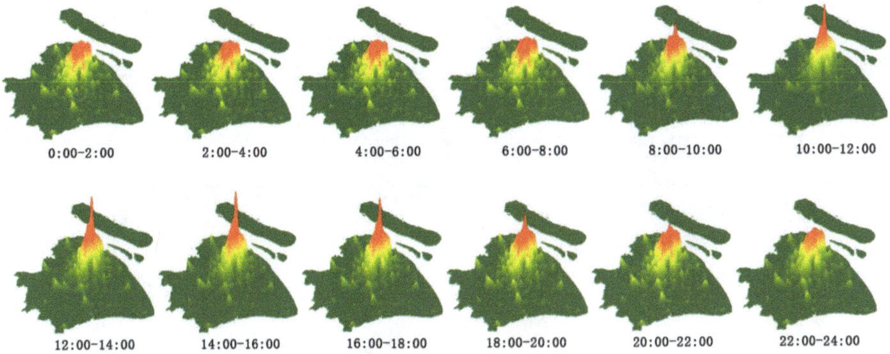

Fig. 13.2 Distribution of population in different periods

deconcentrate during nighttime. Huge gaps are showed between densities of population in city center and suburban areas during daytime and nighttime. From 6 am to 10 am, people go to city center from suburban areas. At 10 am, the density of population in city center reaches the highest. There is a remarkable monocentric structure of distribution of population, which lasts until 6 pm. After 6 pm, people flow to suburban areas, especially from the city center within the inner ring, a district with many workplaces and few residences. On the contrary, people gather in the new towns, lower level residential areas. The suburban towns on the west side or the city have been well connected with the inner city, while three towns on the east side of Shanghai seem more independent. This phenomenon shows the concentration of workplaces and the lack of residences in city center. Additionally, the trend, concentrating during daytime and deconcentrating during nighttime in city center, shows a great number of demands of commuting in morning and evening rush hour.

13.3.2 Characteristic Division

To analyze the distribution of population during daytime and nighttime, the medians of density of population during daytime, the medians of density of population during nighttime, and the medians of ratio of population during daytime and nighttime are used as thresholds to divide 230 districts in Shanghai. Based on analysis, this study uses the data at 10 am to represent the population during daytime and the data at 10 pm to represent population during nighttime. Because the number of cell phone users during daytime is 1.6 times the number of cell phone users during nighttime, this study standardizes the numbers according to the number of total population in Shanghai in 2015, 24.5 million. The ratio of population during daytime and nighttime is the density of population during daytime divided by the density of population during nighttime.

Based on the analysis of density of population, this study finds there are three rings, city center, suburb, and outer suburb, and the density decreases ring by ring.

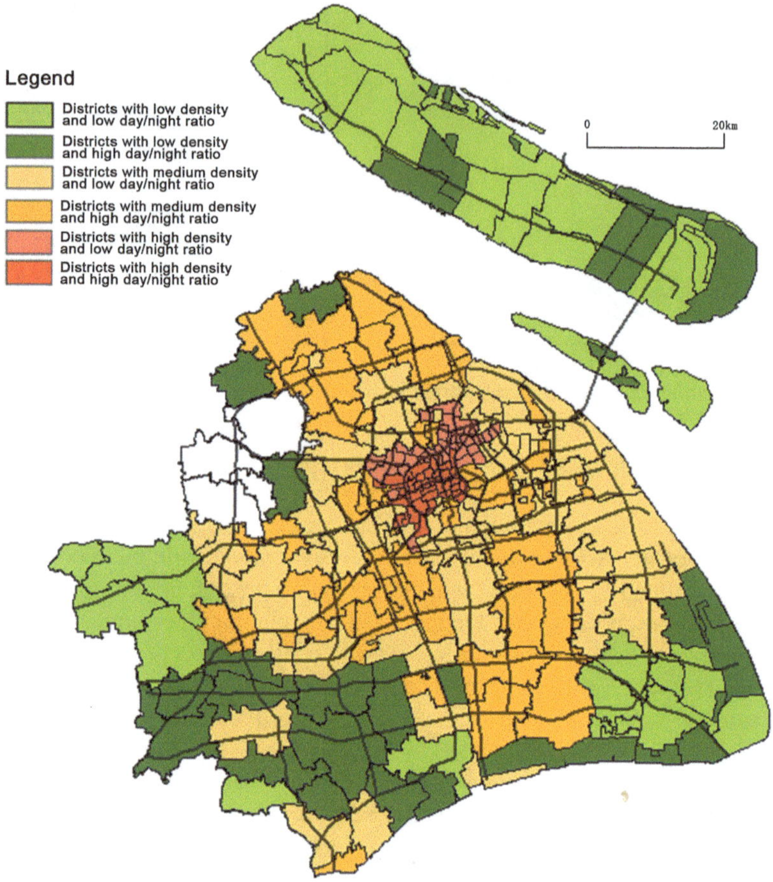

Fig. 13.3 Division based on population density during daytime and nighttime, and their ratio

Combined with ratio of population during daytime and nighttime, this study divides districts in Shanghai into six types. Figure 13.3 shows the division. Districts with low density and low ratio, where intensity of behavior is low, concentrated in the outer suburban rural areas, such as Chongming, Qingpu, Jinshan, Pudong, and so on. Because the types of activities are mixed, the ratio of population during daytime and nighttime is low. Districts with low density and high ratio, where intensity of behavior is low and workplaces concentrate, are in central towns in the outer suburban areas, such as Lv town in Jinshan, Tinglin, Laogang town in Pudong, Nongchang town in Chaoyang, Chengqiao town in Chongming, Shangshi Agricultural Area, and so on. Districts with medium density and low ratio, where types of activities are mixed, concentrate in the suburban areas. Districts with medium density and high ratio concentrate in workplace centers in the suburban areas, such as Xinzhuang, Caohejing, Songjiang industrial areas, Zhangjiang, Jinqiao, Fengxian, and so on. Districts with high density and low ratio, where both workplaces and residences concentrate, are located in the north of the city center, such as Yangpu, Laozhabei, Changning, and so on. Districts with high density and high ratio, where many workplaces and few residences concentrate, locate in the city center within the inner ring.

13.3.3 Difference Between Distribution of Population During Daytime and Nighttime in City Center and Built-Up Areas

After understanding the distribution of population in Shanghai, this study focuses on the distribution of population in city center, areas within the outer ring, and built-up areas to analyze the difference between distributions during daytime and nighttime. Figure 13.4 shows the distribution of city center and built-up areas. Counting the population on weekdays and weekends during daytime and nighttime separately, this study calculates the proportion of population in each period in city center and built-up areas. Table 13.1 shows that the concentration of population in Shanghai is remarkable. Among the total population, about 47.58–50.65%, or 12–12.5 million, concentrate in city center. Similarly, there are about 21.5–22 million, or 86.37–89.04% of the population, concentrating in built-up areas.

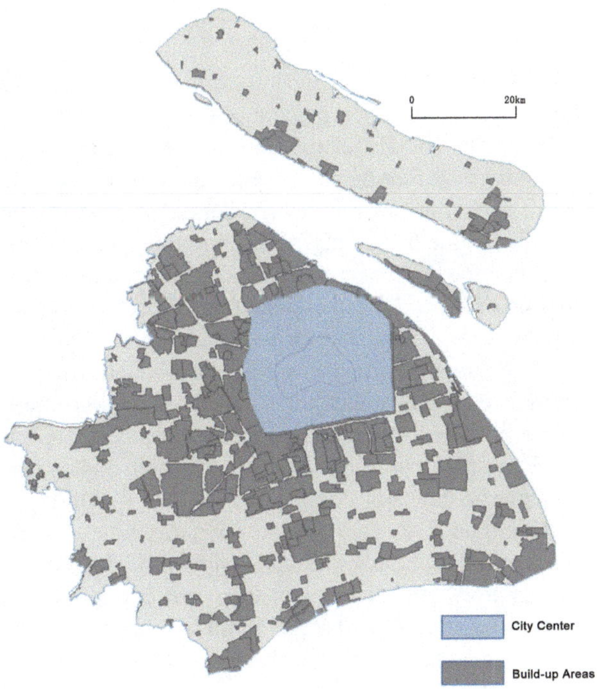

Fig. 13.4 Distribution of city center and built-up areas

Table 13.1 The proportion of daytime and nighttime population in city center and built-up areas on weekdays and weekends

	Weekdays		Weekends	
	Daytime (%)	Nighttime (%)	Daytime (%)	Nighttime (%)
City center	50.65	47.88	49.60	47.58
Built-up areas	89.04	86.78	88.32	86.37%

Because this study analyzes distribution in every 2 h, cell phone users may show in city center and suburbs, or built-up areas and non-built-up areas, in the same time. As a result, the sum of proportion of population in city center and suburbs, or in built-up areas and non-built-up areas, is slightly greater than 100%

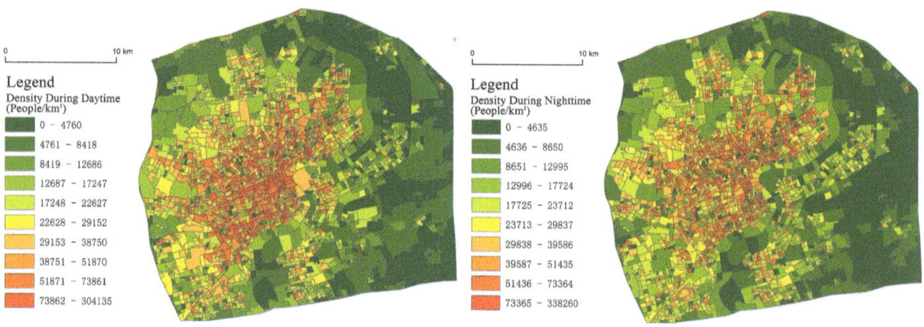

Fig. 13.5 Spatial distribution of density during daytime and nighttime in the city center at census tracts level

More than 86% of the population concentrate in built-up areas, which account for 44.62% of total areas. Forty-seven percent of the population concentrate in city center, which accounts for 10.51% of total areas. The degree of concentration of population in city center is higher than the degree in built-up areas. In addition, the degree of concentration changes with time. The degree of concentration during daytime is higher than the degree during nighttime, and the degree of concentration on weekdays is higher than the degree on weekends. As a result, the degree of concentration during daytime on weekdays in city center is most remarkable.

Next, this study analyzes the distribution of population in city center, where large population concentrate, at census tract level. Figure 13.5 shows the density in city center during daytime is much higher than that during nighttime. Although city center is the single core where population concentrate, there are huge differences between densities of population in each census tract within city center. The densities of population in the census tracts in Puxi are higher than that in Pudong, especially during nighttime. Huangpu River is a natural divider, which causes the huge gap. People still concentrate in Puxi, the traditional central city. Based on the observation of density during daytime, this study finds there are contiguous areas with high density of population in the center of Puxi. In addition, these areas, concentrated within the inner ring, extend to the periphery, including Wujiaochang, Zhenru, Hongqiao, Caohejing, and Lujiazui, that shows fingerlike structure. Other census tracts with high density, like Sanlin in Pudong, and Zhangmiao in Baoshan, become

Fig. 13.6 The ratio of population during daytime and nighttime in city center at census tracts level

the core where population in suburbs concentrate. Based on the analysis of density during nighttime, this study finds the number of census tracts with high density during nighttime decreases and all of them concentrate in Puxi. Due to the lack of residences, the density of population in the census tracts with high density during daytime outside Puxi decrease sharply, such as Lujiazui and Sanlin. The hugest differences between densities during daytime and nighttime in census tracts in Puxi are those census tracts near Hunan Road and Tianping Road in Jingan.

Figure 13.6 shows the ratio of population during daytime and nighttime at census tracts level. First, census tracts with high ratio are continuous within the inner ring. Second, census tracts with low ratio surround the inner ring. And some census tracts with high ratio are dotted with peripheral areas. The ratios in the census tracts within the inner ring, where a large number of business offices concentrate, are the highest. At the same time, the ratios in census tracts beyond those areas are low. People who work inside the inner ring live in these areas. There are no continuous census tracts with high or low ratio in the peripheral areas. On the contrary, some census tracts with high ratio, such as Wujiaochang, sub center, Waigaoqiao Free Trade Zone in Pudong, Jinqiao Export Processing Zone, Zhangjiang Hi-Tech Park, and Caohejing Hi-Tech Park, are dotted with peripheral areas. The distribution of census tracts with low ratio, residential zones is dispersed, such as Dahua districts.

13.4 Analysis of Spatial Behavior

It is people's behaviors that cause the distribution of population change. Behaviors include regular and frequent commuting, irregular and infrequent leisure and consumptions, and the spatial changes of workplaces and residences. As a result, this study makes rules to recognize behaviors based on the research on people's behaviors

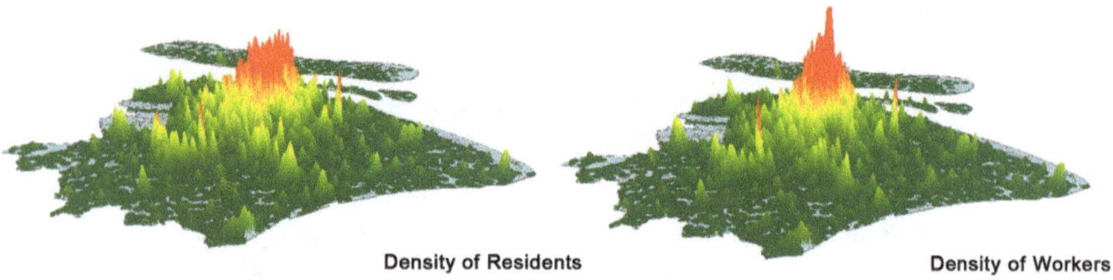

Fig. 13.7 Spatial distribution of residents and workers

and find out workplaces, residences, and the destination of other behaviors.[2] From the point of view of probability and statistics, the population data, multiplied by the corresponding coefficient, is similar to the actual management of the population data and also is certain in the spatial distribution of the reliability.

13.4.1 Distribution of Workplaces and Residences

According to the recognized workplaces and residences, this study analyzes the density of residences and workplaces. Figure 13.7 shows the three-dimensional distribution of density of residents and workers. Both of the distributions are monocentric distribution. Compared with distribution of density of residences, the degree of concentration of workplaces is much higher. When residences move outside, workplaces do not move outside. On the contrary, the degree of concentration of workplaces increases, which raises commuting demands. As a result, when land use is adjusted, workplaces and residences should be adjusted at the same time to reduce the traffic to city center and traffic jams.

13.4.2 Commuting Trips

The average commuting distance of residents, whose workplaces and residences are recognized, is 3571.42 m. Considering the distance between the China Mobile phone base stations, the recognized workplaces and residences have a certain error. In addition, residents do not include people whose workplaces and residents are not recognized. Although it is a little different with traffic survey, this recognized result shows the general commuting situation of residents in Shanghai. Figure 13.8 shows

[2] Compared with sixth national census, the distribution of recognized population is highly relevant to the number of population at districts level.

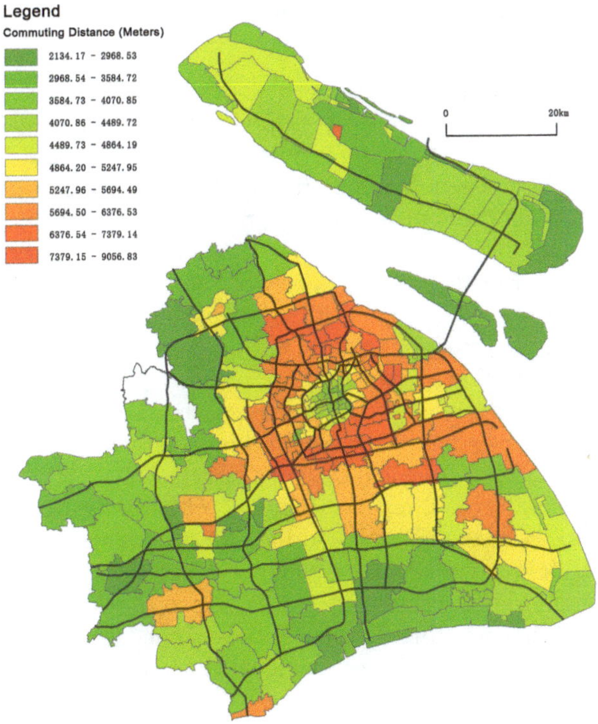

Fig. 13.8 Average commuting distance

that change pattern of average commuting distance inside the outer ring is complete opposite to the change pattern outside the outer ring. From city center to the outer ring, the average commuting distance changes from short, to long. On the contrary, from the outer ring to the periphery of Shanghai, the average commuting distance changes from long, to short. People who have the longest commuting distance are in the districts between city center and periphery, around the outer ring. Combined with land use, this study finds that there are a large number of residential parcels and few workplaces concentrating around the outer ring. As a result, a large number of residents in this area have a long commuting distance. On the other hand, people who have the shortest commuting distance are in the districts inside the inner ring or in the outer suburbs. A great quantity of public facilities and workplaces are inside the inner ring, where traffic developed. People are able to access workplaces with short commuting distance. In the outer suburbs, although there are few workplaces, people would not like to travel so far with inconvenient transportation to access workplaces. Compared with the districts in Puxi, people have longer commuting distance in the districts in Pudong. The reason is that people move to Pudong, where living condition is better, while they do not change their workplaces.

13.4.3 Leisure and Shopping Trips

Besides commuting, leisure and shopping are other main behaviors. Understanding the situation of leisure and shopping is helpful for urban planning. However, cell phone signaling data is not able to show purposes of trips. Because the frequencies of people having leisure and shopping trips are low, and people would stay at the destinations for a while, this study considers the locations where a cell phone user shows at most twice in 2 weeks and stays at least 2 h as the destinations of leisure and shopping trips.[3] Next, the distances between each cell phone user's home and destinations of leisure and shopping trips are calculated. Figure 13.9 shows the average leisure and shopping trips distance.

Similar with commuting, people who live in city center, Songjiang new towns, and Huinan new towns have the shortest leisure and shopping trip distance. The conclusion can be drawn that public facilities in these areas are located well that people are able to access leisure and shopping centers with a short-distance trip. As leisure and shopping centers, especially high-level leisure and shopping centers, concentrating in city center, people who live in Chongming, Jinshang, Fengxian, and districts besides sea in Pudong have longer leisure and shopping trip distance. In addition, compared with people living in Puxi, people who live in Pudong have longer leisure and shopping trips distance. It can be found that leisure and shopping still concentrate in Puxi, which is the traditional downtown of Shanghai.

13.4.4 Dependence on Central City

In former research, this study finds out that the spatial distribution experiences the central aggregation process during daytime and the decentralization to suburban area during nighttime and the densities of workplaces is much higher than the densities of residences in city center. The conclusion could be drawn that a great quantity of population commute to city center from suburban areas. In the same time, because of high-level leisure and shopping centers concentrating in city center, people who live in the outer suburbs have long-distance leisure and shopping trips. As a result, this study calculates the proportion of population who live outside the outer ring and have commuting, leisure, and shopping trips inside the outer ring to analyze the dependence on central city. Figure 13.10 shows the proportion of commuting, leisure, and shopping trips from suburbs to central city.

The most remarkable feature is that proportion of people who have commuting, leisure, and shopping trips to central city is highly related to the distance between districts and central city. In the nearest districts to the central city, the proportion is the highest. When the distance to central city increases, the proportion decreases.

[3] The most recognized destinations of leisure and shopping trips are shopping centers, parks, or public facilities, which are reliable.

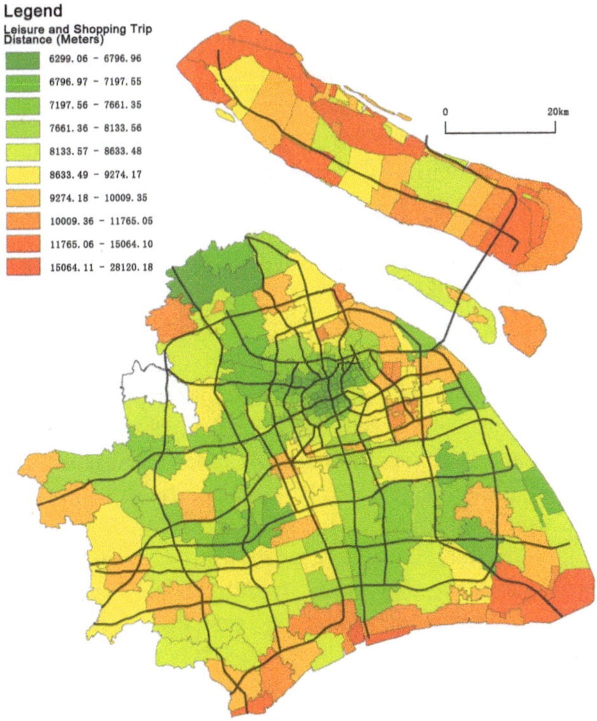

Fig. 13.9 Average leisure and shopping trip distance

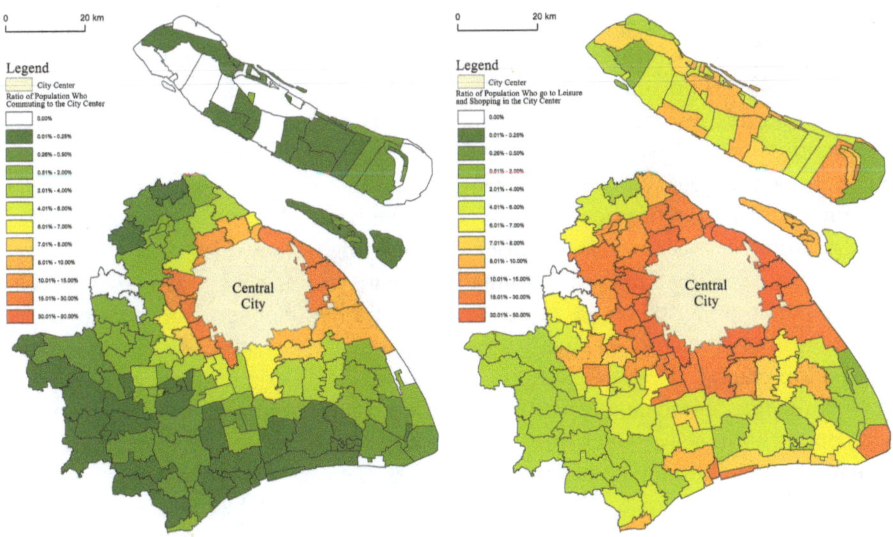

Fig. 13.10 Proportion of commuting, leisure and shopping trips from suburbs to central city at districts level

Compared the number of people who have commuting trips to central city and the number of people who have leisure and shopping trips to central city, this study finds that the degree of dependence of leisure and consumption on central city is higher. In addition, in Chongming, Linggang, and districts besides sea in Fengxian, the proportion of people who have leisure and shopping trips to central city is still high. The public facilities in these districts are inadequate.

13.5 Conclusion and Discussion

Using datasets of cell phone signaling records and related land use data, this study tries to build a framework based on internal relations among population, time, and behavior, which cause distribution of population change, to analyze spatial and temporal dynamic distribution of population in Shanghai. The results show as follows: (1) the density of population appears to be monocentric distribution, and this characteristic is more significant during the day compared to that at night. The spatial distribution experiences the central aggregation process during daytime and the decentralization to suburban area during nighttime. (2) People's behaviors (like commuting, leisure, and consumption) could cause the spatial and temporal distribution of population change. The spatial mismatch between residences and workplaces and the high dependence on the central area result in uneven distribution of population and formation of the central aggregating pattern. (3) The dependence level of consumption and leisure on central area is significantly higher than that of employment, especially in the suburban areas adjacent to the central area. In addition, the urban spatial structure of Shanghai, difference between city center and suburban areas, and other characteristics recognized by this study with cell phone signaling data are helpful for urban planning, managements of cities, and implementation of urban planning.

Based on this research, it can be found that using cell phone signaling data has an important meaning and value in the field of spatial and temporal dynamic distribution of population. With cell phone signaling data, details of spatial and temporal dynamic distribution of population, which is the basis of recognizing the mobility of population, assessing urban environment, building framework to deal with emergencies, and creating a digital urban management system, can be recognized in real time. In addition, based on the frequency and duration, planners can recognize different types of behaviors and analyze how these behaviors cause the distribution of population change. Understanding those and features of behaviors, planners can explore the interaction between urban environment and people's spatial behaviors in a comprehensive way, which provides scientific basis for urban planning, urban construction, and urban management. However, the accuracy of cell phone signaling data needs to be improved. First, there is no way to avoid the space error depending on the strength of base station's signal. Second, this study makes some rules to select cell phone users, which may affect the results. Other data should be consulted to improve the reliability. Based on the existing data, this study analyzes the spatial and temporal dynamic distribution of population and access preliminary results. With more data acquisition, building a better framework, optimizing, and applying models to analyze the spatial and temporal dynamic of distribution of population will be the focus of future research.

References

Ahas, R., & Mark, Ü. (2005). Location based service-new challenges for planning and public administration. *Futures, 37*(6), 547–561.

Ahas, R., Aasa, A., Silm, S., & Tiru, M. (2010). Daily rhythms of suburban commuters' movements in the Tallinn metropolitan area: Case study with mobile positioning data. *Transportation Research Part C: Emerging Technologies, 18*(1), 45–54.

Akkerman, A. (1995). The urban household pattern of daytime population change. *The Annals of Regional Science, 29*(1), 1–16.

Bian, X., Chen, H., & Cao, G. (2013). Patterns of regional urbanization and its implications: An empirical study of the Yangtze River Delta region. *Geographical Research, 32*(12), 2281–2291.

Calabrese, F., Colonna, M., Lovisolo, P., Parata, D., & Ratti, C. (2011). Real-time urban monitoring using cell phones: A case study in Rome. *IEEE Transactions on Intelligent Transportation Systems, 12*(1), 141–151.

Calabrese, F., Mi, D., Lorenzo, G., Ferreira, J., & Ratti, C. (2013). Understanding individual mobility patterns from urban sensing data: A mobile phone trace example. *Transportation Research Part C: Emerging Technologies, 26*, 301–313.

Deville, P., Linard, C., Martin, S., et al. (2014). Dynamic population mapping using mobile phone data. *Proceedings of the National Academy of Sciences, 111*(45), 15888–15893.

Diao, M., Zhu, Y., Ferreira, J., & Ratti, C. (2015). Inferring individual daily activities from mobile phone traces: A Boston example. *Environment and Planning B: Planning and Design, 43*(5), 1–21.

Ding, L., Niu, X., & Song, X. (2015). Identifying the commuting area of Shanghai central city using mobile phone data. *City Planning Review, 39*(9), 100–106.

Elvidge, C. D., Imhoff, M., Baugh, K. E., et al. (2001). Night-time lights of the world: 1994–1995. *ISPRS Journal of Photogrammetry and Remote Sensing, 56*, 81–99.

Foley, D. L. (1952). The daily movement of population into central business districts. *American Sociological Review, 17*(5), 538–543.

Foley, D. L. (1954). Urban daytime population: A field for demographic ecological analysis. *Social Forces, 32*, 323–330.

Guo, C., Zhen, F., & Zhu, S. (2014). Progress and prospect of the application of smart phone LBS data in urban researches. *Human Geography, 29*(6), 18–23.

Li, Y., Liu, H., & Tang, Q. (2015). Spatial-temporal patterns of China's interprovincial migration during 1985–2010. *Geographical Research, 34*(6), 1135–1148.

Liu, Y., Xiao, Y., Gao, S., Kang, C., & Wang, Y. (2011). A review of human mobility research based on location aware devices. *Geography and Geo-Information Science, 27*(4), 8–13+3+2.

Liu, Y., Liu, X., Gao, S., et al. (2015). Social sensing: A new approach to understanding our socioeconomic environments. *Annals of the Association of American Geographers, 105*(3), 512–530.

Long, Y., Shen, Z., & Mao, Q. (2011). Retrieving individual attributes from aggregate dataset for urban micro-simulation: A preliminary exploration. *Acta Geographica Sinica, 66*(3), 416–426.

Long, Y., Mao, M., Mao, Q., Shen, Z., & Zhang, Y. (2014). Fine-scale urban modelling and its opportunities in the "Big data" era: Methods, data and empirical studies. *Human Geography, 29*(3), 7–13.

Niu, X., & Ding, L. (2015). Analyzing job-housing spatial relationship in Shanghai using mobile phone data: Some conclusions and discussions. *Shanghai Urban Planning Review*, (121), 39–43.

Niu, X., Ding, L., & Song, X. (2014). Understanding urban spatial structure of Shanghai central city based on mobile phone data. *Urban Planning Forum, 6*, 61–67.

Pan, Q., Jin, X., & Zhou, Y. (2013). Population change and spatiotemporal distribution of China in recent 300 years. *Geographical Research, 32*, 1291–1302.

Qin, X., Wei, Y., Chen, W., & Duan, X. (2013). Population expansion and polycentric development of Nanjing city in a period of hyper-growth. *Geographical Research, 32*(4), 711–719.

Rao, Y., Song, J., & Yu, W. (2015). Spatial pattern and mechanism of population growth in metropolitan Beijing. *Geographical Research, 34*(1), 149–156.

Reades, J., Calabrese, F., & Ratti, C. (2009). Eigenplaces: Analysing cities using the space-time structure of the mobile phone network. *Environment and Planning B: Planning and Design, 36*(5), 824–836.

Roddis, S. M., & Richardson, A. J. (1998). *Construction of daytime activity profiles from households travel survey data*. Washington DC: National Research Council.

Sleeter, R., & Wood, N. (2006). Estimating daytime and nighttime population density for coastal communities in Oregon. *44th Urban and Regional Information Systems Association Annual Conference*. Columbia, British.

The Floating Population Service Management Department of national population and Family Planning Commission. (2011). *Floating population in China: 2011*. Beijing: China population publishing house.

Vieira, M., Frias-Martinez, V., Oliver, N., & Frias-Martinez, E. (2010). Characterizing Dense Urban Areas from Mobile Phone-Call Data: Discovery and Social Dynamics, In *Proceedings of 2010 IEEE Second International Conference* (pp. 241–248). Minneapolis.

Wang, L., Yang, Y., & Feng, Z. (2014). Prediction of China's population in 2020 and 2030 on county scale. *Geographical Research, 33*, 310–322.

Wang, D., Wang, C., Xie, D., et al. (2015a). Comparison of retail trade areas of retail centers with different Hierarchical levels: A case study of East Nanjing Road, Wujiaochang, Anshan Road in Shanghai. *Urban Planning Forum*, (223), 51–61.

Wang, D., Zhong, W., Xie, D., & Ye, H. (2015b). The application of cell phone signaling data in the assessment of urban built environment: A case study of Baoshan district in Shanghai. *Urban Planning Forum*, (225), 82–90.

Wu, Z., Chai, Y., Dang, A., et al. (2015). Geography interact with big data: Dialogue and reflection. *Geographical Research, 34*, 2207–2221.

Yang, C., & Ning, Y. (2015). Evolution of spatial pattern of inter-provincial migration and its impacts on urbanization in China. *Geographical Research, 34*(8), 1492–1506.

Yang, Z., Long, Y., & Douay, N. (2015). Opportunities and limitations of big data applications to human and economic geography: The state of the art. *Progress in Geography, 34*(4), 410–417.

Yuan, Y., Raubal, M., & Liu, Y. (2012). Correlating mobile phone usage and travel behavior: A case study of Harbin, China. *Computers, Environment and Urban Systems, 36*(2), 118–130.

Zhang, X., Chai, Y., Chen, Z., & Tan, Y. (2016). Analysis of spatial and temporal patterns of daily activities of suburban residents based on GPS data: A case study of the Shangdi-Qinghe area of Beijing. *International Review for Spatial Planning and Sustainable Development, 4*, 4–16.

Zhen, F., & Wang, B. (2015). Rethinking human geography in the age of big data. *Geographical Research, 34*(5), 803–811.

Zhen, F., Qin, X., & Xi, G. (2015). The innovation of geography and human geography in the information era. *Scientia Geographica Sinica, 35*(1), 11–18.

Chapter 14
Application of Big Data in the Study of Urban Spatial Structures

Yi Shi and Junyan Yang

Abstract Communication devices such as cell phones not only capture large amounts of human behavioral data but also provide information about the structure of cities and their dynamic properties. In this chapter, we focus on the latter aspects by studying mobile phone signaling data recorded during 8 days in Shanghai. We first define a behavioral density parameter that measures the intensity of individuals' behaviors, allowing us to highlight the centers of urban structure. We then propose a polycentric model to test the polycentric structure of Shanghai. The result shows that Shanghai has a polycentric urban structure. We study the behavioral density of the polycentric urban structure and show in particular that signaling data is not only effective in detecting behavioral density but also allows us to distinguish different categories of urban centers. These results point to the possibility of identifying the urban spatial structure using mobile phone signaling data.

Keywords Urban spatial structure • Urban centers • Big data • Phone signaling data • Individual behavior density

14.1 Introduction: Research Progress of Urban Space Based on Mobile Phone Signaling Data

Individual spatial and temporal behavior is the spatial position of the individual at different times when the behavior is carried out. The aggregation intensity of individuals in a specific space during a specific period is defined as the individual behavior density. If one considers "Mobile phone landscape: the use of mobile phone positioning data of the city analysis" (Ratti et al. 2006) as the beginning of bringing mobile phone data into urban planning research, the relationship between individual behavior density and urban location is the first question that concerns

Y. Shi (✉) • J. Yang
School of Architecture, Southeast University, Nanjing, China
e-mail: shiyi@seu.edu.cn

Z. Shen, M. Li (eds.), *Big Data Support of Urban Planning and Management*,
Advances in Geographic Information Science, DOI 10.1007/978-3-319-51929-6_14

planning scholars. Based on the study of cell phone data, it has been found that call detail records (CDRs) of mobile phones reflect the difference in population density in different city regions in cities such as Milan (Ratti et al. 2006; Pulselli et al. 2008), Rome (Reades et al. 2007; Girardin et al. 2008), New York (Girardin et al. 2009; Isaacman et al. 2010), Boston (Calabrese et al. 2010) and Los Angeles (Isaacman et al. 2010). As a further validation, the density of mobile phone users in Portugal, France (Deville et al. 2014), and New York (Becker et al. 2011) was compared with census data. It was found that the population density distribution based on mobile phone CDR data and short message service (SMS) records was highly consistent with the Census Bureau's census data. The research shows that behavior density estimation based on mobile phone data has high reliability in urban individual distribution analysis.

Because mobile phone signaling data has a much higher sampling rate, which can record positioning signals of all the mobile devices in a power-on state, and a higher update rate (positioning every 0.5 h on average) than CDR data or SMG data, mobile phone signaling data is a more accurate data source for behavioral density estimation in urban research. Krisp (2010) estimates the spatial distribution of a residential population for the planning of fire and first aid facilities based on the spatial density of mobile users at night. Niu et al. (2014), Ding et al. (2015) identifies urban spatial structure using mobile phone user density based on mobile phone signaling data in Shanghai. Wang et al. (2015) compared consumers' spatial distribution in three commercial centers in Shanghai, Nanjing Dong Lu, Wujiaochang, and Anshanlu, using mobile phone signaling data. The behavioral density analysis based on mobile phone signaling data is not only a technical exploration of new data types but has become an important method to apply to urban structure and behavioral research.

14.2 Data Processing and Analysis: Calculation of Individual Behavior Density in 3D Perspective

14.2.1 Mobile Phone Data

This study uses Shanghai's mobile phone data from the biggest communication service operator in China, including base station coordinates and mobile signaling data. The signaling data uses an encrypted phone identification number that has been removed from the user's attributes and does not involve any personal information belonging to the mobile user. The mobile phone signaling data was recorded over a total of 8 days in 2013, including 1 weekday and 1 weekend day in every season. During the above recording time, mobile phone signaling data from 9578 base stations in Shanghai recorded about 20 million different mobile phone identification numbers, representing roughly the same amount of 2G mobile phone users.

14.2.2 Mobile Phone Data Analysis Method

Through mobile phone signaling data, the population spatial aggregation situation at a particular time can be simulated. This study presents a method for calculating behavior density based on three-dimensional space. Through this algorithm, mobile phone signaling data collected at units based on the location of mobile phone base stations can be transformed into mobile phone user data with urban lot as the unit. An important feature of the algorithm is that it introduces the distribution weight of the 3D activity area, which improves the accuracy of the algorithm in the microscopic scale of the city (Figs. 14.1 and 14.2).

14.3 Dynamic Characteristics of Behavioral Density

Urban spatial structure results from the aggregation of different types of facilities in a city. Commercial service facilities gather to form the city center, and the gathering of nonpublic facilities such as residences forms the urban community. This paper examines the relationship between urban spatial structure and behavioral density distribution from the perspective of public and nonpublic facilities.

14.3.1 Relevant Differences Between Daytime and Nighttime Behavior Density and Facility Floor Area Ratio

We chose the floor area ratio (FAR) of public service facilities as the research object for public service facility layout. As public service agencies are the core feature of the city center, the FAR of public service facilities can be used as a measure of the development of urban lots from small commercial land to urban centers.

A linear regression equation was used to study the diurnal behavior density change relative to the central system layout. The public service facility FAR (FARa) of each plot was taken as the dependent variable in the linear regression model. The FAR of all lots and the FAR of non-common service facilities (FARb) are used as reference dependent variables. The average hourly behavioral density from 1:00 to 24:00 on working days was taken as the independent variable, which was recorded as (H1, H2, H3, ..., H24). The coefficient of determining R2 represents the proportion of the portion of the total data sample that can be explained by the regression equation. The higher the R2, the greater the proportion of the portion of the total data sample that can be explained by the regression equation.

Table 14.1 shows the regression coefficient of the hourly behavior density on weekdays from 0:00 to 24:00 and the FAR for all lots, FARa and FARb. Among the coefficients of determination between FAR and hourly behavioral density, H8 has

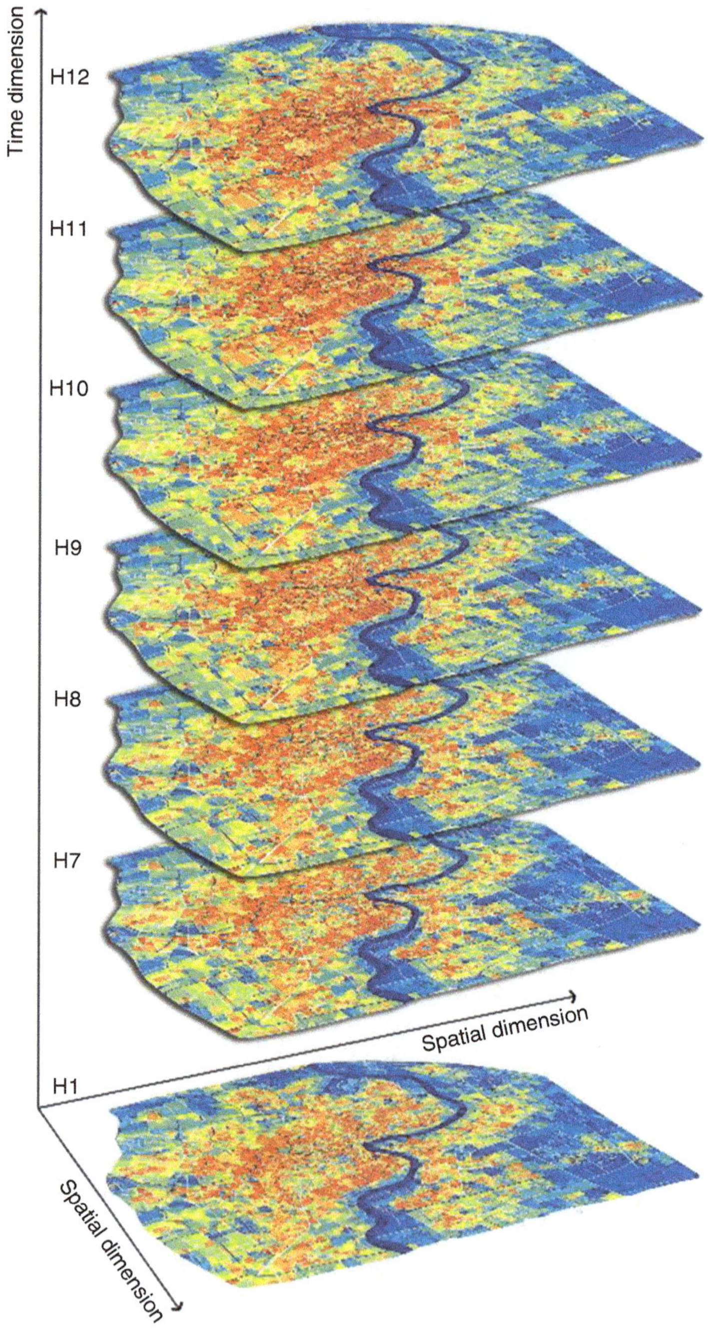

Fig. 14.1 Behavior density at different times

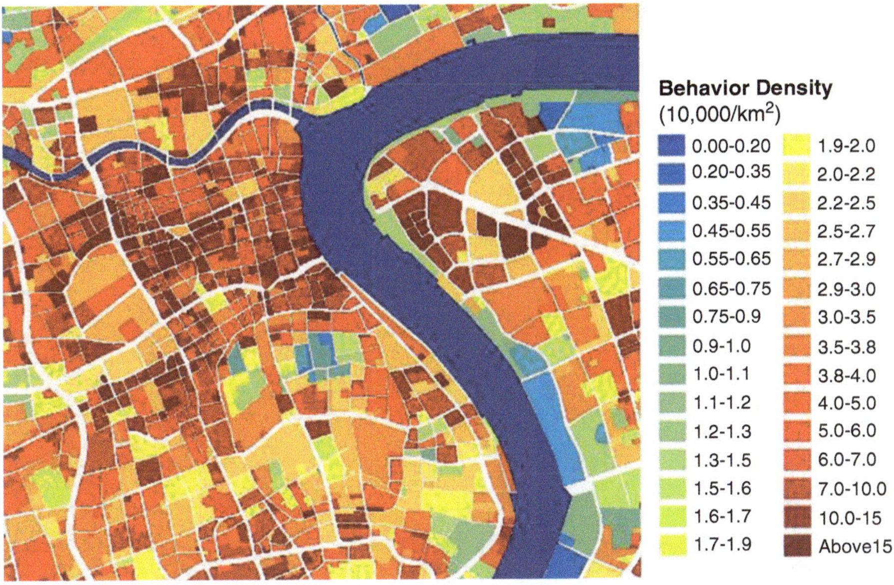

Fig. 14.2 Behavior density calculation results using lot as unit

the highest R2, with a value of 0.636, and H15 has the lowest R2, with a value of 0.603. The difference between the highest R2 and the lowest R2 is small. This indicates that there is no significant difference in the effect of changes in FAR between daytime and nighttime behavioral density.

The FAR of public service facilities (FARa) has the highest correlation with H12, with an R2 of 0.706 and a B coefficient of 0.840, and the lowest correlation with H5, with an R2 of 0.439 and a B coefficient of 0.663. On the contrary, the FAR of nonpublic service facilities has the highest correlation with H1 and H2, with both R2 coefficients measuring as 0.640 and both B coefficients measuring 0.800, and the lowest correlation with H15, with an R2 coefficient of 0.577 and a B coefficient of 0.760.

The B coefficients of the FAR, FARa, and FARb and the behavioral density from 1:00 to 24:00 on weekdays are plotted on the graph and connected (Fig. 14.3). It can be seen from the figure that the regression coefficient of the FARa varies significantly during the day and night cycle on weekdays. The B coefficient shows a low level from 4:00 to 5:00, and it rises sharply after 6:00. The B coefficient maintains a level above 0.800 from 9:00 to 18:00 and then rapidly declines to 0.693 from 18:00 to 24:00, maintaining a low level between 0.650 and 0.700 from 24:00 to 0.700 the next day.

The regression coefficient curve of FARb also shows a significant fluctuation in the day and night cycle on weekdays. The B coefficient is maintained at a high level of about 0.800 from 23:00 to 8:00, decreasing to 0.782 at 9:00. The coefficient from 9:00 to 17:00 is maintained at a low level around 0.760 and returns to 0.800 from 17:00 to 24:00.

Table 14.1 Regression coefficient of behavior density and FAR at different times

	FAR	FAR_a		FAR_b		
	R2	Beta	R2	Beta	R2	Beta
H1	0.629	0.793	0.464	0.681	0.640	0.800
H2	0.628	0.793	0.455	0.674	0.640	0.800
H3	0.627	0.792	0.447	0.669	0.639	0.800
H4	0.625	0.791	0.441	0.664	0.639	0.799
H5	0.625	0.790	0.439	0.663	0.638	0.799
H6	0.625	0.791	0.441	0.664	0.638	0.799
H7	0.629	0.793	0.468	0.684	0.639	0.799
H8	0.636	0.798	0.539	0.734	0.635	0.797
H9	0.634	0.796	0.651	0.807	0.611	0.782
H10	0.622	0.788	0.694	0.833	0.591	0.769
H11	0.616	0.785	0.705	0.840	0.586	0.766
H12	0.613	0.783	0.706	0.840	0.585	0.765
H13	0.609	0.781	0.699	0.836	0.583	0.764
H14	0.604	0.777	0.695	0.834	0.579	0.761
H15	0.603	0.776	0.696	0.834	0.577	0.760
H16	0.605	0.778	0.697	0.835	0.578	0.761
H17	0.610	0.781	0.697	0.835	0.583	0.764
H18	0.613	0.783	0.679	0.824	0.586	0.766
H19	0.617	0.785	0.617	0.785	0.598	0.773
H20	0.625	0.791	0.585	0.767	0.612	0.782
H21	0.632	0.795	0.572	0.757	0.623	0.790
H22	0.634	0.797	0.544	0.738	0.631	0.795
H23	0.633	0.796	0.513	0.716	0.636	0.798
H24	0.631	0.794	0.480	0.693	0.639	0.799

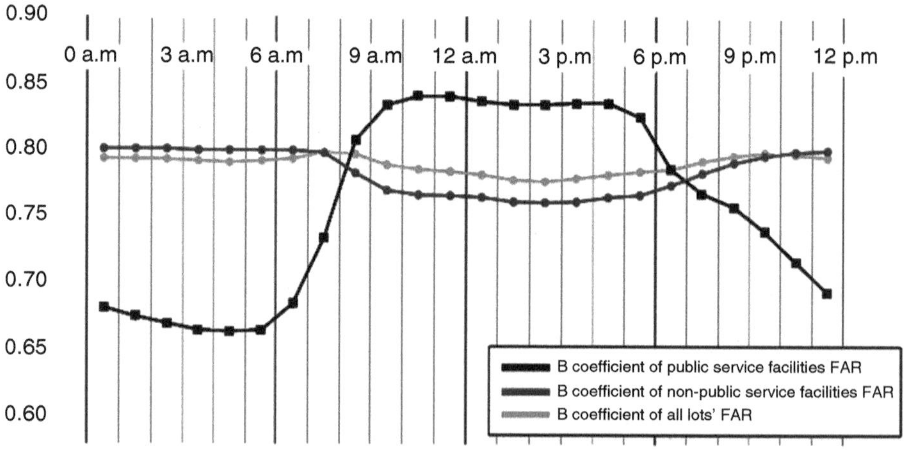

Fig. 14.3 B coefficient of behavior density and FAR, FAR_a, FAR_b at different times

For FARa and FARb, there is a relatively stable peak range between 9:00 and 17:00, and from 24:00 to 6:00, which is highly consistent with the daily pattern of workplace and residence behavior. In the working hours from 9:00 to 17:00, individuals mainly conduct public behaviors such as working, eating, and spending time in recreational activities. Therefore, the distribution of behavior density in this period is significantly related to the public service facilities FARa. The time period from 24:00 to 6:00 constitutes non-working hours when behavior is mainly private, such as staying home, resting, and sleeping, so the correlation with FARa decreases, and correlation with FARb significantly increases. The time period from 5:00 to 9:00 and from 17:00 to 24:00 corresponds to the transition period between the working and sleeping periods; thus, the regression coefficient is also fluctuating. Thus, the change in behavioral density between day and night reflects, in fact, the dynamic relationship between public behavior, nonpublic behavior, and urban facility layout: public behavior density and public service facility layout have a significant correlation. However, the density of nonpublic behavior is closely related to the layout of nonpublic service facilities, including residential facilities. The correlation between the two is a progressive cycle with changing day and night dynamics. The dynamic correlation between facilities and the behavioral density proves that the factors that promote the formation of urban centers in a specific area of the city do not include the gathering of the population, but the aggregation of specific behaviors such as employment, recreation, and commuting.

14.3.2 Relevance Differences Between Weekday and Weekend Behavior Density and FAR

When the study of individual behavior density is placed on a longer time scale, it is possible to compare the relevant differences between the weekday and weekend behavior density and the layout of urban facilities. The linear regression model is built with the public service facility FAR (FARa) as the dependent variable. The behavioral density of weekdays and weekend days from 0:00 to 24:00 is the independent variable, denoted as H1, H2, H3, ..., H24.

Table 14.2 lists the B coefficients of behavioral density from 0:00 to 24:00 on weekdays/weekends with FARa. It is found that the B coefficients of weekdays are higher than that of weekends for each hour. The difference between the two is most significant from 9:00 to 19:00 and becomes smaller from 0:00 to 8:00. This is because the working day and holiday behavior patterns are mainly reflected in the time interval from 9:00 to 19:00, due to more people leaving their residence to participate in public behaviors on weekdays than on weekends. Therefore, the correlation between behavior density on weekdays and public service facilities is more significant than it is on the weekend (Fig. 14.4).

Comparing the B coefficients of weekdays with those of the weekend, another observed difference is that the weekend B coefficients drop more slowly after

Table 14.2 Regression coefficient of behavior density and FAR on weekdays/weekends

Weekday	FAR_a		Weekend	FAR_a	
	R2	Beta		R2	Beta
H1	0.464	0.681	H1	0.445	0.667
H2	0.455	0.674	H2	0.437	0.661
H3	0.447	0.669	H3	0.430	0.656
H4	0.441	0.664	H4	0.426	0.652
H5	0.439	0.663	H5	0.423	0.651
H6	0.441	0.664	H6	0.425	0.652
H7	0.468	0.684	H7	0.448	0.670
H8	0.539	0.734	H8	0.504	0.710
H9	0.651	0.807	H9	0.603	0.776
H10	0.694	0.833	H10	0.656	0.810
H11	0.705	0.840	H11	0.667	0.817
H12	0.706	0.840	H12	0.667	0.817
H13	0.699	0.836	H13	0.652	0.810
H14	0.695	0.834	H14	0.651	0.807
H15	0.696	0.834	H15	0.647	0.804
H16	0.697	0.835	H16	0.643	0.802
H17	0.697	0.835	H17	0.641	0.801
H18	0.679	0.824	H18	0.626	0.791
H19	0.617	0.785	H19	0.581	0.763
H20	0.585	0.767	H20	0.553	0.744
H21	0.572	0.757	H21	0.531	0.729
H22	0.544	0.738	H22	0.515	0.718
H23	0.513	0.716	H23	0.490	0.700
H24	0.480	0.693	H24	0.460	0.678

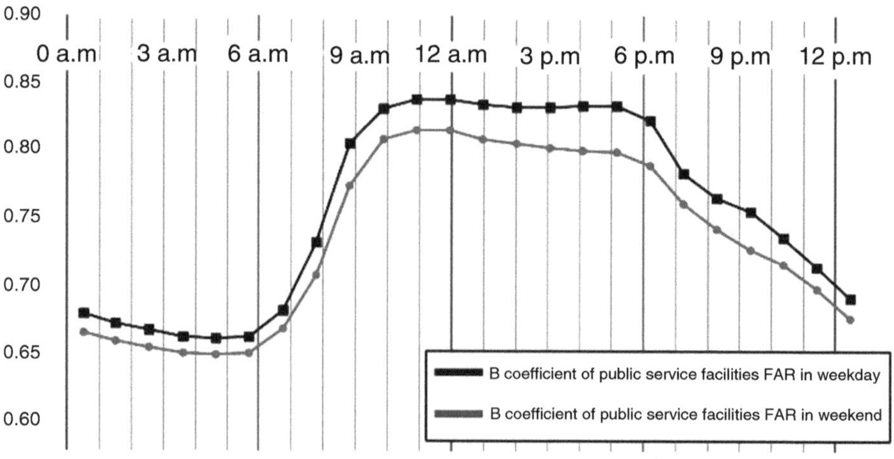

Fig. 14.4 B coefficient of behavior density and FAR_a on weekdays/weekends

19:00 in the evening. This is because public activities on the weekend are not strongly affected by working hours; thus, public activities at night are more numerous and diversified.

Although the content and form of individual behaviors on the weekdays and weekends are very different, the circadian fluctuation pattern of the correlation between the public service facility FAR and the individual behavioral density does not change, which indicates the change in public behavior content such as work, shopping, and recreation will not affect the existence of dynamic association rules between public service facility layout and individual behavior density.

14.3.3 Relevance Differences in Seasonal Behavior Density and FAR

When individual behavior density is observed over a longer period, the relevant differences between the seasonal behavior density and the layout of urban facilities can be compared. The linear regression model was built using the public service facility FAR (FARa). The behavioral densities of the weekdays from 0:00 to 24:00 in spring, summer, autumn, and winter are the independent variables, denoted as H1, H2, H3, ..., H24. All these days were sunny, with a maximum outdoor temperature of 15 °C (spring), 35 °C (summer), 25 °C (autumn), and 5 °C (winter).

Table 14.3 lists the B coefficients of behavior density from 0:00 to 24:00 in different seasons with FARa. The minimum value of the B coefficients in each season both appeared in the morning, between 4:00 and 5:00. The individual behavior density in the fall has the highest correlation with the FARa at all times, especially at the highest levels, which were around 0.85 in the period from 9:00 to 17:00. This period also contains the highest point in the B coefficient curve during the spring and winter, which are 0.83 and 0.80, respectively. The B coefficient of summer in the early morning and evening is similar to that of autumn but shows a significant decrease at noon, unlike in other seasons. Combining the temperature of each sampling date, the difference of the B coefficient of the behavioral density between different seasons shows the influence of the season on public activity. Autumn, with a maximum temperature of 25 °C, is most suitable, so the correlation between the behavioral density and public service facilities FARa is most significant. Spring, with a daily maximum temperature of 15 °C in the daytime, is more suitable for public activities, so the B coefficient is only slightly lower than the fall. However, nights and mornings in spring are relatively cold, so the B coefficient is low. Since the temperature at noon in summer is too hot (with a maximum temperature of 35 ° C), the intensity of public activity at that time is suppressed. In winter, with the highest daily temperature of 5 °C, the willingness to participate in social activities decreased due to the lowest temperatures of the year, so the correlation between behavioral density and public service facilities was significantly lower than that of other seasons (Fig. 14.5).

Table 14.3 Regression coefficient of behavior density and FAR in different seasons

	FAR_a (spring)		FAR_a (summer)		FAR_a (autumn)		FAR_a (winter)	
	R2	Beta	R2	Beta	R2	Beta	R2	Beta
H1	0.425	0.652	0.442	0.665	0.466	0.682	0.442	0.665
H2	0.418	0.647	0.433	0.658	0.457	0.676	0.433	0.658
H3	0.410	0.641	0.428	0.654	0.451	0.672	0.424	0.651
H4	0.407	0.638	0.424	0.651	0.447	0.669	0.414	0.643
H5	0.405	0.636	0.422	0.649	0.445	0.667	0.412	0.642
H6	0.408	0.639	0.424	0.651	0.447	0.669	0.412	0.642
H7	0.435	0.660	0.452	0.673	0.473	0.688	0.434	0.659
H8	0.507	0.712	0.522	0.723	0.532	0.729	0.501	0.706
H9	0.622	0.789	0.637	0.798	0.639	0.799	0.607	0.779
H10	0.666	0.816	0.663	0.814	0.694	0.833	0.630	0.793
H11	0.675	0.821	0.649	0.806	0.711	0.843	0.638	0.799
H12	0.676	0.822	0.630	0.794	0.715	0.845	0.636	0.797
H13	0.670	0.819	0.618	0.786	0.714	0.845	0.630	0.794
H14	0.668	0.817	0.616	0.785	0.712	0.844	0.626	0.792
H15	0.669	0.818	0.618	0.786	0.712	0.844	0.625	0.791
H16	0.671	0.819	0.626	0.791	0.712	0.844	0.624	0.790
H17	0.672	0.820	0.637	0.798	0.712	0.844	0.623	0.789
H18	0.658	0.811	0.638	0.799	0.688	0.829	0.613	0.783
H19	0.606	0.778	0.584	0.765	0.620	0.787	0.562	0.750
H20	0.577	0.760	0.552	0.743	0.586	0.766	0.536	0.732
H21	0.545	0.738	0.537	0.733	0.580	0.761	0.532	0.729
H22	0.509	0.714	0.515	0.718	0.550	0.742	0.508	0.713
H23	0.473	0.688	0.492	0.701	0.510	0.714	0.485	0.697
H24	0.440	0.663	0.457	0.676	0.480	0.693	0.459	0.677

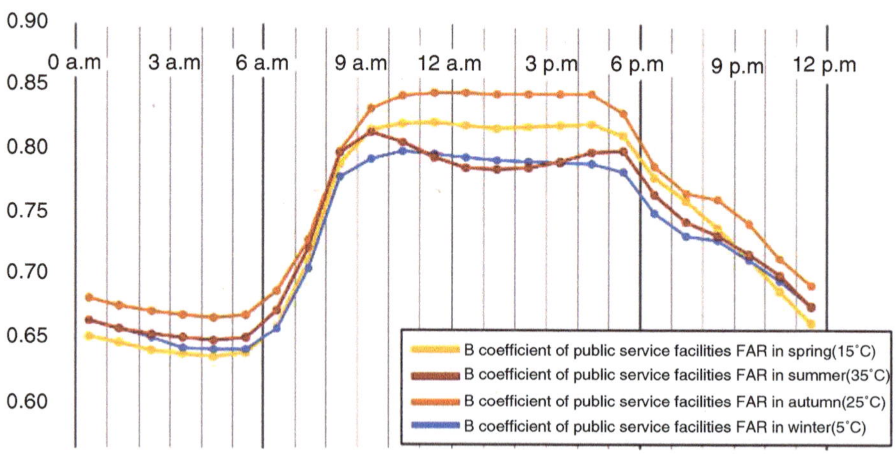

Fig. 14.5 B coefficient of behavior density and FAR_a in different seasons

As the seasons change the behavior pattern of the population, the correlation between the behavioral density of different seasons and the public facilities fluctuates. However, the volatility difference of the correlation is consistent with the influence of the urban environment on travel, commuting, and recreational activities under seasonal influence, thus proving the stability of the association between the urban facility layout and the individual behavior density over longer periods.

14.4 Identification of Shanghai Polycentric Structure

Examining the urban spatial structure from a behavioral perspective, the urban center is found to be a high-intensity gathering area of public activity in the city. Therefore, the identification of urban polycentric structure needs to calculate the public behavior density in different urban lots. This paper uses the distribution of mobile phone users during the daytime (from 9:00 to 18:00), which indicates the distribution of urban social activities and uses the density of mobile users in the abovementioned period as the baseline for urban polycentric structure (Fig. 14.6).

First, the regions where daytime behavior density is highly concentrated needed to be identified. The daily mean density of each plot was calculated from 9:00 am to

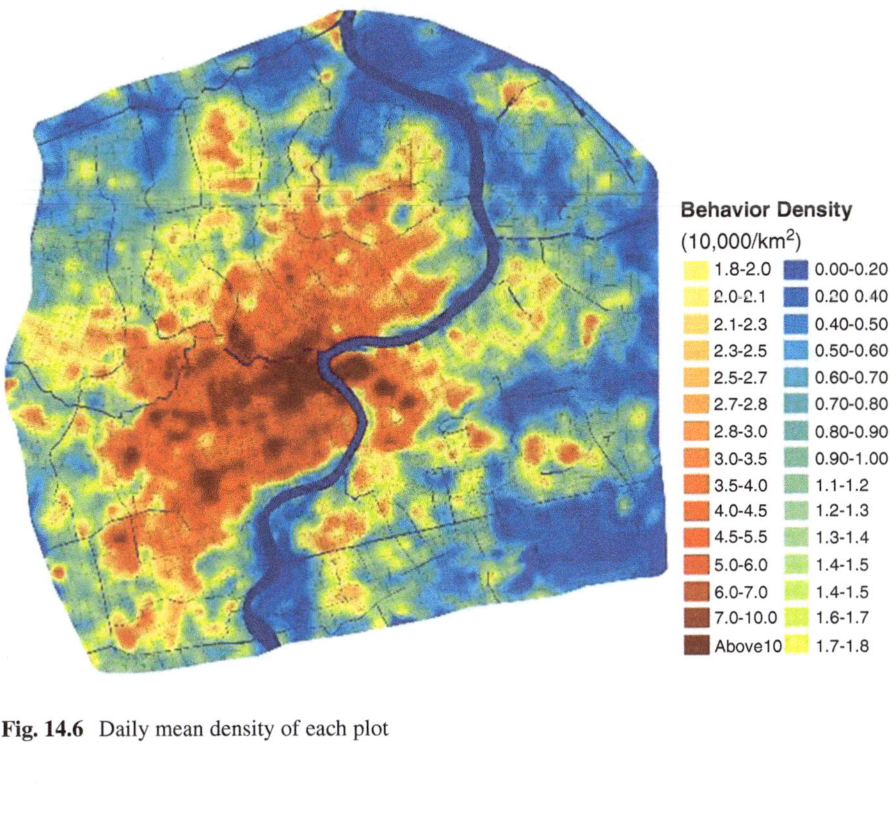

Fig. 14.6 Daily mean density of each plot

6:00 pm on working days. The mean value distribution of daytime behavior density was plotted using the method of nuclear density calculation. The kernel polynomial interpolation uses local polynomial interpolation, plotting the contours to divide the city into areas of different behavior densities, with a contour spacing of 10,000 people/km^2. By comparing the range of different contours, the 60,000/km^2 contour is used as the standard for identifying the city center. An enclosed area greater than 0.2 km^2 located within a range of 60,000 km^2 or greater is identified as an urban center.

According to the above methods, People's Square, Lujiazui, Xujiahui, Shanghai Railway Station, Dapuqiao, Wujiaochang, Hongkou Football Stadium, Changshou Road, Hongqiao, Riverside Road, Tangqiao, Zhangjiang, and Caohejing and other lots are identified as urban centers (Fig. 14.7). The spatial extent of the central area is defined using its high behavior density blocks. The total area of the 12 urban centers is 45.2 km^2. According to the daily behavior density, the number of mobile phone users in the central area is about 1.85 million, which is about 10% of the mobile phone users in Shanghai.

Fig. 14.7 Identification of Shanghai polycentric structure

14.5 Classification of Shanghai Polycentric Structure

14.5.1 Based on Behavioral Density

The concentration of individual behavior reflects its level as an urban center. The average daily population density in each central area is estimated and standardized based on the number of mobile phone users in the area of 100,000 people/km^2. For comparison, the urban center is divided into four grades by a natural breaks classification, which represents the first-, second-, third-, and fourth-level centers. People's Square is a first-level center; second-level centers include Lujiazui, Shanghai Railway Station, Xujiahui, Hongqiao, Dapuqiao, and Changshoulu; third-level centers include Caohejing, Wujiaochang, Hongkou Stadium, and Linping Road; fourth-level centers include Tangqiao and Zhangjiang (Fig. 14.8).

14.5.2 Based on Influence Distance

The influence distance of urban centers also reflects differences in the center level. The residence of an urban center space user can be identified based on mobile phone signaling data, as well as the distance between the user's residence and the urban center. The level of urban center classification reflects these factors.

The phone base station coordinates of all central area users were screened from 0:00 to 5:00 on weekdays if users were located at the same base station for at least half of the time (3 h or more). The area where the base station is located is identified as the user's residence. According to the above method, a total of 97,000 people were identified from the continuous signaling data of 200,000 mobile users, and the coordinates of the user's residence were identified. For example, there were 6017 mobile phone users identified who had been active in the central area of the People's Square Centre during the day. These users are distributed around 72 mobile phone base stations at the center of the People's Square during the day and spread out to 1519 mobile phone base stations at night. Linking these residences to the center reflects the influence range of the People's Square Centre.

The average distance between an urban center and its users' residence is defined as that center's influence distance. People's Square, which has the largest influence distance, reaching 9.0 km. Wujiaochang, Lujiazui, Xujiahui, Hongqiao, Zhangjiang, and the railway station have influence distances between 7.0 and 8.0 km. Dapuqiao, Changshoulu, Tangqiao, and Linpinglu have influence distances between 5.0 and 6.0 km. Caohejing and Hongkou stadium have the lowest distance, which are 4.9 km and 4.6 km, respectively. After the standardization, the urban centers are divided into four grades by the natural breaks classification. The People's Square is a first-level center. The second-level centers are Wujiaochang, Lujiazui, Xujiahui, Hongqiao, Zhangjiang, and the train station. The third-level centers are Dapuqiao, Changshoulu, Tangqiao, and Linpinglu. The fourth-level centers are Caohejing and Hongkou Stadium (Fig. 14.9, Table 14.4).

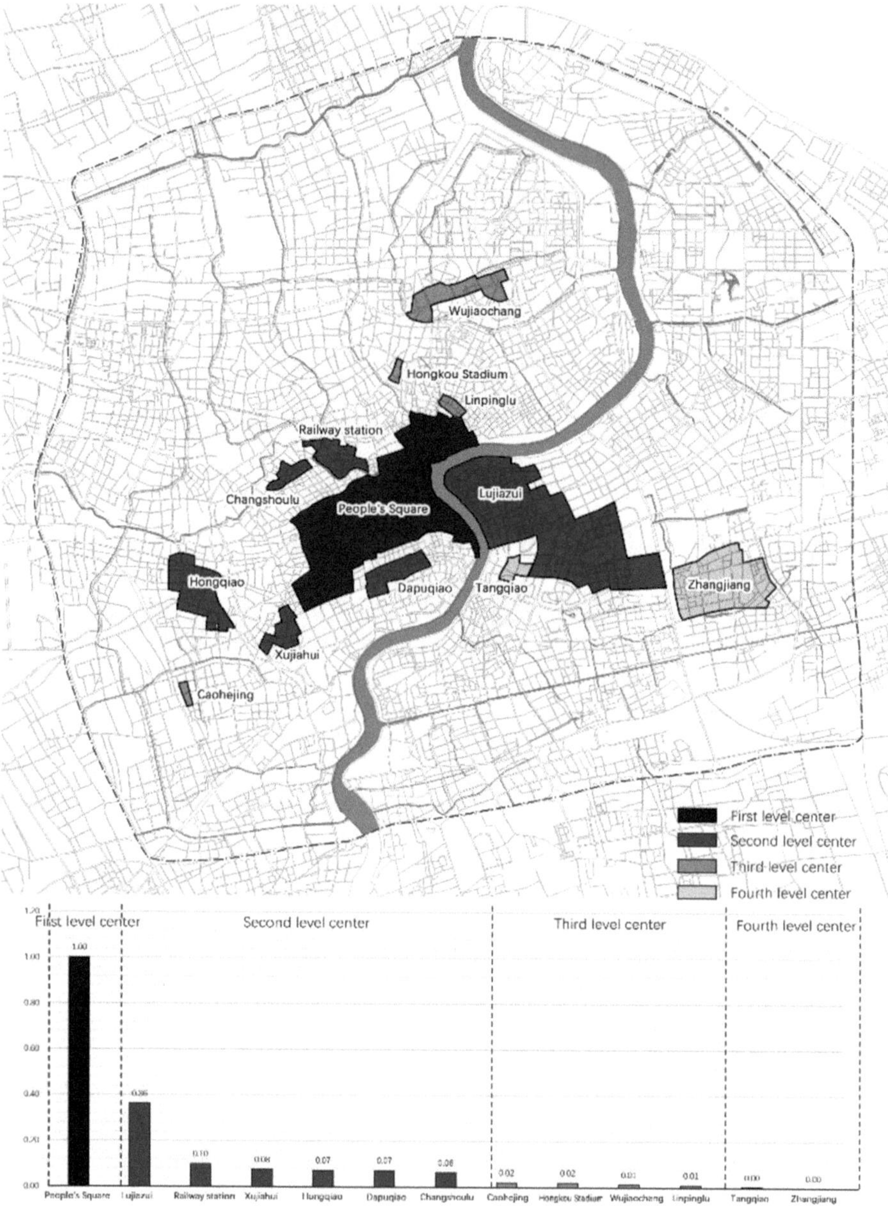

Fig. 14.8 Classification of Shanghai polycentric structure based on behavioral density

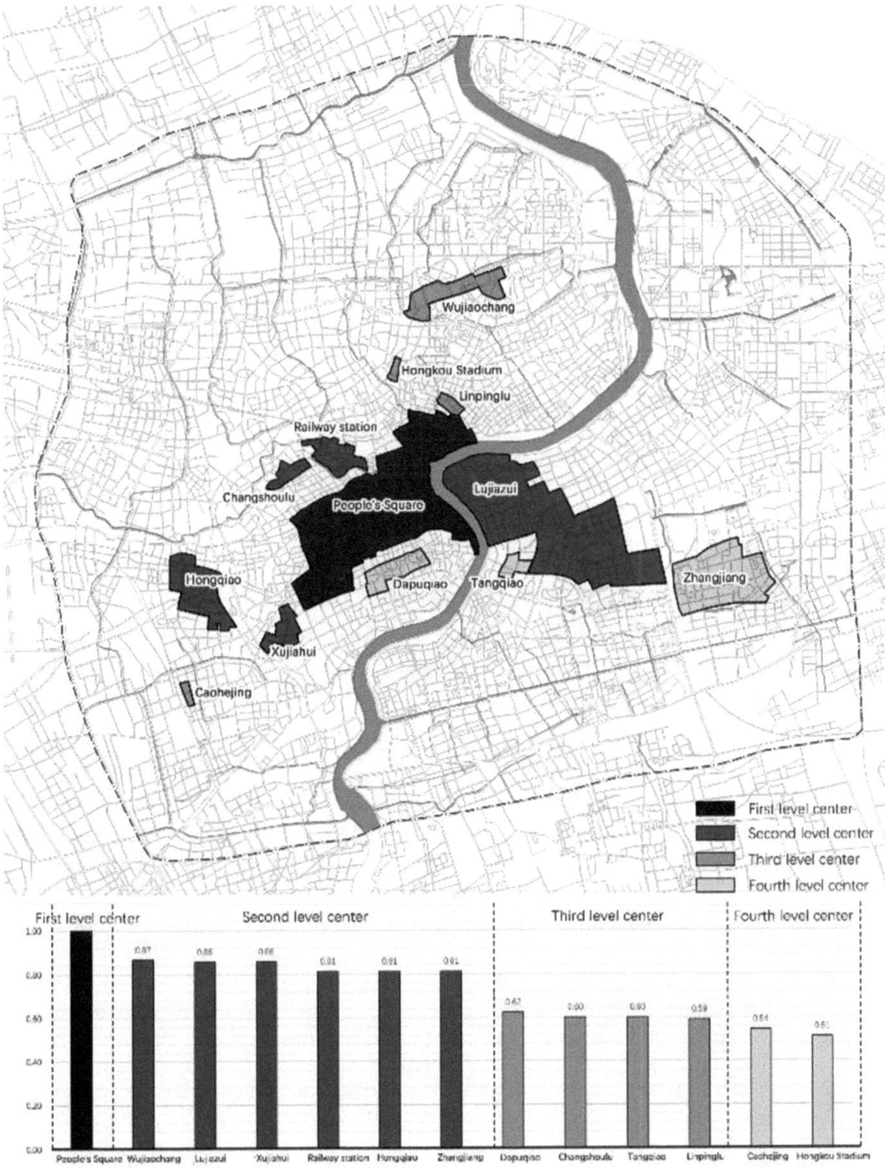

Fig. 14.9 Classification of Shanghai polycentric structure based on influence distance

Table 14.4 Urban centers' influence distance

Name	Influence distance (km)	Standardization
People's Square	9	1.00
Wujiaochang	7.8	0.87
Lujiazui	7.7	0.86
Xujiahui	7.7	0.86
Train station	7.3	0.81
Hongqiao	7.3	0.81
Zhangjiang	7.3	0.81
Dapuqiao	5.6	0.62
Changshoulu	5.4	0.60
Tangqiao	5.4	0.60
Linpinglu	5.3	0.59
Caohejing	4.9	0.54
Hongkou	4.6	0.51

14.5.3 Comparison of Two Methods of Classification

In this paper, we propose two ways to classify urban centers using mobile phone signaling data: one is based on behavior density, and the other is based on influence distance. The urban center classification based on behavior density represents the centers' attraction of public activity. The urban center classification based on influence distance represents the service range of the centers. Using SPSS software to calculate the correlation of the classification results under the two methods, the correlation coefficient is found to be 0.627, which shows that the stronger the centers' attraction to activity, the greater the scope of its urban impact.

By comparing the classification results of the central area under the two methods, we find that:

1. Based on the mobile phone signaling data, the classification of urban centers can be effectively divided into different levels in the central area.

The two classification methods based on mobile phone signaling data can clearly divide the urban centers into four categories. People's Square, regardless of the density of behavior or the impact of distance, is at the highest level, significantly ahead of other central areas. The railway station, Xujiahui, Wujiaochang, Hongqiao, and other second-level centers have similar level characteristics. These types of urban centers are characterized by a greater range of attractions and greater impact on public activities than low-level centers but are still significantly below the first level. Lujiazui's center in particular, which appeared in the late twentieth century, is in a transitional stage between the two categories. Lujiazui center's behavior density is much higher than other second-level centers, but it lags behind traditional city centers, such as the People's Square, in influence distance. The third- and fourth-level centers are below the above two types of centers, both in terms of density of behavior and influence distance.

2. The classification method results show the difference between the urban center levels.

Although the behavior density and the influence distance both reflect the grade of the urban centers, their expression is different. The behavior density reflects the ability of the urban center to attract public activity, while the influence distance reflects the spatial influence of the centers. By comparing the results of the center classification from the two methods, we find the difference of the two methods in the classification.

In the calculation of the behavior density and influence distance, the values for the first-level center Shanghai People's Square are standardized as 1. Based on the influence distance measurement, the standardized values of for Lujiazui, the railway station, Xujiahui, and other second-level centers reach 0.8–09. However, based on the behavior density calculation, the standardized values of the second-level centers only reach 0.06–0.36. This illustrates significant differences between the first- and second-level centers in their ability to attract public activity. The influence distance of the second-level centers is close to that of the first-level center, but between them there is a large gap in behavior density, as the first-level center corresponds to the municipal main center, while the second-level center corresponds to the municipal subcenter. These two kinds of centers serve the entire city rather than the district they are located in; thus, the scope of their impact is similar. The first-level centers provide more comprehensive services, whereas the second-level centers are mainly responsible for one of commercial, business, administrative, or other specific services. It is because of the specificity in service content of the second-level centers that there is a small gap between them and the first-level centers in influence distance, but large gap in the ability to attract public activity (Table 14.5).

The situation is the opposite for second-level centers and third-level centers. According to the results of influence distance measurement, the standardized numerical gap between second-level centers and third-level centers is about 0.19–0.27.

Table 14.5 Comparison of two methods of classification

Urban center name	Behavioral density	Influence distance
People's Square	1.00	1.00
Lujiazui	0.36	0.86
Shanghai train station	0.10	0.81
Xujiahui	0.08	0.86
Hongqiao	0.07	0.81
Wujiaochang	0.06	0.87
Dapuqiao	0.07	0.62
Changshoulu	0.06	0.60
Linpinglu	0.05	0.59
Zhangjiang	0.02	0.81
Caohejing	0.02	0.54
Hongkou Stadium	0.02	0.51
Tangqiao	0.01	0.60

According to the results of behavior density measurement, the standardized numerical gap between second-level centers and third-level centers is only about 0.02–0.09. This illustrates the difference between second-level centers and third-level centers. These two kinds of centers are closer in behavior density but show a wide gap in influence distance. The second-level centers mainly correspond to the municipal subcenters, while the third-level centers mainly correspond to the district centers. The non-equivalence of the results of the two methods proves that although the second-level centers and third-level centers are closer in the behavior activity intensity, the second-level centers are significantly larger than the third-level centers.

Acknowledgments This research is supported by a grant from the National Natural Science Foundation of China (No.51708103).

References

Becker, R. A., Caceres, R., Hanson, K., Loh, J. M., Urbanek, S., Varshavsky, A., & Volinsky, C. (2011). A tale of one city: Using cellular network data for urban planning. *IEEE Pervasive Computing, 10*(4), 18–26.

Calabrese, F., Pereira, F. C., Di Lorenzo, G., Liu, L., & Ratti, C. (2010, May). The geography of taste: Analyzing cell-phone mobility and social events. In *International Conference on Pervasive Computing* (pp. 22–37). Berlin/Heidelberg: Springer.

Deville, P., Linard, C., Martin, S., Gilbert, M., Stevens, F. R., Gaughan, A. E., et al. (2014). Dynamic population mapping using mobile phone data. *Proceedings of the National Academy of Sciences, 111*(45), 15888–15893.

Girardin, F., Calabrese, F., Dal Fiore, F., Ratti, C., & Blat, J. (2008). Digital footprinting: Uncovering tourists with user-generated content. *IEEE Pervasive Computing, 7*(4), 36–43.

Girardin, F., Vaccari, A., Gerber, A., Biderman, A., & Ratti, C. (2009). Quantifying urban attractiveness from the distribution and density of digital footprints. *International Journal of Spatial Data Infrastructure Research, 4*, 175–200.

Isaacman, S., Becker, R., Cáceres, R., Kobourov, S., Rowland, J., & Varshavsky, A. (2010, February). A tale of two cities. In *Proceedings of the Eleventh Workshop on Mobile Computing Systems & Applications* (pp. 19–24). ACM.

Krisp, J. M. (2010). Planning fire and rescue services by visualizing mobile phone density. *Journal of Urban Technology, 17*(1), 61–69.

Niu, X., Ding, L., & Song, X. (2014). Urban spatial structure of Shanghai central city based on mobile phone data. *Journal of Urban Planning, 6*, 61–67.

Ding, L., Niu, X., & Song, X. (2015). Identifying the commuting area of shanghai central city using mobile phone data. *Journal of Urban Planning, 9*, 100–106.

Pulselli, R., Ramono, P., Ratti, C., & Tiezzi, E. (2008). Computing urban mobile landscapes through monitoring population density based on cellphone chatting. *International Journal of Design and Nature and Ecodynamics, 3*(2), 121–134.

Ratti, C., Frenchman, D., Pulselli, R. M., & Williams, S. (2006). Mobile landscapes: Using location data from cell phones for urban analysis. *Environment and Planning B: Planning and Design, 33*(5), 727–748.

Reades, J., Calabrese, F., Sevtsuk, A., & Ratti, C. (2007). Cellular census: Explorations in urban data collection. *IEEE Pervasive Computing, 6*(3), 30–38.

Wang D., Wang C., et al. (2015). Comparison of business classes of different levels of business centers in Shanghai based on mobile phone signaling data - A case study of Nanjing East Road and Wujiaochang City, Shanghai, China. Anshan Road as an example. *Urban Planning Journal, 3*, 008.

Chapter 15
Application of Cellular Data in Traffic Planning

Jianhui Lai, Yanyan Chen, Zijun Wu, Guang Yuan, and Miaoyi Li

Abstract Traffic planning is a very important tool that enables planners to determine solutions for traffic congestion. In practice, there are many difficulties, including a lack of data, that have prevented traffic planners from fully understanding urban travel characteristics, and thus the implementation of some traffic planning projects has not met the travel demands of urban residents. The traditional method is to estimate travel demand by combining the nature of land use and the intensity of development. However, for traffic planning, the prediction of factors, such as the travel distance, commuting distance, and other indicators, relies on a "four-stage model," which needs large amounts of difficult to obtain modeling data, and the results do not have a high degree of precision. In addition it is not practical for all applications. In this study, we used cellular data to extract the distribution of people, including the distribution of their jobs, travel, and other parameters, and established a city scape based only on job distribution, with spatial location data used to model the travel characteristics. Based on the parameters of population density, job density, balance of jobs and residents, distance to the city center, and distance to the workplace, the trip density, the daily trip distance, and the distance commuted, a predictive model for different urban areas was established by multiple linear regression. The correlation coefficients of the model indicated that the accuracy of trip density prediction was very high, while the accuracy of daily trip distance and distance commuted was lower, but the model was considered suitable for practical application.

Keywords Traffic planning • Cellular data

J. Lai (✉) • Z. Wu
Beijing Institute for Scientific and Engineering Computing, Beijing, China
e-mail: lseelight@163.com

Y. Chen • G. Yuan
College of Metropolitan Transportation Beijing University of Technology, Beijing University of Technology, Beijing, China

M. Li
Joint International FZUKU Lab SPSD, Fuzhou University, Fuzhou City, China

Z. Shen, M. Li (eds.), *Big Data Support of Urban Planning and Management*,
Advances in Geographic Information Science, DOI 10.1007/978-3-319-51929-6_15

15.1 Introduction

The concept of excess or wasteful commuting was initially introduced by Hamilton and has become a well-established line of inquiry in transportation research in recent decades. Banister (1997) reported that the average commuting distances in European and North American metropolitan areas were increasing each year. There is no doubt that the continual expansion of cities has played a role in the growth of the distances traveled. However, it appears that the potential to travel over increasingly larger distances has systematically resulted in the physical separation of functions. In particular the separation of residential areas and the workplace has played an important role in increasing the commuting distance.

In this study, cellular data was used to study the distribution of urban jobs and traffic links between workplace trips and to establish a model to forecast travel intensity and travel distance.

The density and distribution of urban housing and jobs are not only the most fundamental indicators of the intensity of urban economic structure and regional activities but are also the most important input data of any traffic demand model. Studies at home and abroad have indicated that these factors have a close relationship with the trip characteristics of urban residents. For example, the higher the density of jobs is, the greater the trip intensity is. The higher the degree of mixing between residential areas and workplaces, the shorter the trip distance is, and the greater the possibility of using public transport is. The distribution of residential areas and jobs in cities can be used to determine urban population mobility and establish effective travel regulation while providing a reference for urban management and planning. Many studies of the characteristics of urban housing and employment have been conducted; for example, Zhang et al. (2013) and Liu and Fanbu (2011) studied the Beijing metropolitan area to determine the relationship between employment and residential locations. Using data from the second national census of basic units of Beijing and the fifth population census of Beijing, the researchers established a geographical database using a geographic information system (GIS) to identify the spatial distributions of employment and residences. Zhu (2013), Liang et al. (2003), and Du et al. (2007) used population census data or artificial surveys to study the relation between employment and residential areas in big cities.

Investigating the trips and household information of urban residents is still the most common method to acquire residential and employment data. The locations and travel time of respondent are recorded in detail each day by questionnaire investigations. However, the method has two main disadvantages. First, the cost of such investigations is high and can be very difficult, with professional investigators required. Second, the sampling rate is low, and with the expansion of cities is continually decreasing, which makes the uniformity of samples very difficult to achieve. The development of techniques such as GPS and wireless communication networks has provided an effective method for tracking population mobility. The positioning accuracy of GPS data is high (about 10 cm) and the sampling frequency is easy to set.

However, dedicated GPS receiving equipment needs to be configured for each respondent, and GPS signals cannot be acquired within a building. It is easy to use wireless communication networks to record the location of urban residents.

15.2 Cellular Data

In the process of an individual using a cellular phone under conditions with an overlay of signals, several actions can occur, which the wireless network is triggered to record. These include the Cell-ID of the phone, the connecting base station, location area code (LAC), and the time of the trigger action (see Table 15.1). As reported by Lai (2013), Sasaki and Nishii (2010), the process is called a trigger event, with various event types recognized and listed below. We refer to this as cellular (or signaling) data:

- Calling and called
- Receiving and sending messages
- Turning on/off phone
- Periodic location update
- Normal location update

The position of a base station is used to represent the location of cellular phone users. The position is not the actual position of a phone user. In a city, the density of base stations is high, but in terms of the cellular phone's actual position, there is an error of 300–500 m, while for a countryside, the error is about tens of meters. This can be used to track urban population mobility (Fig 15.1).

15.2.1 Data Size

When a cellular phone user triggers an event, one location point is recorded. The more location points that are recorded, the easier it is to track their position. Path matching was conducted by tracing the location points and potential routes and was

Table 15.1 Cellular phone data

Field name	Data type	Information content
IMSI	VARCHAR2	Phone identifier is stored in the subscriber identity module (SIM), which can differentiate between individual phone users
TIME	DATE	Timestamp is a time record, which is added by the manufacturers to the signal process, and is accurate to 1 s
LAC	NUMBER (8)	Location area code
Cell-ID	NUMBER (8)	Cell area identity
Event-ID	NUMBER (8)	Event types: (1) periodic location update, (2) turning off phone, (3) turning on phone, (4) normal location update

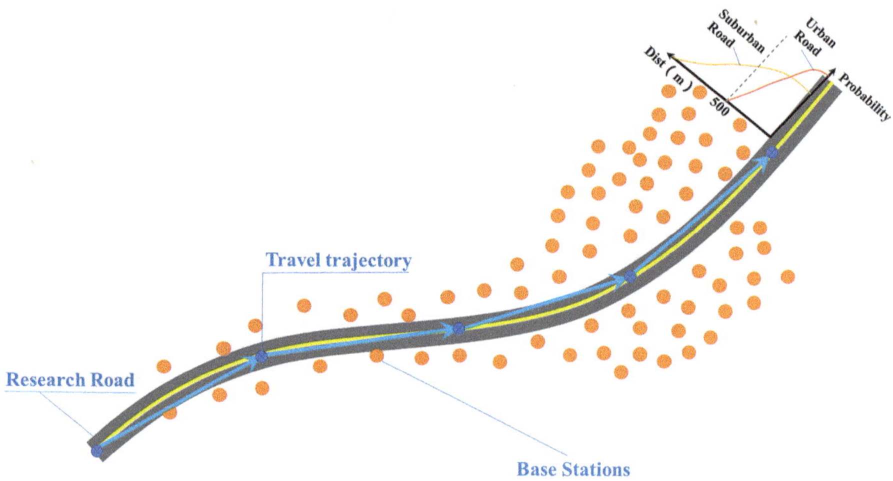

Fig. 15.1 The trajectory of cellular data

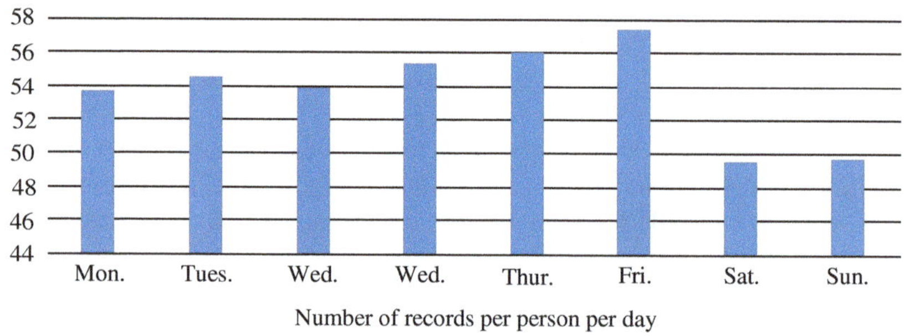

Fig. 15.2 The average number of communication events for cellular phone users

used to analyze the speed of a cellular phone user's movement. Prioritizing the analysis of a cellular phone user's trace features in call detail records is beneficial for adopting the correct methods to build an analysis model. Data from the China Mobile network in Beijing was tested. The location of a phone on this network in Beijing area is periodically updated every 2 h. The number of communication events for a cellular phone user is shown in Fig. 15.2. The effective data volume (i.e., event records) per person per day was 53.8, with about 55.2 event records on weekdays and 49.7 on weekend days. This is because people make more calls and send more short messages on weekdays triggering more events.

After analyzing the number of events triggered by cellular phone users at different periods on 1 day (Fig. 15.3a and Fig. 15.4a), it was found that each user triggered an average of 2.1 events per hour. The number of events triggered in each hour of 1 day differed greatly. For example, the number of events triggered each hour from 0.00 to 06.00 h was <1, because cellular phone users sleep at night and rarely use their phones. The information collected for this period was mainly the data pro-

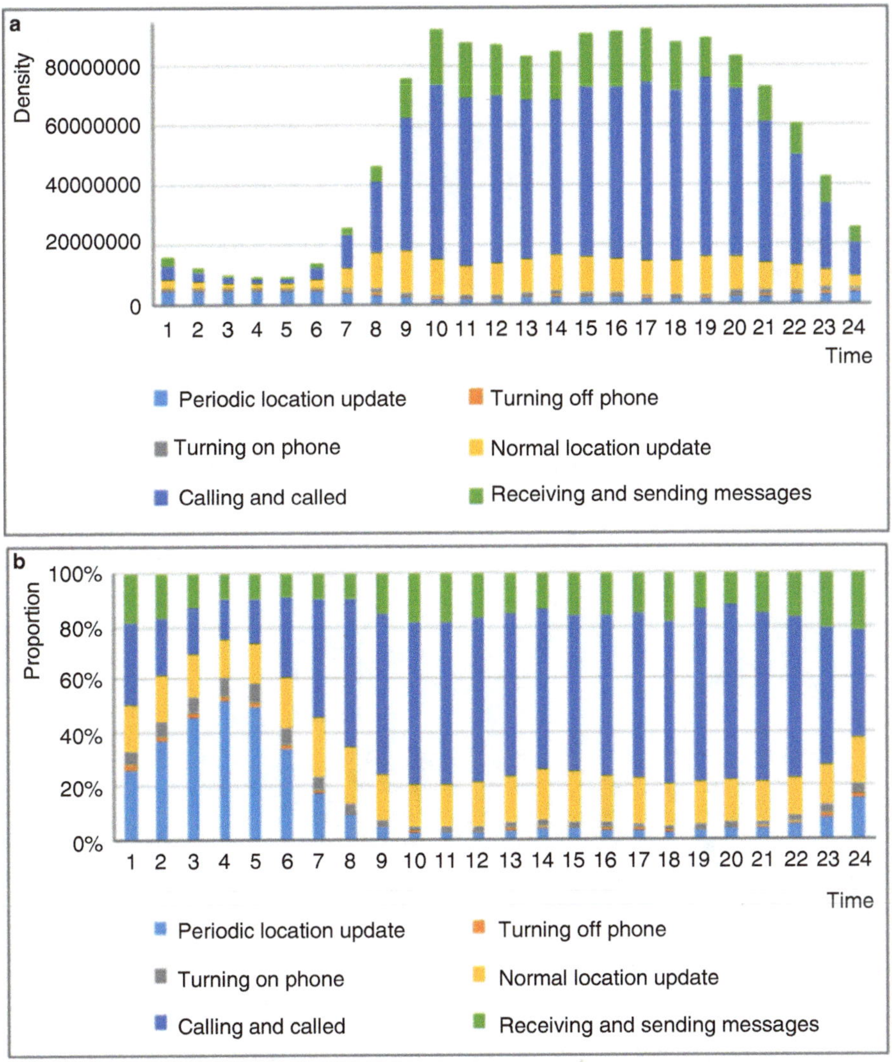

Fig. 15.3 The type of communication events for cellular phone users (weekdays). (**a**) The density of communication events. (**b**) The proportion of communication events

duced by periodic position updates, but the average number of communication events triggered each hour from 08.00 to 18.00 h was about 3 (i.e., one data message every 20 min).

The proportion and density of different types of events throughout the day differed substantially. The density of communication events in the day and at night varied, with six events per person per hour in the daytime and less than one event per person per hour after dawn. The densities of normal position updates and call making and answering events differed substantially. Cellular phone users undertook

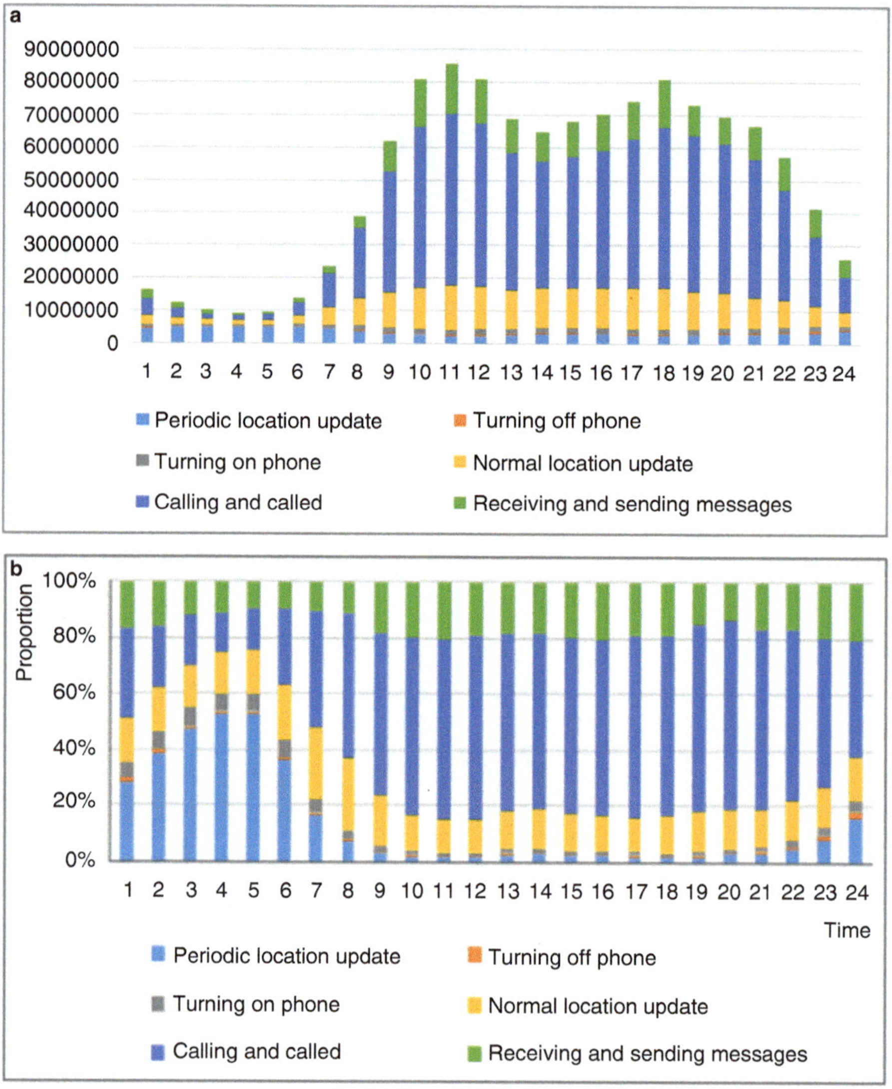

Fig. 15.4 The type of communication events for cellular phone users (weekend). (**a**) The density of communication events. (**b**) The proportion of communication events

more cross-regional activities in the daytime, and the density of normal position updates was about 1 per person per hour, while the density of call making and answering was about 4 per person per hour. Most cellular phone users were asleep at night, and therefore the density of normal position updates fell to 0.1 per person per hour, and the density of call making and answering was about 0.2 per person per hour. Among all of the various communication events, the proportion of periodic position updates and call making and answering differed substantially over 1 day.

In the daytime, more than 60% of communication events were call making and answering, while at night, about 40% of communication events were periodic position updates. Starting up and shutting down events accounted for a relatively low proportion of all events in the daytime and at night.

On weekends (Fig. 15.4), communication events happened less frequently than on weekdays and the proportion of communication events differed slightly.

Therefore, for single cellular phone users, the data collected was affected by long time intervals between data, no fixed term of collection, and large differences between day and night time data.

15.2.2 Time Interval

As shown by the above analysis, the collection of call record details did not occur over a fixed term. The time intervals between adjacent records in different periods were recorded and are shown in Fig. 15.5. From 7:00 to 19:00, the average time interval between data production was 1800 s, with a standard difference of about 3000 s. During the night, the average time interval was about 4000 s and the standard difference was 4500 s. This indicates that the average time interval between call detail records during the daytime was shorter than at night. The fluctuation between adjacent time intervals was relatively large and was larger at night than during the daytime.

15.3 Calculation of Traffic Status

15.3.1 Residential Areas and Workplaces

Urban residents move between their place of residence and workplace by commuting. When both are in the same area, the commuting distance is shorter. The first position was used as the starting point, the distance from the starting point to the position point was the abscissa (ignoring the direction), and time was the ordinate axis. The resulting figure was referred to as a trip space-time diagram. From the diagram, the relationship of time and space for a cellular phone user could be observed; for example, Fig. 15.6 shows the trip space-time diagram of a cellular phone user over a week. The user is generally at home in the evening and moves to another location during the day. The shape of the space-time diagram on Monday, Tuesday, and Thursday was similar, with no movement. There was movement over a short distance at 12.00 h on Wednesday. The cellular phone user was not in the same location in the morning and afternoon on Friday. They went shopping at 10.00 h on Saturday and returned to their place of residence at 15.00 h. They made no trips on Sunday. By comparing the time intervals and duration that the cellular

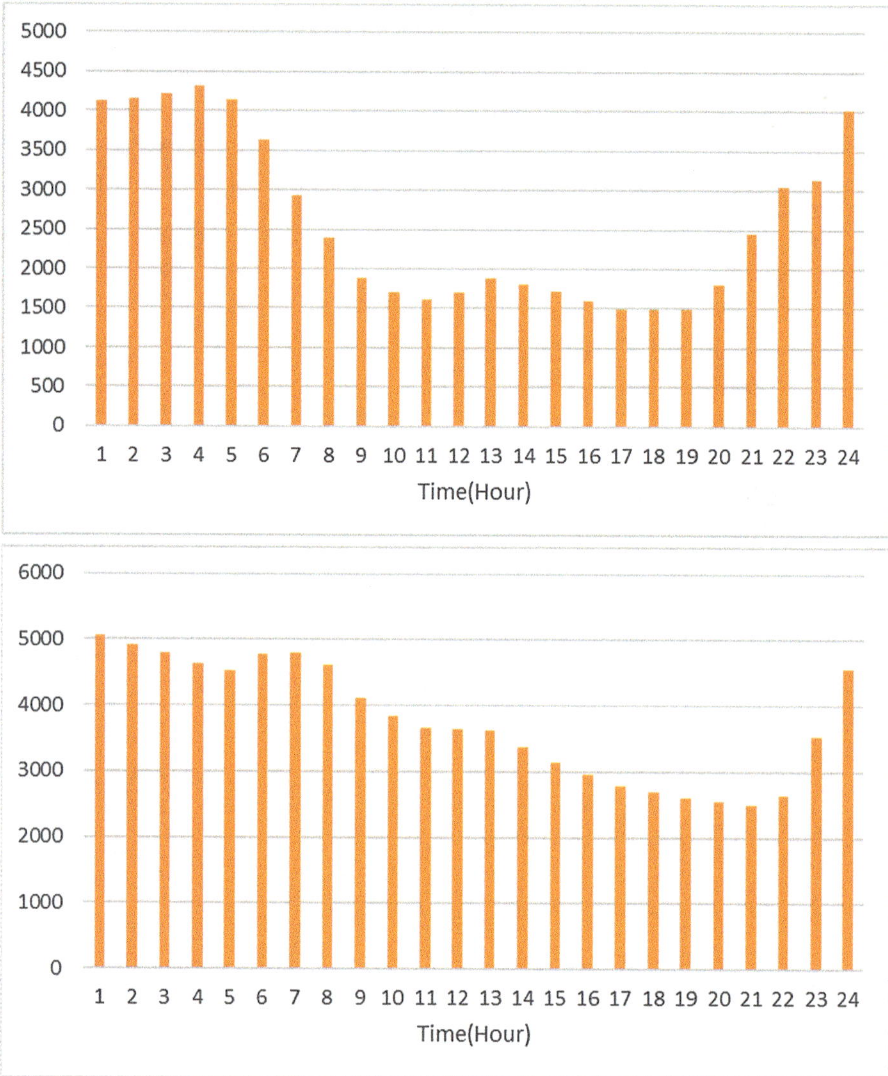

Fig. 15.5 The time interval between call record details. (**a**) Average value of time interval (s). (**b**) Standard deviation of time interval (s)

phone user remained in the area, we could determine if their place of residence and workplace were within the same area. If the cellular phone user remained in a particular area during the evening for several days, it was considered to be their place of residence.

When a cellular phone user remains in the same position, with no movement, the connection with the base station theoretically remains the same. In reality, signals from cellular phones may move back and forth (i.e., ping-ponging) between different base stations, and the corresponding cellular phone user's position was then recorded

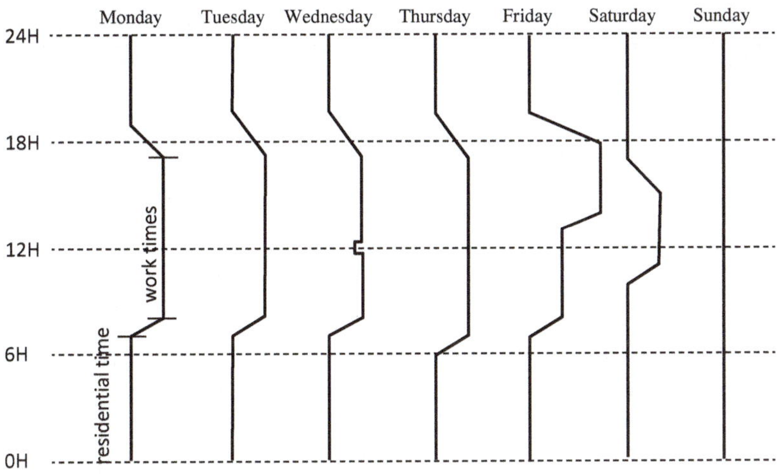

Fig. 15.6 Trip space-time diagram of a cellular phone user

Fig. 15.7 Distribution of a cellular phone user's base station connection

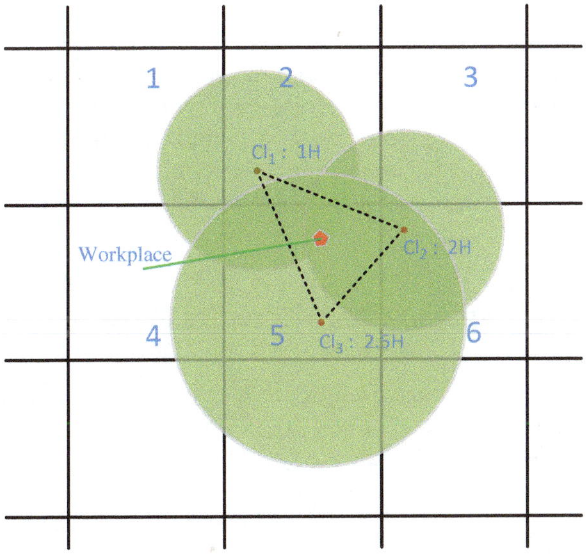

as a "ping-ponging handover." When the overlay of a signal in a particular location was weak, this feature was more evident. In Fig. 15.7, the position of the red star indicates the actual workplace of a cellular phone user. It connected to three base stations CI_1 , CI_2 , CI_3. The duration of connection was 1, 2, and 2.5 h, respectively. The three base stations were located in areas 2, 6, and 5. Therefore, the area of any connected base station may be the actual area in which the cellular phone user was located. To reduce the influence of error on the results, the data for several days was compared using the weighted average position of the three base stations to represent the position of a single base station.

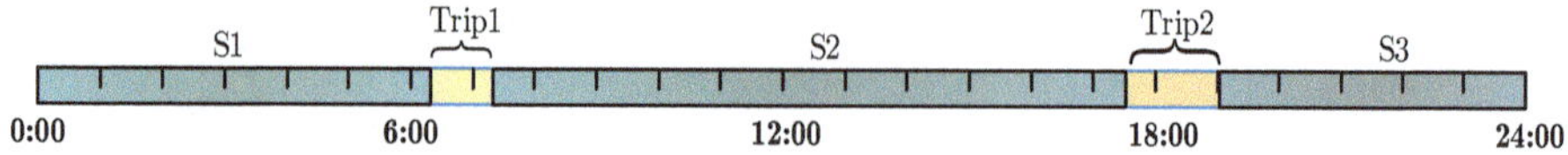

Fig. 15.8 Relationship between trips and stop points

15.3.2 Trips

Cellular data can track the entire process of an urban resident's activity, and therefore it can be used to calculate traffic travel information, such as travel origin-destination (OD) pairs, travel distance, and travel time. Before calculating each of these indicators, we first determined the cellular phone user's state (i.e., moving or stationary), by assessing if they were undertaking continuous movement. If the cellular phone user was in the process of traveling, the time interval between the start and the end of the movement was the time consumed during the trip, and the middle track chain was the distance traveled. If the time of two adjacent trips was the same, it may indicate where the cellular phone user was living or working (Fig. 15.8).

The process of assessing the state of movement of a cellular phone user was based on a naive Bayes classifier, which uses a certain number of known states, combined with characterization parameters and the user's category (resident population or floating population) to determine the state of movement of a cellular phone user at any point in time. The procedure is shown in Fig 15.9.

15.4 Spatial Characteristics of the Traffic Demand in Beijing

Cellular data was used to collect travel data between residential areas and workplaces and to study the relationship between the balance of time spent at work and home and travel in the city.

15.4.1 Distribution of Workplaces and Residential Areas

In the study, five indices were selected to describe the spatial characteristics of Beijing: residential density, job density, degree of balance between time spent at work and home, distance to the city center, and distance to the workplace. The first three indices were derived from the cellular network. The last two indices were derived from a GIS platform. We conducted a statistical analysis to show the spatial characteristics of land use in Beijing.

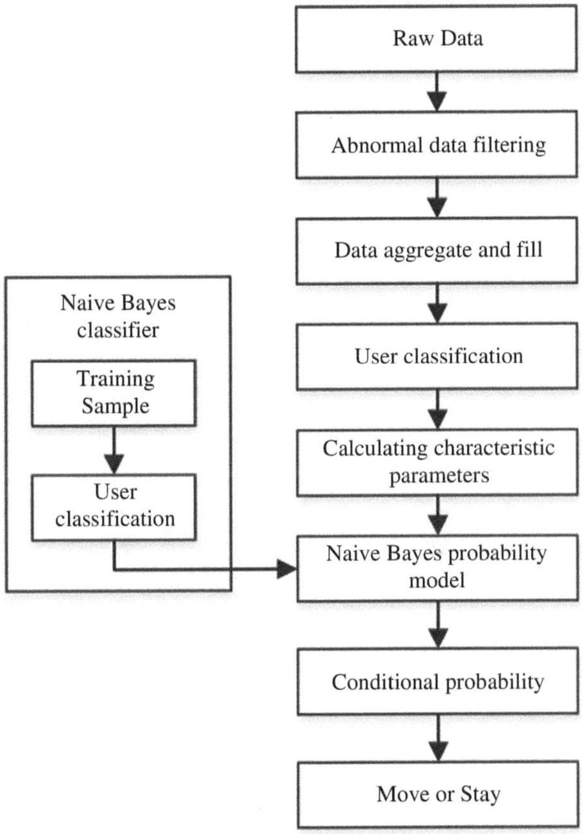

Fig. 15.9 The method used to determine the state of movement of a cellular phone user

15.4.1.1 Population Density

Population density is the sum of residents in one zone of the city divided by the area of the zone. Figure 15.10 shows the spatial distribution of residential density used for studying traffic flows in the city.

The residential density was high within the Fourth Ring Road. It was found that 38% of the city's residents lived within the Third Ring Road. The average residential density was 24022.2 residents per km^2, with a standard deviation of 20996.9. Figure 15.11 shows the residential density for the area within the Second Ring Road, Second-Third Ring Road, and Third-Fourth Ring Road and beyond the Fourth Ring Road. The orange and blue columns show the residential density for the southern and northern areas of the city, respectively. It can be seen that beyond the Third Ring Road, the residential density of the northern region was higher than that of southern region.

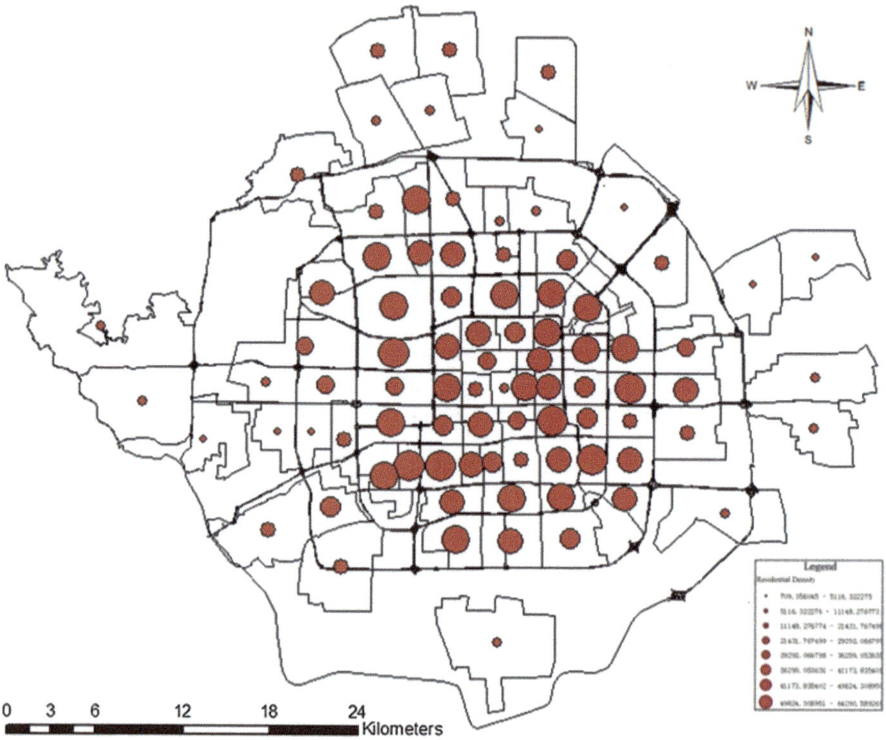

Fig. 15.10 Spatial distribution of residential density in the surveyed zones of Beijing (unit: residents per square km)

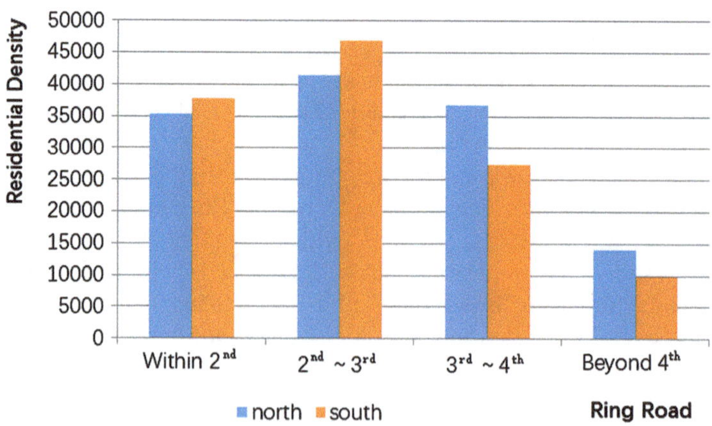

Fig. 15.11 Residential density in Beijing (unit: residents per square km)

15.4.1.2 Job Density

Job density is the sum of jobs in one zone divided by the area of the zone, i.e., the same calculation method as for population density. Figure 15.12 shows the spatial distribution of job density used for studying traffic flows in the city.

The job density was high within the Fourth Ring Road and north of the area between the Fourth Ring Road and Fifth Ring Road. It was found that 49% of residents lived within the Third Ring Road. The average job density was 20,026 per km^2, and standard deviation is 21,065, which means the distribution of job density was more uneven than the distribution of residential density. Figure 15.13 shows the job density for the area within the Second Ring Road, Second-Third Ring Road, and Third-Fourth Ring Road and beyond the Fourth Ring Road. The orange and blue columns show the residential density for the southern and northern areas of the city, respectively. The job density in the northern area was higher than in the southern area.

Fig. 15.12 Spatial distribution of job density in the surveyed zones of Beijing (unit: residents per square km)

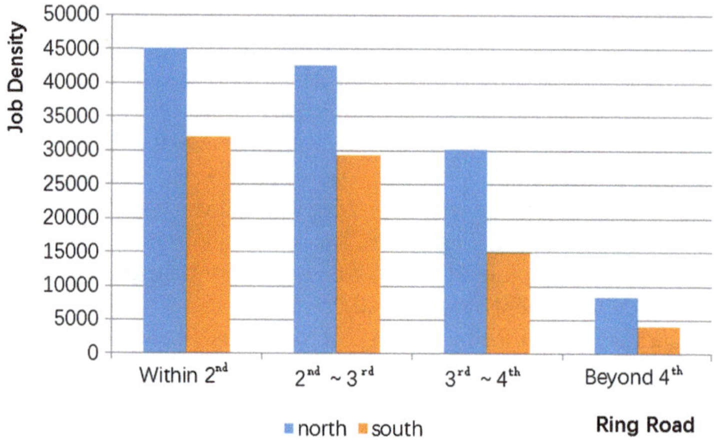

Fig. 15.13 Job density in Beijing (unit: per square km)

15.4.1.3 Balance of Time Spent at Work and Home

The balance between time spent at home and at work (*H*) was calculated by the following formula:

$$H = \frac{W}{R \cdot \alpha} \tag{1}$$

where *W*: job count
R: resident count
α: the ratio of the working population to the total population

W and *R* are total numbers of people working and living in a zone. In Beijing, the working age is 16–60 for males and 16–55 for females. The population whose age falls into the working age range is regarded to represent the workable population. For Beijing, α is 72.8%.

W and *R* are total numbers of people working and living in a zone.

The closer *H* is to 1, the more balanced is the time spent at work and home (Fig 15.14).

Figure 15.15 shows job density for the area within the Second Ring Road, Second-Third Ring Road, and Third-Fourth Ring Road and beyond the Fourth Ring Road. The gray and green columns show the statistics for the southern and northern areas, respectively. Within the Second Ring Road, the number of available jobs was greater than the number of workable residents. Within the Second Ring Road and Third Ring Road, and in the northern area beyond the Fourth Ring Road, the number of available jobs was nearly equal to the number of residents. In the southern area beyond the Fourth Ring Road, the number of workable residents was greater than the number of jobs. The balance of time spent at work and home in the southern area of the city was more skewed than in the northern area. These results indicate

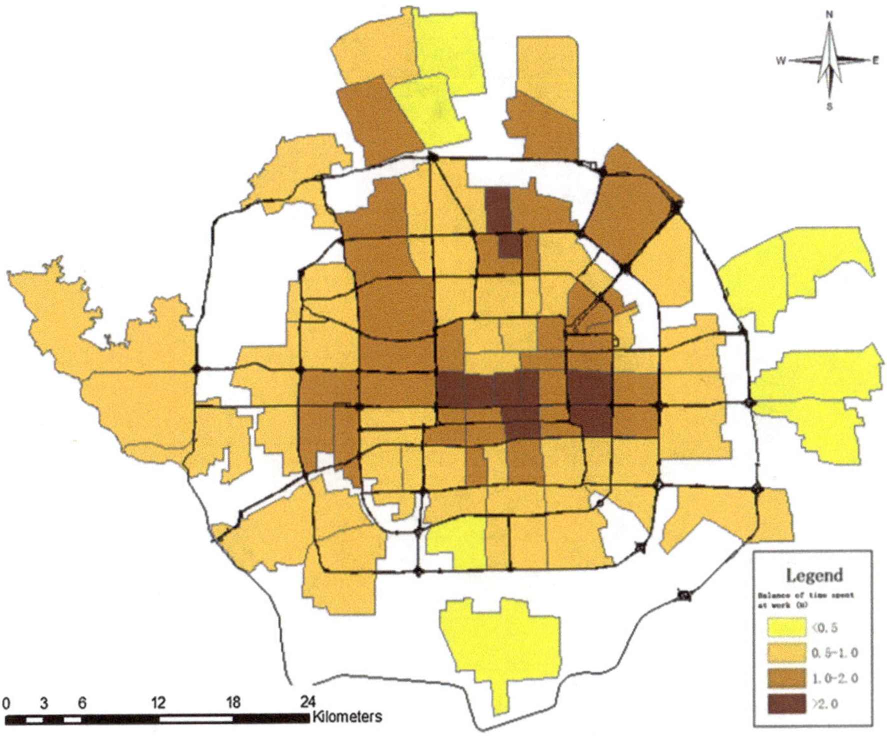

Fig. 15.14 Spatial distribution of the balance of time spent at work and home in the surveyed zones of Beijing

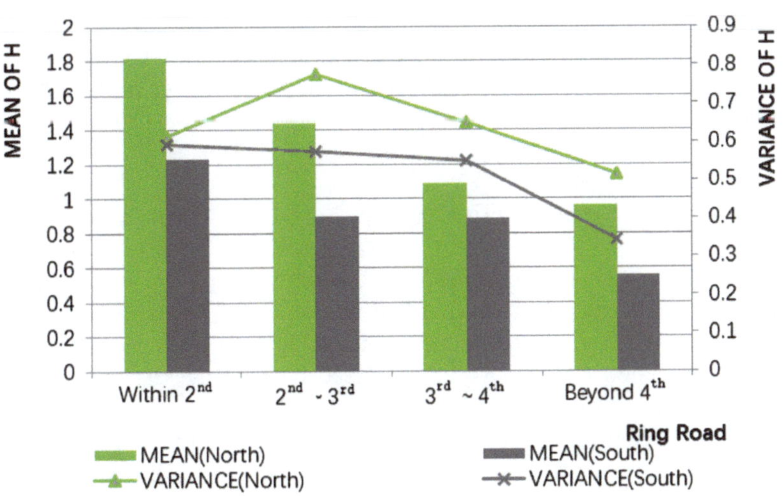

Fig. 15.15 Statistics regarding the balance between time spent at home and at work (*H*)

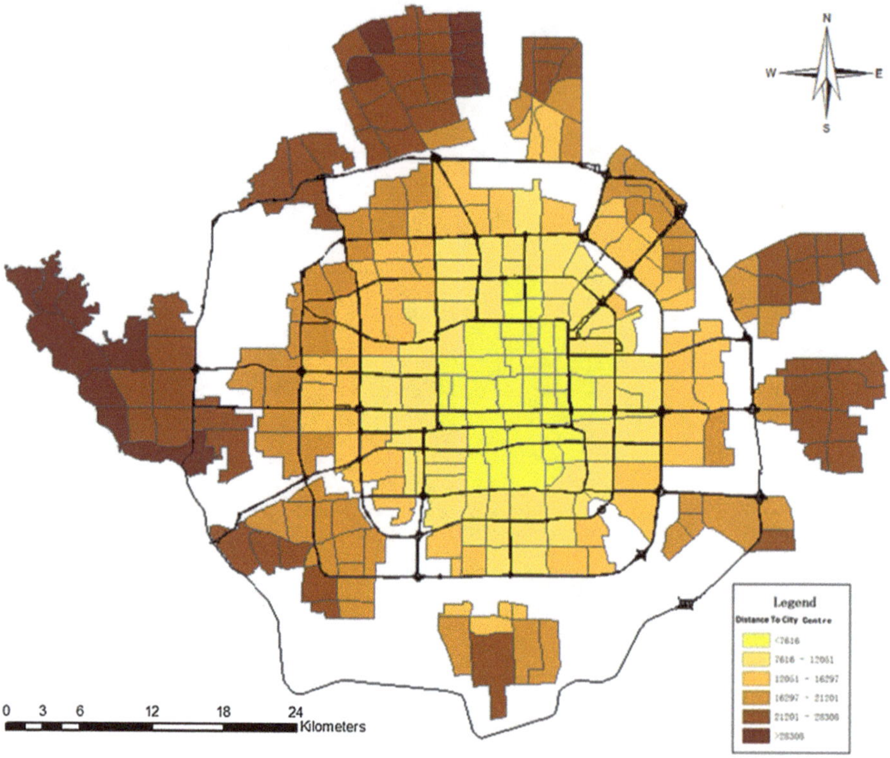

Fig. 15.16 Spatial distribution of the distance to the city center in the surveyed zones of Beijing

that more workplaces were located in the center and north of the city. The balance of time spent at work and home was skewed in the northern area of the old city and southern area beyond the Fourth Ring Road.

15.4.1.4 Distance to the City Center

Tiananmen Square, at the south gate of the Forbidden City, has always been considered the city center of Beijing and was selected as the geometric center for this study. We used GIS to calculate the distance from each surveyed traffic zone to this geometric center and considered it to be the distance to the city center (dc).

Figure 15.16 shows the distribution of the distance to city center. The average distance to the city center for the studied traffic zones was 15.9 km and the standard deviation was 7.452 km.

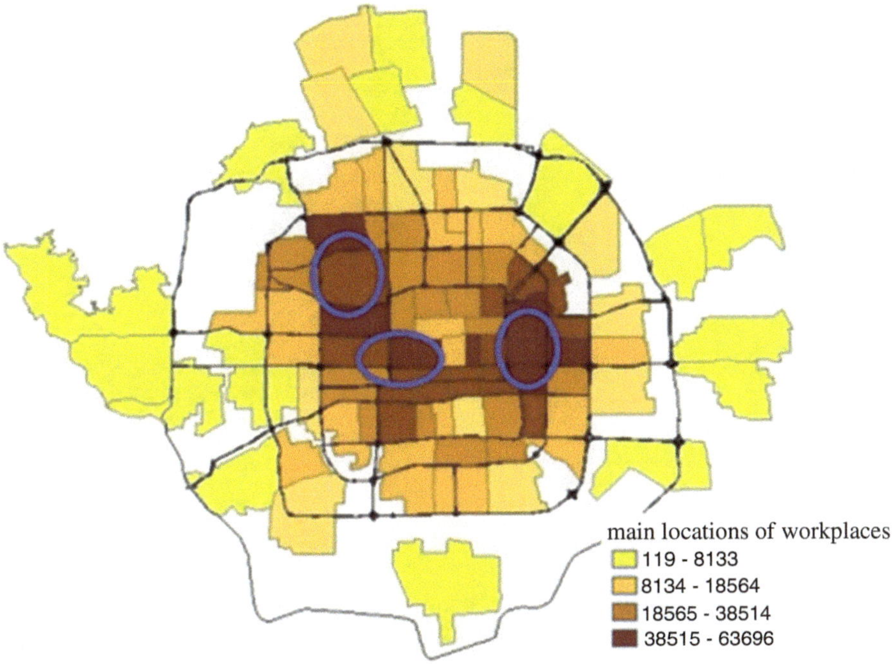

Fig. 15.17 The main locations of workplaces in the in the surveyed zones of Beijing

15.4.1.5 Distance to the Workplace

From the job density data in the surveyed traffic zones, three districts were identified as the main locations of workplaces: the central business district (CBD), financial district, and a high-tech development district. Figure 15.17 shows the location of these three districts (marked by blue circles) and indicates that they were generally situated on the Second Ring Road, Third Ring Road, and Fourth Ring Road. We then used GIS to calculate the distances from each surveyed traffic zone to the centroid of the nearest workplace location, which was considered to be the distance to the workplace (d_w). Figure 15.18 shows the distribution of the distance to the workplace. The average distance was 16,260 m, and the standard deviation was 5722 m.

15.4.2 Trip Distribution

Three indices were used to describe the spatial characteristics of land use in Beijing, trip density, trip distance, and distance commuted. We conducted a statistical analysis to determine the spatial characteristics of land use in Beijing.

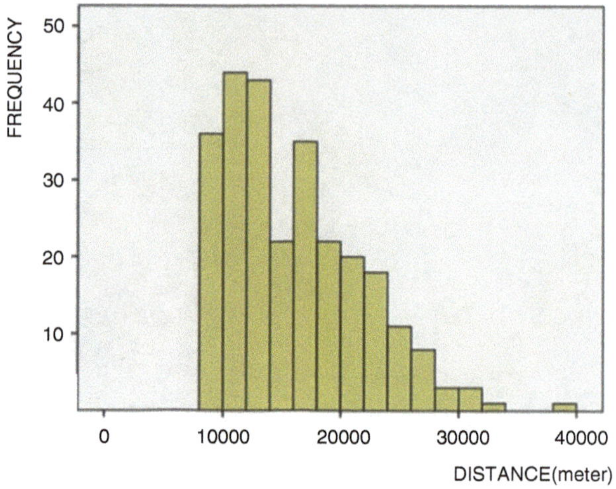

Fig. 15.18 The distance to workplaces

15.4.2.1 Trip Density

Trip density (F) is the sum of all trips in one zone, i.e., trip generation (G) divided by the area of the zone, where G is the sum of trips within the zone.

Figures 15.19 and 15.20 show the spatial distribution and trip density statistics in the surveyed zones of Beijing.

The average trip density in the core built-up area was 10,414 per km^2. For almost 90% of the area, the trip density was lower than 220,000 per km^2. About 10% of the area had a high trip density, with the highest density being 400,000 trips per km^2. The zones with a high trip density were mainly along the Second Ring Road and Third Ring Road, as well as Chang'an Street, which is the longest east-west-oriented main road.

The average trip number for each resident was 2.3425 trips per person.

15.4.2.2 Daily Trip Distance and Distance Commuted

Trip distance is the daily average trip distance for all the trips that occurred in a zone.

The distance commuted is the average trip distance from home to the workplace for the residents in a particular zone.

Figures 15.21 and 15.22 show the spatial distribution and statistics of trip distance and distance commuted.

Both the daily trip distance and distance commuted had a similar spatial pattern, with the average distance commuted being 12.84 km and the average trip distance being 12.59 km. The proportion of short commutes (no longer than 5 km) among all trips was larger than among daily trips.

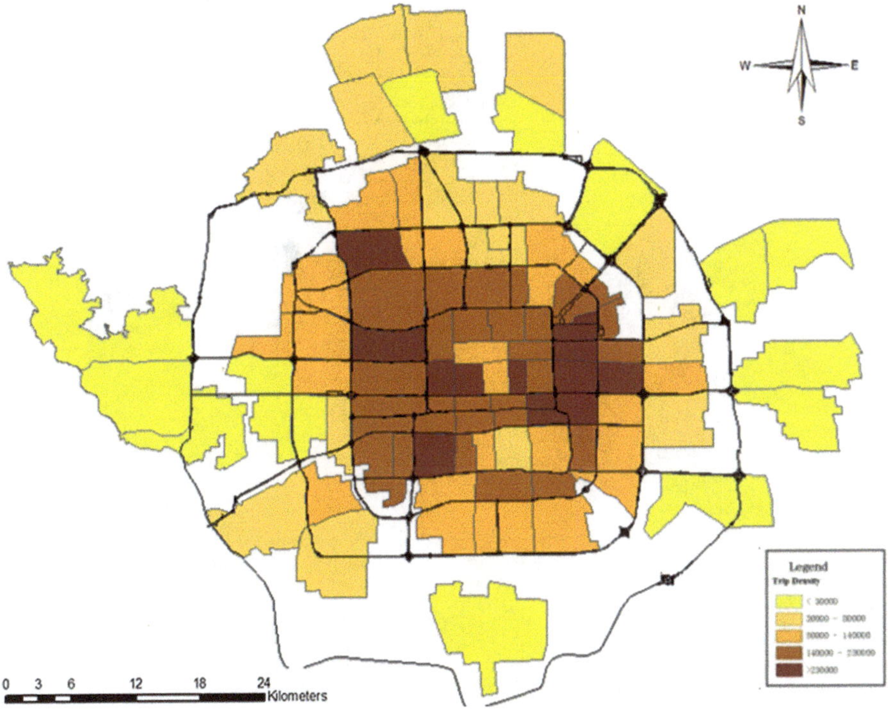

Fig. 15.19 Spatial distribution of trip density (F: trips per km^2) in the surveyed zones of Beijing

Fig. 15.20 Cumulative frequency curve of trip density (F)

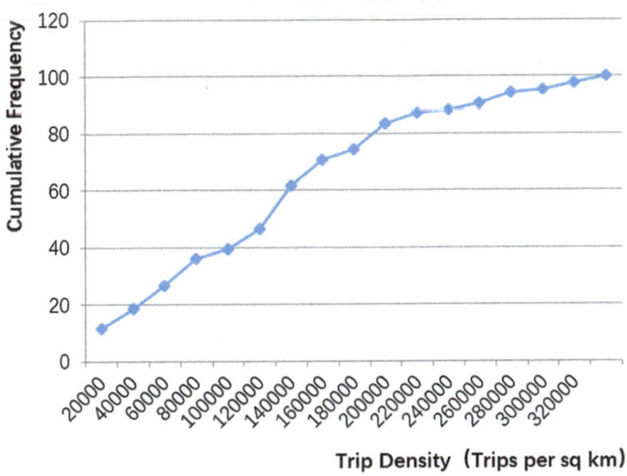

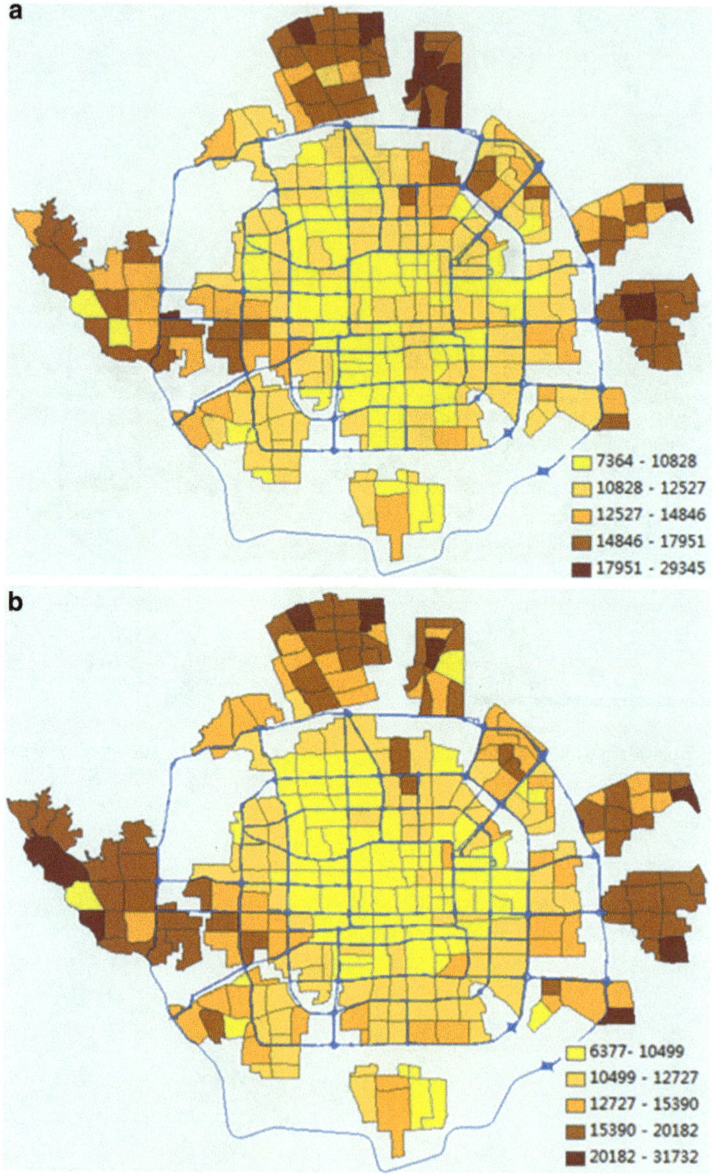

Fig. 15.21 Spatial distribution of daily trip distance and distance commuted. (**a**) Daily trip distance (m). (**b**) Distance commuted (m)